VISCOELASTIC AND VISCOPLASTIC MATERIALS

Edited by **Mohamed Fathy El-Amin**

Viscoelastic and Viscoplastic Materials
http://dx.doi.org/10.5772/61921
Edited by Mohamed Fathy El-Amin

Contributors

Bing Wei, Roozbeh Naemi, Nachiappan Chockalingam, Sara Behforootan, Panagiotis Chatzistergos, Silvia Ioan, Yosuke Nishitani, Benjamin Ramirez-Wong, Nestor Ponce-Garcia, Anayansi Escalante-Aburto, Patricia Isábel Torres-Chávez, Juan De Dios Figueroa-Cardenas, Dan Andrei Şerban, Fernando Cortés, Jon García-Barruetabeña, Naomi Chesler, Zhijie Wang, Mark Golob, Salim Messaoudi, Sandra Carillo, Claudio Giorgi, Antonio Marcos Gonçalves De Lima, Mario Lázaro, César F. Casanova, Carlos Lázaro, Mikael Hedenqvist, Henrik Ullsten, Mikael Gällstedt

Published by InTech

Janeza Trdine 9, 51000 Rijeka, Croatia

Publishing Process Manager Andrea Koric

Technical Editor SPi Global

Cover InTech Design team

First published August, 2016

Printed in Croatia

Additional hard copies can be obtained from orders@intechopen.com

Viscoelastic and Viscoplastic Materials, Edited by Mohamed Fathy El-Amin

p. cm.

Print ISBN 978-953-51-2602-7

Online ISBN 978-953-51-2603-4

For EU product safety concerns:
IN TECH d.o.o., Prolaz Marije Krucifikse Kozulić 3, 51000 Rijeka, Croatia,
info@intechopen.com or visit our website at intechopen.com.

Meet the editor

Prof. Mohamed F. El-Amin is a full-time professor of applied mathematics and computational sciences at the Effat University, Saudi Arabia. Also, he is a visiting professor at the King Abdullah University of Science and Technology (KAUST), and is a full-time professor at the Aswan University, Egypt. As a mathematician, he has over 20 years of research experience in the field of computational sciences, applied mathematics, heat and mass transfer, fluid dynamics, turbulence, petroleum engineering, etc. After obtaining his PhD in 2001, he held research positions in several universities, including South Valley University (Egypt), Stuttgart University (Germany) and Kyushu University (Japan). The research of Dr. El-Amin has resulted in over 130 publications.

Contents

Preface

This book introduces numerous selected advanced topics in viscoelastic and viscoplastic materials. The book effectively blends theoretical, numerical, modeling and experimental aspects of viscoelastic and viscoplastic materials that are usually encountered in many research areas such as chemical, mechanical and petroleum engineering. The book consists of 14 chapters that can serve as an important reference for researchers and engineers working in the field of viscoelastic and viscoplastic materials.

The first chapter provides a synopsis of polymer flooding as an emerging and advantageous EOR process and also provides a guideline for pilot design. The objective of the second chapter is to evaluate mechanical properties, including the viscoelasticity of the wheat kernels, using compression tests. The third chapter studies the viscoelastic behaviour of liquid crystal polymers in composite systems. Thermal properties of Hemp fiber–reinforced plants-derived polyamide biomass composites and their dynamic viscoelastic properties in molten state are presented in Chapter 4. The fifth chapter studies a practical approach to select plasticizers for proteins such as thermoformed wheat gluten, which involved 30 plasticizer candidates. Chapter 6 investigates dynamic viscoelastic properties of carbon nanofiber–filled polyamide composites and the blend of these composites and thermoplastic elastomer in molten state. The aim of the seventh chapter is to review the current state of knowledge regarding the viscoelastic behavior of cardiovascular tissues, whereas Chapter 8 is focuses on the study of the r nonviscous modes of a nonviscously damped vibrating system. In Chapter 9, the viscoplastic behavior of a polyamide-based thermoplastic polymer is investigated by performing numerous tests that highlighted the influence of time and temperature on the mechanical behavior. The viscoelasticity in foot-ground interaction is studied in Chapter 10, such that an insight into the mechanical behavior of plantar soft tissue during loading with specific emphasis on heel pad, which is the first point of contact during normal gait. Chapter 11 presents a review in the field of viscoelastic adhesive applications for noise and vibration reduction. Chapter 12 concerns with the general stability problem in viscoelastic, namely the motion of a viscoelastic body and the effect of the dissipation induced by the viscoelastic (integral) term on the solution. The aim of Chapter 13 is to present a collection of nonstandard viscoelastic kernels, with special emphasis on singular and time-dependent kernels. Finally, finite element modeling and experiments of systems with viscoelastic materials for vibration attenuation are covered by Chapter 14.

Professor Mohamed F. El-Amin, Ph.D
Effat University,
Jeddah, Kingdom of Saudi Arabia

Chapter 1

Advances in Polymer Flooding

Bing Wei

Additional information is available at the end of the chapter

http://dx.doi.org/10.5772/64069

Abstract

Polymer flooding, which has been successfully used in numerous enhanced oil recovery (EOR) projects in the world, was able to promote oil recovery by 12 to 15% of original oil in place (OOIP). When a reservoir is flooded with viscous polymer solution, the mobility ratio between water and oil becomes more favorable relative to conventional waterflooding, which leads to a significant increase in the volumetric sweep efficiency. Furthermore, recent research based on laboratory studies and pilot field testing has proved that the displacement efficiency (at pore scale) can also be improved due to the elasticity of polymer solution. Therefore, this chapter first introduces the recovery mechanisms that have been proposed to explain oil displacement by polymer flooding within oil reservoirs. The development of EOR polymers from chemical structure to physical property is also reviewed. The experience and learning of polymer flooding accumulated in the last 20 years in the Daqing Oilfield, China, which is the most successful oilfield in the world implementing polymer flooding technique, are summarized and discussed. The aim of this chapter is to provide a synopsis of polymer flooding as an emerging and advantageous EOR process and also provide a guideline for pilot design.

Keywords: enhanced oil recovery, polymer flooding, mechanism, polymer type, field application, experience

1. Introduction

As early as the 1960s, polymer flooding had been suggested as an oil recovery process to further increase oil recovery after waterflooding; the main purpose of adding polymer was to increase the viscosity of the displacing fluid, which is commonly water [1]. This work was followed by wide research attention to recognize the benefits of polymer flooding in oil recovery applications [2–4].

For most oil reservoirs, especially heterogeneous formations, at least half of the reserved oil still leaves behind after extensive waterflooding due to the unfavorable mobility ratio between water and oil. Once a preferential flow path is formed between injector and producer, the subsequently injected water would flow straight to the production well bypassing the oil bearing zones, which ultimately causes a low sweep efficiency and oil recovery. In order to cover the bypassed oil zones, polymer is usually used to thicken the injection water and makes the mobility of water and oil comparable. Through polymer flooding, the poor mobility ratio encountered in conventional waterflooding is corrected, and consequently the volumetric sweep efficiency of the water-flooded reservoirs can be significantly improved. Among all the EOR methods, polymer flooding is considered as one of the most promising technologies because of its technical and commercial feasibility. The worldwide interests in polymer flooding applications were further stimulated recently by the exciting field reports from the scaled use in Daqing oilfield in China, with incremental oil productions of up to 300,000 barrels per day [5].

The overall oil recovery efficiency in oil production processes is generally governed by two sub efficiencies, *i.e.*, macroscopic and microscopic recovery efficiency. The macroscopic recovery efficiency refers to the volume that the flooding agents are able to sweep; while microscopic recovery efficiency is a measure of the effectiveness of the displacing fluid(s) in mobilizing the oil trapped at pore scale by capillary forces. In other words, any mechanism that can improve either macroscale or microscale oil recovery efficiency is beneficial for increasing oil production [6].

The mechanisms of polymer flooding have been pursued since it incepted [7]. The well-established relationship between capillary number and oil recovery indicates that a substantial increase in oil recovery at the pore level (microscale) can be obtained only when the capillary number is increased by several thousand times. However, for polymer flooding, the capillary number is normally increased less than 100 times [8, 9]. Therefore, it was previously suggested that polymer flooding can only improve the volumetric sweep efficiency without any effect on the microscopic displacement efficiency [10]. However, in Daqing oilfield, China, the oil recovery factor using polymer flooding was increased by up to 13% OOIP, and this value seemed unachieved only relying on sweep efficiency improvement. This fact made researchers to revisit the oil recovery mechanisms occurred in polymer flooding. After almost 15 years' efforts, it was demonstrated that the incremental oil recovery by polymer flooding could also be explained by the simultaneous increased microscopic displacement efficiency due to the distinctive flow characteristic of polymer solutions.

Regarding EOR polymers, hydrolyzed polyacrylamides (HPAMs) are still the most widely used polymer to date in oilfields because of their availability in large quantity with customized properties (molecular weight, hydrolysis degree, etc.) and low manufacturing cost [11, 12]. However, it is known that polyacrylamides are very susceptible to chemical, mechanical, thermal, and microbial degradation, and this issue might affect its acceptance in type II and type III reservoirs having high temperature and salinity. As alternatives to HPAM, many novel polymers with tough properties have been proposed in the past several years, such as hydrophobically modified polymers and biopolymers [13].

Therefore, this chapter first summarizes all the oil recovery mechanisms involved in polymer flooding related to the sweep efficiency and displacement efficiency in order to assist people rapidly understand this popular and emerging EOR technique. Further, EOR polymers from chemical structure to physical properties are discussed. In the end, field experience and learning of polymer flooding in China are presented.

2. Oil recovery mechanisms of polymer flooding

2.1. Macroscopic sweep efficiency improvement

2.1.1. Mobility control

The primary objective of polymer flooding is to control the mobility of the aqueous phase. Mobility is defined as the ratio of the relative permeability to the displacing phase divided by the viscosity of the displacing phase. The dissolution of polymer to water reduces its mobility by thickening the aqueous phase and significantly diminishes the formation of viscous fingerings and/or channels [10, 14, 15]. Therefore, polymer flooding is very effective in improving the volumetric sweep efficiency.

Mobility ratio (*M*) is the key parameter to quantify the mobility contrast between aqueous phase and oil phase. The mobility ratio for waterflooding is given by the following expression (Eq. 1) [14]:

$$M = {}^{\lambda_w}\!/_{\lambda_o} = \frac{{}^{k_{rw}}\!/_{\mu_w}}{{}^{k_{ro}}\!/_{\mu_o}} = \frac{k_{rw}\mu_o}{k_{ro}\mu_w} \qquad (1)$$

where λ is the fluid mobility, k_r is the relative permeability, and λ is the fluid viscosity; the subscripts *w* and *o* denote water phase and oil phase, respectively.

The conventional concept of mobility ratio (*M*) distinguishes "favorable" mobility conditions when $M \leq 1$ and "unfavorable" mobility conditions when $M > 1$ [16, 17]. From this point, it can be easily understood that polymer flooding enables the correct mobility ratio (*M*) to a favorable level by increasing the viscosity of water (μ_w).

2.1.2. Disproportionate permeability reduction (DPR)

In addition to viscosity increase, another mechanism, DPR, can also facilitate polymer flooding to improve the macroscopic sweep efficiency. Disproportionate permeability reduction means polymer considerably reduces the water relative permeability (k_{rw}) while producing minimum reduction in the oil relative permeability [18–20].

The reasons causing DPR in polymer flooding include: 1) Segregation of flow pathways (water and oil); 2) Shrinking/swelling of polymer; 3) Layer formation on pore wall by adsorbed polymer; and 4) Wettability alteration [21–25]. However, it was believed that polymer adsorption and segregation of flow pathways were the dominant mechanisms compared to other two mechanisms [26–29].

2.1.3. Flow resistance induced by polymer elasticity

Flow resistance is the third mechanism that can improve volumetric or macroscopic sweep efficiency during polymer flooding. The effect of viscoelastic fluid on the formation of an "internal cake" (frictional pressure drop) was reported by Dehghanpour and Kuru [30]. They observed that polymer solution with higher elasticity yielded significantly higher pressure drop during flow through porous media. A similar trend was obtained by Urbissinova et al. [31], who evaluated the contribution of polymer's elasticity in EOR. Likewise, they observed that the polymer solution having higher elasticity corresponds to higher flow resistance to flow through porous media than the polymer with lower elasticity, even though their shear viscosities were identical, which in turn improved the macroscopic sweep efficiency and oil recovery (**Figure 1**) [32, 33]. This mechanism may also improve the microscopic displacement efficiency by displacing residual oil immobilized by capillary forces and rock configuration [8]. Furthermore, the elasticity can also maintain the stability of the propagating front and minimize fingers as shown in **Figure 2** [34].

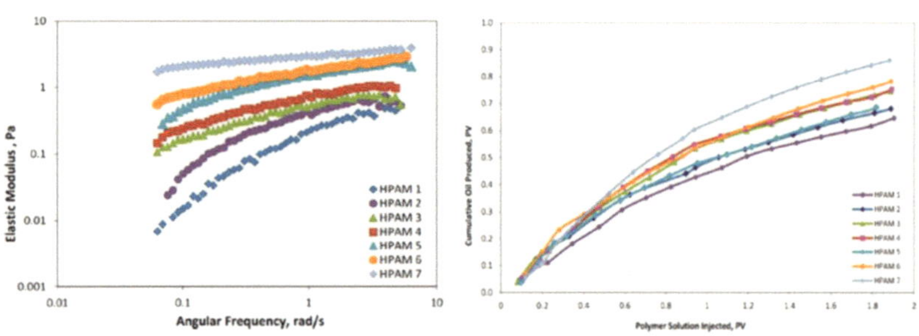

Figure 1. The effect of polymer elasticity on oil recovery (similar viscosity). (Source: [33])

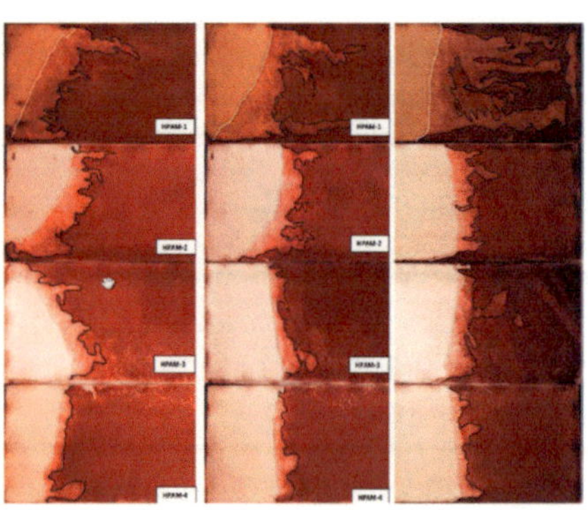

Figure 2. Propagating front of polymer solution. (Source: [34]) (Elasticity: HPAM-4>HPAM-3>HPAM-2>HPAM-1)

2.2. Microscopic displacement efficiency improvement

In the last decade, it has been confirmed that polymer flooding can increase the microscopic displacement efficiency by mobilizing and displacing residual oil saturation (trapped by capillary forces). This phenomenon was attributed to the elasticity of the polymer solutions.

Wang et al. [8] first studied this subject by evaluating the effectiveness of polymer flooding in displacing "residual oil" after waterflooding under different conditions as follows. (1) Residual oil in "dead ends"; (2) Residual oil films on rock; (3) Residual oil in pore throats trapped by capillary forces; (4) Residual oil un-swept in microscale heterogeneous portions of the porous media (**Figure 3**). In all cases, it was observed that residual oil was further reduced after polymer flooding. The proposed mechanisms improving the microscopic displacement efficiency during polymer flooding include pulling effect, stripping effect, oil thread, and shear thickening effect, which are discussed in the subsequent paragraphs.

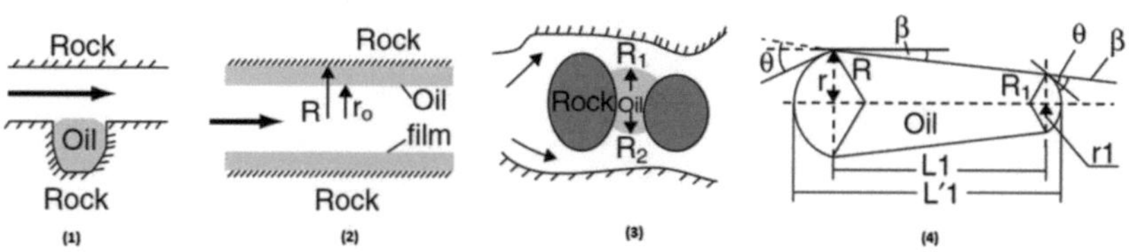

Figure 3. Simplified models of residual oil distribution after waterflooding. (Source: [35])

2.2.1. *Pulling effect*

It was found that if an elastic fluid flows over dead ends, normal stresses between oil and polymer solution are generated in addition to the shear stresses resulting from the long molecular chains [36, 37]. Therefore, polymer molecules impose a larger force on oil droplets and thus pull them out of dead ends. The amount of residual oil pulled out from dead ends seems proportional to the elasticity of the driving fluid (**Figure 4**). The viscoelastic polymer pushes the fluid ahead and pulls the fluids beside and behind, while the nonelastic fluids (water and glycerin) as shown in **Figure 4** are also capable of pushing the fluids ahead but cannot "pull" out oil from the dead end.

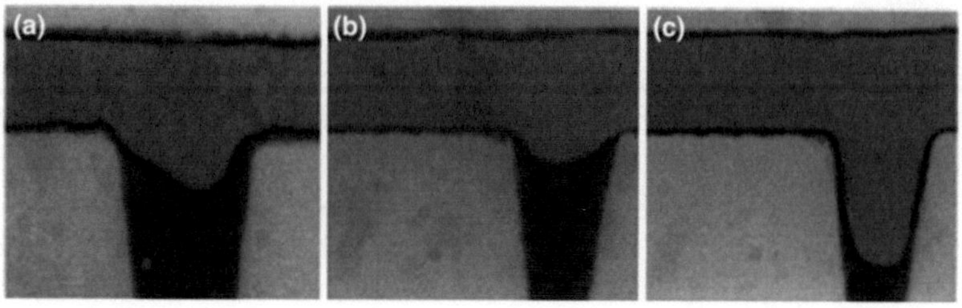

Figure 4. Residual oil in the dead ends after a: water; b: glycerin (without elasticity); c: HPAM flooding (source: [5])

This effect has been further investigated through numerical simulation of polymer flooding over dead ends [38, 39]. A simplified physical model with dead end geometry was used (**Figure 5**). A modified upper-convected Maxwell (MUCM) model which considered viscosity, elasticity, and also non-Newtonian power lower behaviors of polymer solution was established. Using this model, the contours of velocity and stream function can be drawn and the microscopic efficiency quantified (**Figure 6**).

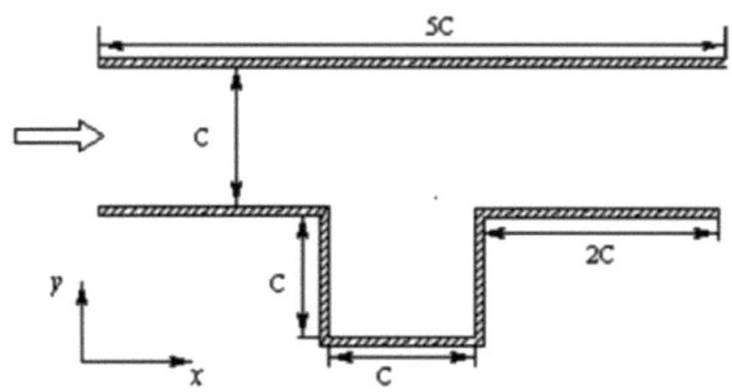

Figure 5. Micropore with dead end model. (Source: [38])

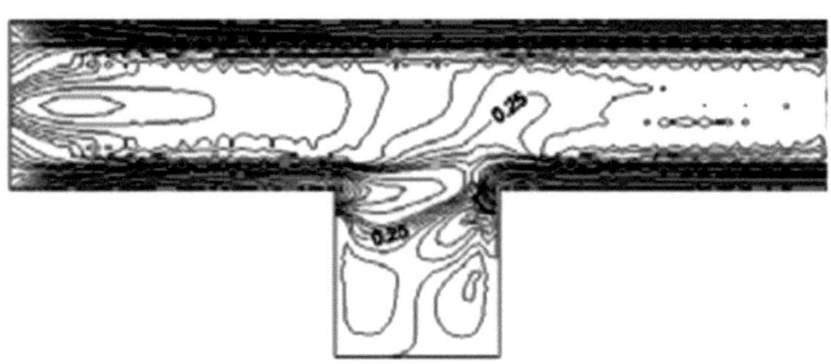

Figure 6. An example of the stress contour determined by the MUCM model. (Source: [38])

2.2.2. Stripping effect

For oil-wet porous media, residual oil is attached on the rock surface in the form of a continuous oil film (**Figure 3(2)**). Wang et al. [8, 36] compared the velocity profile of a Newtonian and a non-Newtonian fluid in a capillary and observed that the velocity gradient near the capillary wall for the elastic fluid is noticeably higher than that for the Newtonian fluid (**Figure 7**). Therefore, a stronger force is produced during flow of polymer solutions compared to water, which thus facilitates stripping the oil films off rock surface and eventually promotes the microscopic displacement efficiency [39–42]. The alteration in wettability from oil-wet to more water-wet due to the stripping effect would also increase oil recovery.

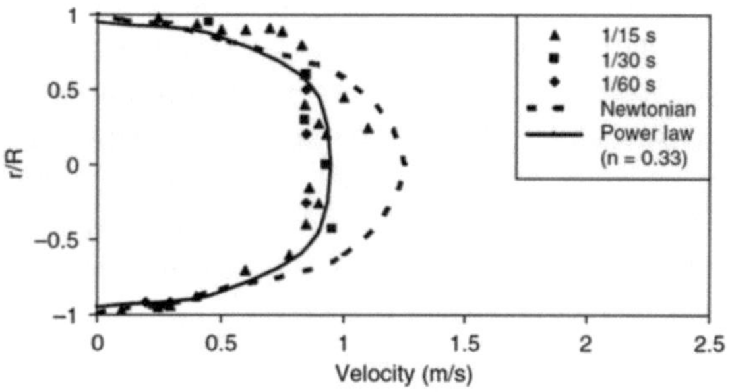

Figure 7. Velocity profiles of water and an HPAM solution in a capillary. (Source: [36])

2.2.3. *Oil thread*

The third possible mechanism related to the elastic property of polymers is oil thread. Oil is pulled out by polymers into oil columns and then forms oil threads as they aggregate with the residual oil downstream [5, 43]. However, due to the high interfacial tension between oil and polymer solution, the long oil column would be destabilized and broken into oil droplets, the oil droplets will be probably re-entrapped by capillary forces [44]. Elastic polymer solutions are able to stabilize oil threads as a result of the normal stress. It is believed that the normal force bearing on the convex surface of oil thread is larger than the normal force bearing on the concave surface. The essential consequence of this effect is to stabilize oil threads and prevent them from deforming, which therefore improves the displacement efficiency [41]. Furthermore, this force increases with the Deborah number (N_{Deh}), which is a dimensionless number commonly employed to describe the elasticity of polymer solutions in porous media (**Figure 8**).

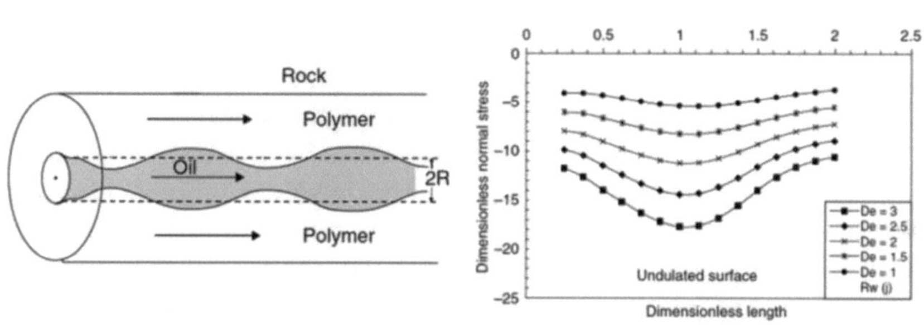

Figure 8. Illustration of the oil thread mechanism. (Source: [43])

2.3.4. *Shear thickening*

The flow behavior of polymers as a function of shear rate (flow velocity gradient) in porous media can be generally divided into three regions: Newtonian, shear thinning, and shear thickening as shown in **Figure 9**. When polymer molecules flow through a series of pore bodies and pore throats, the flow field experiences elongation and contraction. Shear thickening phenomenon would occur when the flow velocity is so high that the polymer molecules do not have sufficient relaxation time to re-coil to adapt the flow geometry. The stretched polymer chains result in a high apparent viscosity (shear thickening). This effect can help the driving fluid to rapidly displace the mobile but hard-to-displace oil or to displace the bypassed oil in small-scale heterogeneities more effectively [45]. Therefore, the shear thickening effect is considered as one of the possible mechanisms for microscopic displacement efficiency improvement in polymer flooding [46]. Moreover, the high apparent viscosity is also beneficial for promoting the macroscopic sweep efficiency [47].

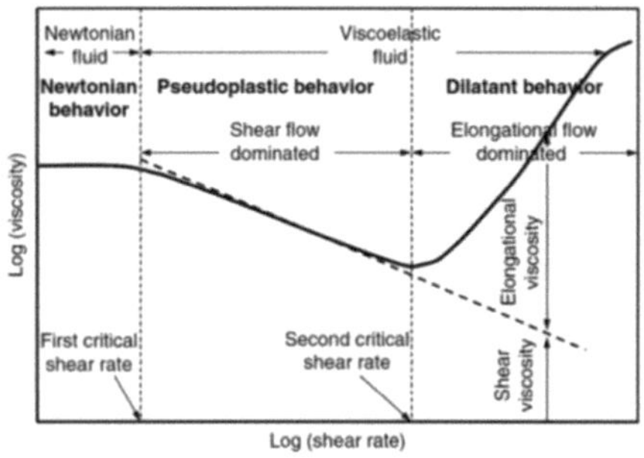

Figure 9. Flow behavior of viscoelastic fluid. (Source: [48])

3. EOR polymers

3.1. Synthetic Polymers

HPAM, partially hydrolyzed polyacrylamide, is by far the most used polymer in EOR application, which is a copolymer of PAM and PAA obtained by partial hydrolysis of PAM or copolymerization of sodium acrylate and acrylamide [49]. The chemical structure of HPAM is shown in **Figure 10**. For most cases, the degree of hydrolysis of the acrylamide monomers ranges from 25 to 35%. The fact that the monomeric units have to be hydrolyzed is probably related to the formation of the corresponding salt. The negative charges loaded onto polymer backbone due to the hydrolysis are responsible for prominent stretching (electric repulsion) of polymer chains, which results in viscosity increase. On the other hand, the degree of hydrolysis cannot be too high, otherwise, HPAM becomes too sensitive to salinity and hardness of brine (shielding effect) [50].

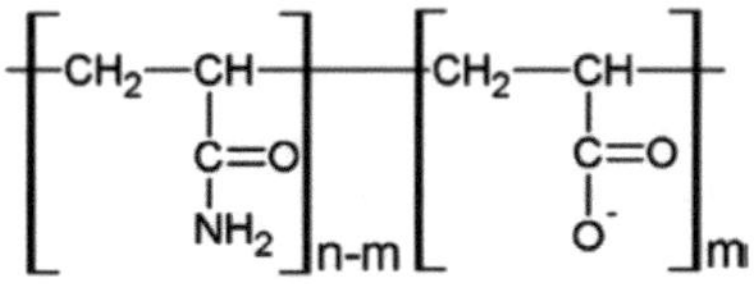

Figure 10. Chemical structure of HPAM.

However, given the harsh conditions present in most of oil reservoirs, problems and limitation arise with the use of conventional HPAM because of the significant chemical and mechanical degradation. Therefore, other tolerant polymers such as hydrophobically associating polymers and biopolymers have been proposed in the market [51].

The design of hydrophobically associating polymers is based on the concept of association between hydrophobic groups that are incorporated in the backbone of the polymers. These hydrophobic groups could be randomly or block-like distributed, and coupled at one or both ends. When the polymer is dissolved in water, the hydrophobic groups associate forming micro-domains, which leads to increase in the hydrodynamic volume and accordingly yields an improved thickening capacity [52]. **Figure 11** shows the structure of a typical hydrophobically associating polymer.

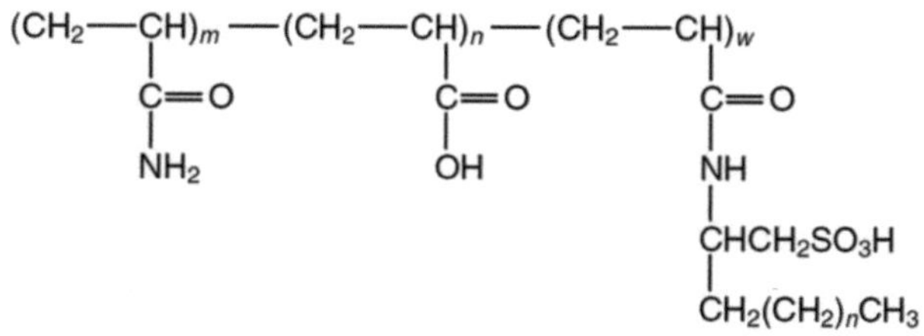

Figure 11. Structure of a hydrophobically associating polymer. (Source: [48])

Recently, Zaitoun et al. reported a new class of novel EOR polymers, which contain nVP (n-Vinyl-Pyrrolidone) and ATBS (Acrylamido tert-Butyl Sulfonate). The presence of nVP can enhance the thermal stability of polymers, while the function of ATBS is to improve both thermal stability and salinity tolerance. **Figure 12** shows the chemical structure of nVP and ATBS as well as the commercial names of the polymers.

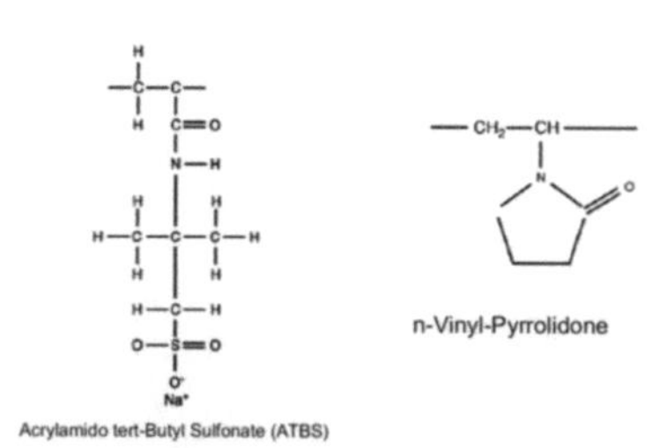

Commercial Name	Acrylamide	%AAH (Acrylate)	%ATBS	% nVP	Molecular Weight (10^6 g/mol)	Concentration (ppm)	Viscosity (mPa·s)
FA920	100				8	2700	17.57
3630 S	72.2	27.8			18.5	850	18.69
AN 105	95		5		6	2200	18.84
SAV505	50			50	6–8	4800	17.3
3430 S	67.2	32.8			12	1400	18.22
AN 125	75		25		8	2000	18.29

Figure 12. Chemical composition of the polymers. (Source: [53])

KYPAM is a commercial EOR polymer, which is produced by Beijing Hengju, China and is being widely used in many oilfields such as Daqing and Shengli, etc. It is a typical salinity-tolerant EOR polymer, which is functioned by different types of groups. The basic structure of KYPAM is shown in **Figure 13**.

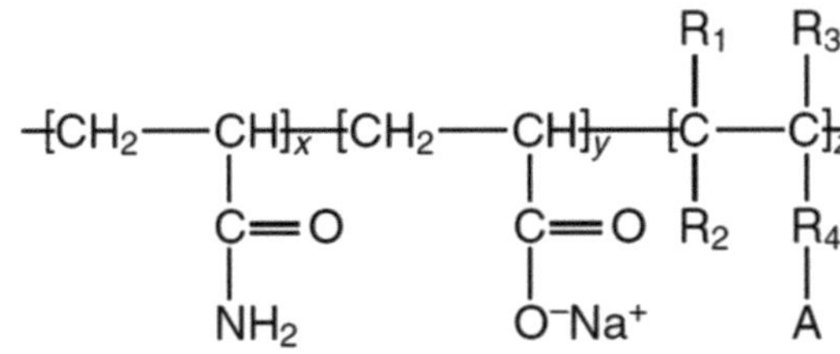

Figure 13. Structure of KYPAM. (Source: [54])

where R_1, R_2,and R_3 could be either H or alkyl. A represents an ionic functional group that is tolerant to Ca^{2+} and/or Mg^{2+}. The elasticity of KYPAM is controlled by R_1, R_2, and R_3, and R_4 determines the salinity tolerance of KYPAM.

KYPAM has been successfully used in China; a good example is Lamadian Field, Daqing. In 2001, the northwestern block of Lamadian field carried out a field test using KYPAM. Polymer solution which was prepared with produced water was injected in May 2001. **Figure 14** illustrates the average water cut change during KYPAM test and HPAM test in a neighbor block. It should be emphasized that the HPAM solution was prepared with fresh water. It was observed that the average water cut was reduced by 15% by KYPAM compared to conventional HPAM [55].

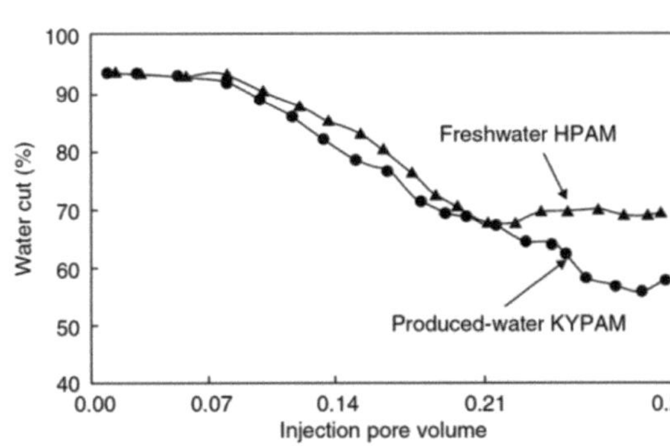

Figure 14. Effect of HPAM and KYPAM on water cut. (Source: [55])

3.2. Biopolymers

In addition to the synthetic polymers, another kind of polymer frequently used is biopolymer, such as xanthan gum. Xanthan gum is a polymersaccharide which is produced by fermentation of glucose or fructose by different bacteria. The chemical structure of xanthan gum as shown in **Figure 15** illustrates the presence of two glucose units, two mannose units, and one glucuronic acid unit with a molar ratio of 2.8-2-2 [56]. X-ray diffraction studies proved that xanthan backbone has a helical structure where the side chains fold down along the helix. The average molecular weight of xanthan gum used in EOR processes is from 1–15 million.

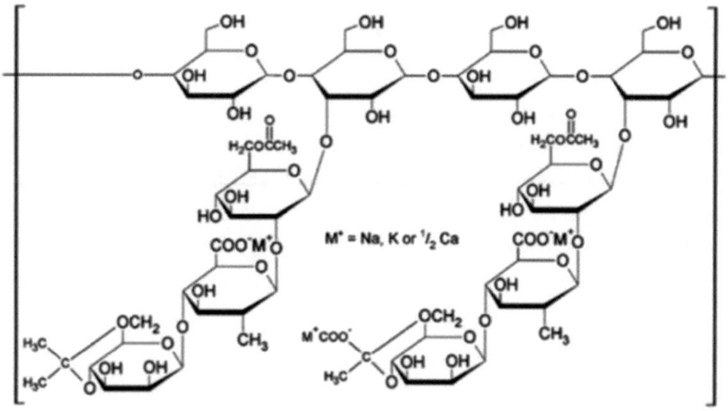

Figure 15. Chemical structure of xanthan gum. (Source: [57])

Due to its unique structure, xanthan gum is quite resistant to mechanical shear and salinity. The chemical and physical properties of xanthan gum related to EOR use have been systematically investigated in our previous reports [58–61]. It was demonstrated that xanthan gum experienced less viscosity loss when exposed to harsh reservoir conditions compared with HPAM and hydrophobic polymers. Therefore, a piston-like flow pattern was established during xanthan gum propagation in porous media.

A successful pilot test using xanthan gum was conducted in the 7th block in Shengli oilfield, China (65°C, 5500TDS). The injection of xanthan gum solution was launched after the waterflooding was exhausted. Totally, 0.356PV polymer solution with a concentration of 480PV·mg/L was injected into the target formation ($Ng5^5$); the whole process took about 825days. The results indicated that a reduction in water cut from 97 to 92% was observed [13].

Hydroxyethycellulose (HEC) is another biopolymer that has been implemented in oilfields. HEC was produced from insoluble cellulose by chemical modification. Unlike xanthan gum, HEC is a synthetically produced biopolymer and thus does not contain cellular debris, which has been implicated as a cause of formation plugging. As a result of the nonionic feature and rigid backbone, HEC is able to withstand high salt concentration, temperature, and mechanical shear.

HEC is particularly suitable for polymer flooding low temperature but high salinity formations resulting from its tolerance to high salinities, where HPAM and even hydrophobic polymers are discarded due to precipitation. In 1992, a joint field trial of Total and Tatneft was performed in the Romashkino field, Tatarstan. The temperature of this reservoir was 38°C with a salinity of 250g/L. According to the report, the production of 1 million tons of incremental oil was produced due to the injection of 350 tons of HEC [62].

Other potential EOR biopolymers are scleroglucan, simusan, AGBP, welan gum, and so on. However, it seems that these biopolymers have not been implemented in oilfields so far.

4. Experience of polymer flooding and learning in china

Unquestionably, Daqing Oilfield is the most successful example in polymer flooding technology in China and even in the world. Laboratory research investigating the potential of polymer flooding was started in the 1960s. After a few pilot tests, the largest polymer flooding was implemented at Daqing from 2006, which has contributed 22.3% of the total oil production by 2007. At the end of 2007, oil production dedicated by polymer flooding was more than 10 million tons per year [12]. Based on these successful experiences, the use of polymer flooding in the Dagang, Shengli, Karamay oilfields, etc., is being considered or has already been proved.

According to the observations of the response, polymer flooding process can be divided into five stages as shown in **Figure 16** [63].

1. ***Initial stage***. Polymer solution is injected into formations and injection pressure increases, but water cut does not start to decrease yet.
2. ***Water cut decreasing stage***. Injection pressure keeps increasing and the water cut decrease can be seen.
3. ***Stable stage***. Water cut and production are relatively stable.
4. ***Water cut increasing stage***. After a stable period, the water cut starts to rapidly increase. The produced polymer concentration is quite high.
5. ***Post-water drive stage***. A water drive is followed up.

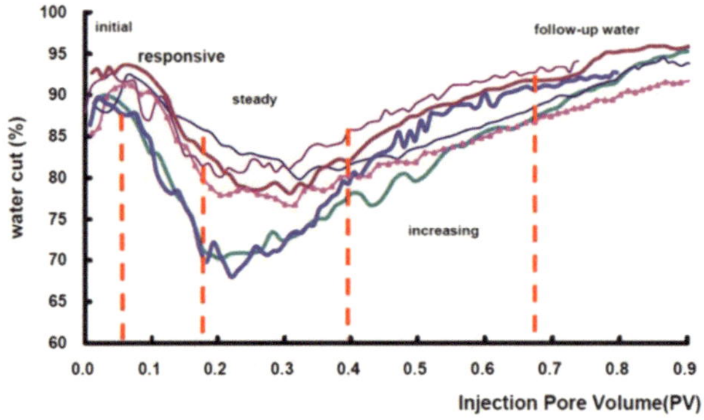

Figure 16. Water cut change during polymer flooding. (Source: [12])

Many key indices including water cut, fluid production capability, injection capability, and injection profile change at different stages of polymer flooding. The most definitive index is water cut, which is increased at the initial stage and also the last stage. Water cut starts to drop at the second stage, which is a response to polymer injection.

After more than 20 years of pilot testing and large-scale commercial applications in polymer flooding in China, a wealth of experience and learning has been accumulated as follows.

4.1. Well pattern design

When designing a polymer flooding project, a number of guidelines should be followed. Designs should accommodate the variations in geological properties for different parts of the field. To optimize the well pattern, the parameters of connectivity factor, permeability ratio, and well spacing should be counted [63].

Connectivity factor is defined as the pore volume accessed by polymer solution divided by the total PV of the oil zone. It was proved that the incremental oil recovery by polymer flooding markedly decreased when this factor is lower than 70%. Therefore, well pattern and polymer molecular weight must be well considered to make this value greater than 70%.

The effect of well pattern on the incremental oil recovery by polymer flooding was compared based on numerical simulations. The results indicated that the magnitudes of incremental oil recovery are very close for the well patterns of line-drive, staggered drive, 4-spot, 5-spot, and 9-spot. However, 5-spot pattern seems attractive for polymer flooding owing to the proper injection pressure and connectivity.

Regarding well spacing, at Daqing, for the oil zones having permeability higher than 400md, net pay thicker than 5m , and V_k greater than 0.65, the well spacing is designed from 200–250 m. However, for the oil zones having lower permeability (>200md), net pay ranging from 1–5m, and V_k less than 0.65, 150–175 m is an reasonable spacing. The prerequisite of any well spacing is to ensure connectivity higher than 70%.

4.2. Polymer injection design

Before a polymer flooding operation, three parameters that are closely related to economic and technical feasibility must be considered. These parameters include polymer molecular weight, polymer concentration, and polymer volume.

The efficiency and the economics of polymer flooding are somehow governed by polymer molecular weight. The selection of the molecular weight is strongly dependent on two factors. High-molecular-weight polymer is normally preferential to minimize the cost from the economic point of view. However, on the other hand, the molecular weight must be small enough to allow polymer to enter and propagate through the porous media. Therefore, for a given rock permeability and pore size, the maximum molecular weight is almost definitized. The ratio of pore throat radius to the root mean square radius of gyration of the polymer must be greater than 5 to eliminate any plugging problem [64]. **Figure 17** plots the maximum molecular weight versus rock permeability.

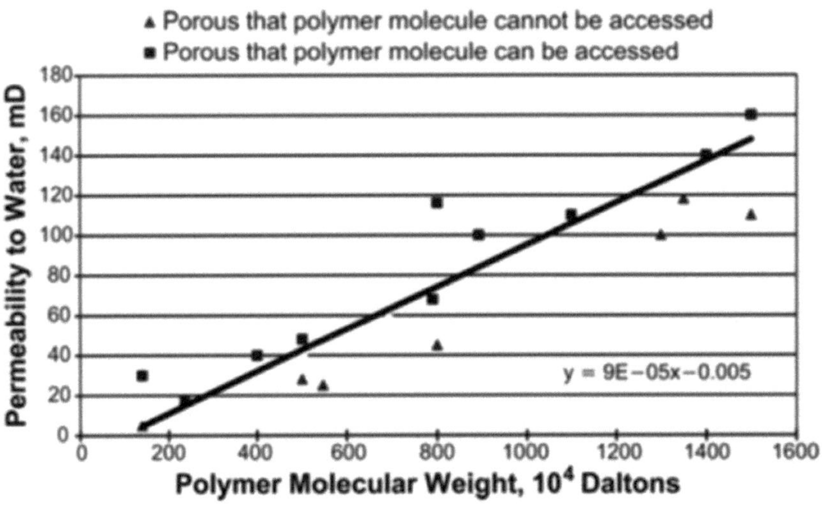

Figure 17. Allowed polymer molecular weight vs permeability. (Source: [65])

The effectiveness of polymer flooding with higher concentration has been technically verified at Daqing. By using higher concentration polymer slugs, the in-depth vertical sweep can be improved in the early stage of polymer flooding. Furthermore, the increased water cut at the third stage of polymer flooding can be significantly mitigated. They found that the water-intake profile became more uniform after injecting 2200 to 2500mg/L polymer solution [66]. This technological trend was also studied in our previous work for heavy oil recovery [59].

According to more than 200 reports of polymer flooding, the average amount of injected polymer solution was from 19 to 150 lb/acre-ft, which is equivalent to 23.3 to 184.2 mg/L·PV if the porosity is 30%. At Daqing, the injected amount of polymer solution usually ranged from 180 to 380 mg/L·PV. However, much larger amounts of polymer solution have been attempted, the largest amount being 771 mg/L·PV, which therefore resulted in a higher incremental oil recovery and improved economics [12].

Acknowledgements

The author wishes to thank all the contributors to the content in this chapter. The copyright owners and authors who permit to re-use the information are acknowledged. I am also grateful to Dr. Laura Romero-Zeron, who led me to this interesting area. My thanks also go to every member in my research group for their advice.

Author details

Bing Wei

Address all correspondence to: bwei@swpu.edu.cn

State Key Laboratory of Oil and Gas Reservoir Geology and Exploitation, Southwest Petroleum University, Chengdu, Sichuan, China

References

[1] Sandiford BB. Laboratory and field studies of water floods using polymer solution to increase oil recovery. Journal of Petroleum Technology. 1964;16:917–922.

[2] Szabo MT. Laboratory investigations of factors influencing polymer flood performance. SPE Journal. 1975;15:338–346.

[3] Dominguez JG, Willhite GP. Retention and flow characteristics of polymer solution in porous media. SPE Journal. 1977;17:111–121.

[4] Gleassure RW. An experimental study of non-Newtonian polymer rheology effects on oil recovery and injectivity. SPE Reservoir Engineering. 1990;5:481–486.

[5] Wang Demin XH, Zhongchun L, Qingyan Y. Study of the mechanism of polymer solution with visco-elastic behavior increasing microscopic oil displacement efficiency and the forming of steady "Oil Thread" flow channels. In: SPE Asia Pacific Oil and Gas Conference and Exhibition; 17–19 April 2001; Jakarta, Indonesia; Society of Petroleum Engineers; SPE 68723.

[6] Romero-Zerón L. Advances in enhanced oil recovery processes. In: Romero-Zerón L, editor. Rijeka, InTech; 2012.

[7] Wei B, Romero-Zerón L, Rodrigue D. Oil displacement mechanisms of viscoelastic polymers in enhanced oil recovery (EOR): a review. Journal of Petroleum Exploration and Production Technology. 2014;4:113–121.

[8] Wang D, Cheng J, Yang Q, Gong W, Li Q. Viscous-elastic polymer can increase microscale displacement efficiency in cores. In: SPE Annual Technical Conference and Exhibition; 1–4 October 2000; Dallas, Texas; Society of Petroleum Engineers; SPE 63227.

[9] Stegemeier GL. Mechanisms of entrapment and mobilization of oil in porous media. Academic Press, New York; 1977.

[10] Du Y, Guan L. Field-scale polymer flooding: Lessons learnt and experiences gained during past 40 years. In: SPE International Petroleum Conference in Mexico; 7–9 November 2004; Puebla Pue, Mexico; Society of Petroleum Engineers; SPE 91787.

[11] Needham RB, Doe PH. Polymer flooding review. Journal of Petroleum Technology. 1987;39:1503–1507.

[12] Dong HZ, Fang SF, Wang DM, Wang JY, Liu Z, Hong WH. Review of practical experience & management by polymer flooding at Daqing. In: SPE Symposium on Improved Oil Recovery; 20–23 April 2008; Tulsa, Oklahoma, USA; Society of Petroleum Engineers; SPE 114342.

[13] Guo XH, Li WD, Tian J, Liu YZ. Pilot Test of Xanthan Gum Flooding in Shengli Oilfield. In: SPE Asia Pacific Improved Oil Recovery Conference; 25–26 October 1999; Kuala Lumpur, Malaysia; Society of Petroleum Engineers; SPE 57294.

[14] Pitts MJ, Campbell TA, Surkalo H, Wyatt K. Polymer flood of the Rapdan pool. SPE Reservoir Engineering. 1995;10:183–186.

[15] Kotlar HK, Selle O, Torsaeter O. Enhanced oil recovery by COMB-flow: Polymer floods revitalized. In: International Symposium on Oilfield Chemistry; 28 February–2 March 2007; Houston, Texas, USA; Society of Petroleum Engineers; SPE 106421.

[16] Lake LW. Enhanced oil recovery. Prentic Hall, New Jersey; 1989.

[17] Dyes AB, Caudle BH, Ericjson RA. Oil production after breakthrough as influenced by mobility ratio. Journal of Petroleum Technology. 1954;6:81–86.

[18] Schneider FN, Owens WW. Steady-state measurements of relative permeability for polymer/oil systems. SPE Journal. 1982;22:79–86.

[19] Niu JG, Chen P, Shao ZB, Wang DM, Sun G, Li Y. Research and development of polymer enhanced oil recovery. Petroleum Industry Press; 2006.

[20] Taber JJ, Martin FD. Technical screening guides for the enhanced recovery of oil. In: SPE Annual Technical Conference and Exhibition; 5–8 October 1983; San Francisco, California; Society of Petroleum Engineers; SPE 12069.

[21] Mennella A, Chiappa L, Bryant SL, Burrafato G. Pore-scale mechanism for selective permeability reduction by polymer injection. In: SPE/DOE Improved Oil Recovery Symposium; 19–22 April 1998 ; Tulsa, Oklahoma; Society of Petroleum Engineers; SPE 39634.

[22] Liang JT, Seright RS. Further investigations of why gels reduce water permeability more than oil permeability. SPE Production and Facilities. 1997;12:225–230.

[23] Nilsson S, Stavland A, Jonsbraten HC. Mechanistic study of disproportionate permeability reduction. In: SPE/DOE Improved Oil Recovery Symposium; 19–22 April 1998; Tulsa, Oklahoma; Society of Petroleum Engineers; SPE 39635.

[24] Barreau P, Bertin H, Lasseux D, Glénat P, Zaitoun A. Water control in producing wells: influence of an adsorbed-polymer layer on relative permeabilities and capillary pressure. SPE Reservoir Engineering. 1997;12:234–239.

[25] Zaitoun A, Kohler N. Two-phase flow through porous media: Effect of an adsorbed polymer layer. In: SPE Annual Technical Conference and Exhibition; 2–5 October 1988; Houston, Texas; Society of Petroleum Engineers; SPE 18085.

[26] Al-Sharji HH, Grattoni CA, Dawe RA, Zimmerman RW. Disproportionate permeability reduction due to polymer adsorption entanglement. In: SPE European Formation Damage Conference; 21–22 May 2001; The Hague, Netherlands; Society of Petroleum Engineers; SPE 68972.

[27] Córdova M, Mogollón JL, Molero H, Navas M. Sorbed polyacrylamides: selective permeability parameters using surface techniques. In: SPE/DOE Improved Oil Recovery Symposium; 13–17 April 2002; Tulsa, Oklahoma; Society of Petroleum Engineers; SPE 75210.

[28] Denys K, Fichen C, Zaitoun A. Bridging adsorption of cationic polyacrylamides in porous media. In: SPE International Symposium on Oilfield Chemistry; 13–16 February 2001; Houston, Texas; Society of Petroleum Engineers; SPE 64984.

[29] Stavland A, Nilsson S. Segregated flow is the governing mechanism of disproportionate permeability reduction in water and gas shutoff. In: SPE Annual Technical Conference and Exhibition; 30 September–3 October 2001; New Orleans, Louisiana; Society of Petroleum Engineers; SEP 71510.

[30] Dehghanpour H, Kuru E. A new look at the viscoelastic fluid flow in porous media – a possible mechanism of internal cake formation and formation damage conrol. In: SPE International Symposium on Oilfield Chemistry; 20–22 April 2009; The Woodlands,Texas; Society of Petroleum Engineers; SPE 121640.

[31] Urbissinova TS, Trivedi JJ, Kuru E. Effect of elasticity during viscoelastic polymer flooding: a possible mechanism of increasing the sweep efficiency. Journal of Canadian Petroleum Technology. 2010;49:49–56.

[32] Veerabhadrappa SK. Study of effects of polymer elasticity on enhanced oil recovery by core flooding and visualization experiments (thesis). University of Alberta; 2012.

[33] Veerabhadrappa SK, Doda A, Trivedi JJ, Kuru E. On the effect of polymer elasticity on secondary and tertiary oil recovery. Industrial and Engineering Chemistry Research. 2013;52:18421–18428.

[34] Veerabhadrappa SK, Trivedi JJ, Kuru E. Visual confirmation of the elasticity dependence of unstable secondary polymer floods. Industrial & Engineering Chemistry Research. 2013;52:6234–6241.

[35] Wang D, Xia H, Liu Z, Anda, Yang Q. Study of the mechanism of polymer solution with visco-elastic behavior increasing microscopic oil displacement efficiency and the forming of steady "oil thread" flow channels. In: SPE Asia Pacific Oil and Gas Conference and Exhibition; 17–19 April 2001; Jakarta, Indonesia; Society of Petroleum Engineers; SEP 68723.

[36] Wang D, Wang G, Wu W, Xia H, Yin H. The influence of viscoelasticity on displacement efficiency From micro – to macroscale. In: SPE Annual Technical Conference and Exhibition; 11–14 November 2007; Anaheim, California, USA; Society of Petroleum Engineers; SEP 109016.

[37] Wang DM, Wang G, Xia HF. Large scale high visco-elastic fluid flooding in the field achieves high recoveries. In: SPE Enhanced Oil Recovery Conference; 19–21 July 2011; Kuala Lumpur, Malaysia; Society of Petroleum Engineers; SPE 144294.

[38] Yin H, Wang D, Zhong H. Study on flow behaviors of viscoelastic polymer solution in micropore with dead end. In: SPE Annual Technical Conference and Exhibition; 24–27 September 2006; San Antonio, Texas, USA; Society of Petroleum Engineers; SPE 101950.

[39] Xia H, Wang D, Wang G, Wu J. Effect of polymer solution viscoelasticity on residual oil. Petroleum Science and Technology. 2008;26:398–412.

[40] Jiang H, Wu W, Wang D, Zeng Y, Zhao S, Nie J. The effect of elasticity on displacement efficiency in the lab and results of high concentration polymer flooding in the field. In: SPE Annual Technical Conference and Exhibition; 21–24 September 2008; Denver, Colorado, USA; Society of Petroleum Engineers; SPE 115315.

[41] Sedaghat MH, Ghazanfari MH, Parvazdavani M, Morshedi S. Experimental investigation of microscopic/macroscopic efficiency of polymer flooding in fractured heavy oil five-spot systems. Journal of Energy Resources Technology. 2013;135:364–392.

[42] Xia H, Ju Y, Kong F, Wu J. Effect of elastic behavior of HPAM solutions on displacement efficiency under mixed wettability conditions. In: SPE Annual Technical Conference and Exhibition; 26–29 September 2004; Houston, Texas; Society of Petroleum Engineers; SPE 90234.

[43] Luo JH, Liu YZ, Zhu P. Polymer solution properties and displacement mechanisms. Petroleum Industry Press, Beijing; 2006.

[44] Huh C, Pope GA. Residual oil saturation from polymer floods: Laboratory measurements and theoretical interpretation. In: SPE Symposium on Improved Oil Recovery, 20–23 April 2008; Tulsa, Oklahoma, USA; Society of Petroleum Engineers; SPE 113417.

[45] Delshad M, Kim DH, Magbagbeola OA, Huh C, Pope GA, Tarahhom F. Mechanistic interpretation and utilization of viscoelastic behavior of polymer solutions for improved polymer-flood efficiency. In: SPE Symposium on Improved Oil Recovery; 20–23 April 2008 Tulsa, Oklahoma, USA; Society of Petroleum Engineers; SPE 113620.

[46] Jones WM. Polymer additives in reservoir flooding for oil recovery: shear thinning or shear thickening? Journal of Physical D Applied Physics. 1980;13:L87–L88.

[47] Seright RS, Fan T, Wavrik K, Balaban RDC. New insights into polymer rheology in porous media. SPE Journal. 2010; 16:130–140.

[48] Sheng JJ. Polymer flooding. Gulf Professional Publishing, Elsevier, Burlington; 2011.

[49] Morgan SE, Mccormick CL. Water-soluble polymers in enhanced oil recovery. Progress in Polymer Science. 1990;15:103–145.

[50] Shupe RD. Chemical stability of polyacrylamide polymers. Journal of Petroleum Technology. 1981;33:1513–1529.

[51] Wever DAZ, Picchioni F, Broekhuis AA. Polymers for enhanced oil recovery: a paradigm for structure-property relationship in aqueous solution. Progress in Polymer Science. 2011;36:1558–1628.

[52] Feng Y, Billon L, Grassl B, Bastiat G, Borisov O, François J. Hydrophobically associating polyacrylamides and their partially hydrolyzed derivatives prepared by post-modification. 2. Properties of non-hydrolyzed polymers in pure water and brine. Polymer. 2005;46:9283–9295.

[53] Zaitoun A, Makakou P, Blin N, Al-Maamari RS, Al-Hashmi AR, Abdel-Goad M, et al. Shear stability of EOR polymers. SPE Journal. 2012;17:335–339.

[54] Luo JH, Bu RY, Wang PM. Properties of KYPA: a salinity-resistant polymer used in EOR. Oilfield Chemistry. 2002;19:64–67.

[55] Wang YP, Luo JH, Bu RY, Wang PM, Liu YZ. Field Test of KYPAM, a comb-shaped and salinity-resistant polymer. Chemical Industry and Engineering Progress. 2003;22:509–511.

[56] GarcíA-Ochoa F, Santos VE, Casas JA, Gómez E. Xanthan gum: Production, recovery, and properties. Biotechnology Advances. 2000;18:549–579.

[57] Green DW, Whillhite GP. Enhanced oil recovery. Society of Petroleum Engineers, Dallas; 1998.

[58] Wei B, Romero-Zerón L, Rodrigue D. Mechanical properties and flow behavior of polymers for enhanced oil recovery. Journal of Macromolecular Science, Part B: Physics. 2014;53:625–644.

[59] Wei B, Romero-Zerón L. The evaluation of a technological trend in polymer flooding for heavy oil recovery. Petroleum Science and Technology. 2014;32:2396–2404.

[60] Bing W, Romero-Zerón L, Rodrigue D. Evaluation of two new self-assembly polymeric systems for enhanced heavy oil recovery. Indengchemres. 2014;53:16600–16611.

[61] Wei B. Flow characteristics of three enhanced oil recovery polymers in porous media. Journal of Applied Polymer Science. 2015;132. DOI: 10.1002/app.41598.

[62] Loppinet L, Lakovlev S, Glent P. Five years of injection of hydroxyethylcellulose: An ecological pure product for enhanced oil recovery in the field of Romashkino Offshore Europe. In: Offshore Europe; 9–12 September 1997; Aberdeen, United Kingdom; Society of Petroleum Engineers; SPE 38464.

[63] Wang D, Dong H, Lv C, Fu X, Nie J. Review of practical experience of polymer flooding at Daqing. SPE Reservoir Evaluation and Engineering. 2009;12:470–476.

[64] Zhang Y, Li C, Wang X. Discussions on the technical of string with backflush for separate layers during the period of polymer flooding. Oil & Gas Ground Engineering. 2004;23:26–27.

[65] Chen P. Method for determining relative molecular mass of polymer in medium-low permeable reservoirs. Petroleum Geology & Oilfield Development Engineering 2005;24:95–96.

[66] Yang F, Wang D, Yang X, Sui X, Chen Q, Zhang I. High concentration polymer flooding is successful. In: SPE Asia Pacific Oil and Gas Conference and Exhibition; 18–20 October 2004; Perth, Australia; Society of Petroleum Engineers; SPE 88454.

Chapter 2

Mechanical Properties in Wheat (*Triticum aestivum*) Kernels Evaluated by Compression Tests: A Review

Néstor Ponce-García, Benjamín Ramírez-Wong,
Anayansi Escalante-Aburto,
Patricia I. Torres-Chávez and
Juan de Dios Figueroa-Cárdenas

Additional information is available at the end of the chapter

http://dx.doi.org/10.5772/64171

Abstract

To assess the quality of food grains, it is necessary to consider the following two aspects: their general characteristics and their intrinsic quality. Analyzing the quality of wheat kernels is complex due to the particularity of wheat proteins and the diversity of products that can be developed. In contrast, basic factors are used to assess quality aspects, with a focus on kernel hardness. This parameter is usually measured by the force that is required to make the grain rupture. The application of force must be controlled, and hence, the grain will exhibit other mechanical attributes and behavioral characteristics that can be used to evaluate it more objectively. This has led to the development of nondestructive evaluation methods based on the mechanical properties of kernels. This review carried out research on grain wheat, in which the main objective was to evaluate mechanical properties, including the viscoelasticity of the wheat kernels, by using compression tests. The study examined different methods of applying those techniques and the parameters they evaluated. Finally, the results obtained by the different investigation groups that applied the compression tests on wheat kernels were discussed.

Keywords: wheat kernels, mechanical properties, viscoelasticity, compression, hardness

1. Introduction

Wheat is one of the cereals with the highest production worldwide, occupying the second place after rice, with an annual production of 600 million tons per year, almost 70% of which is used by the food industry [1–3]. *Triticum aestivum* (genetically hexaploid) is distinguished from other species in commercial terms and includes two grain categories, that is, soft and hard, that are both destined for the production of flours used in the preparation of bakery products. In contrast, *Triticum durum* (genetically tetraploid) is used to obtain semolina to produce pastas [4].

The commercial value of wheat is characterized by its unique property to form dough. This attribute basically depends on the structure and interaction between the storage proteins of the grain and the gluten fraction, which gives this grain an advantage over other cereals [2]. The industrial quality of wheat is evaluated mainly by the physical, chemical and rheological parameters related to its milling and bakery properties [5].

The latter conducted several investigations that focused on developing methods and evaluation techniques to increase the efficiency of the selection, processing, merchandising and end use of grains [6]. The study of the mechanical properties of wheat grains offers acceptable quality criteria during the milling process, specifically the conditioning parameters, the optimization of energy consumption and the quality of the produced flour.

The mechanical properties as "those related with the behavior of the material under the application of forces" [7], which are linked to the structure, physical state and rheological properties of solid and semisolid materials, which are, in turn, also related to the stress and strain parameters [8, 9]. The study of the mechanical properties of biological materials is complicated and is more complex when these properties are determined in kernels that have small dimensions or elaborate geometric shapes [10], materials that possess a heterogenic nature or those that exhibit viscoelastic behavior [11].

The evaluation of mechanical properties is carried out by using uniaxial compression tests, which require a deformation of the material that may or may not be permanent. Specifically for wheat kernels, the most important properties are the fracture strength and hardness, and both parameters are highly correlated with yield calculations and milling efficiency [12]. By applying tests to evaluate the fundamental mechanical properties of a material, it is possible to clarify more detailed aspects about the material that are not necessarily dependent on the sample geometry, loading conditions, or equipment used during the evaluation [13].

The aim of this chapter is to review the studies that have evaluated the advantages and disadvantages of the mechanical properties of intact wheat kernels by using compression tests and their relations to some aspects of quality.

2. Mechanical properties and compression tests on grains

The analyses performed on intact kernels represent a nondestructive, rapid, objective and complementary evaluation alternative to other methods. This method generates interesting information about the original state of raw materials and the possible functionality of the product [14]. The compression tests contribute substantially to the determination of the mechanical properties and other control aspects of the quality of the kernels, for example in grain flow systems. It is important to determine the material mechanical properties as well as the strain on individual kernels [15, 16].

The study of the stress-strain behavior in grains in their natural state is an interesting approach that provides measurements and objective data and could be used to improve specific processes applied to kernels [17]. However, when a kernel is cut or dissected, its structure and mechanical behavior are modified, particularly when the grain is small. To obtain reliable data, the mechanical properties of biomaterials must be evaluated by using intact kernels [11].

The determination of mechanical properties by compression tests comes from the information provided by a force-strain curve. Some concepts, such as the elastic limit, inflection point, yield point, rupture, apparent elasticity modulus, force and maximum resistance, can be realized from the curve. Those concepts are based on the application of quasi-static loads instead of impact loads [13, 18, 19]. Hooke's law does not apply during the compression of agricultural materials because these materials' properties are greatly affected by other factors, such as the moisture content and temperature [20].

The stress-strain response of grains to compression is determined by the following two aspects: (a) the shape and type of the applied compression (space between surfaces, velocity, sample orientation, among others) and (b) botanical differences in the kernel layers at the compression momentum [21]. The type of equipment, specific conditions of sample preparation, sample geometry and the test velocity must be considered during compression assays [19, 22].

The compression tests are commonly performed at a constant velocity or constant strain. The latter implies that strain increases due to the progressive reduction in the longitude (length), which means that if a fracture is present in the material, the stress-strain curve will show a tendency to decline, despite the deformation and the effect of the increase in velocity. On the contrary, if a contact mechanism dominates, the curve will shows an increase on the slope under a higher velocity, implying an increase in strain [23].

The mechanical behavior of kernels under load compression is time dependent, which means that their characterization should consider the principles of the viscoelastic theory. Viscoelastic materials exhibit the stress-relaxation phenomenon, mainly when sufficiently small stresses are applied. This behavior may be represented by the Maxwell's generalized model [7, 11, 18, 24], which is the Debora's number, a key of the dimensional groups [25].

The most efficient way to evaluate the stress-relaxation phenomenon was through the application of constant deformation (commonly by compression) and measuring the stress as a function of time. The tests at a constant deformation velocity allowed the simultaneous determination of the losses of the relaxation properties [26]. Starting from this point, Khazaei and Mann [27] established that the relaxation time determines how the material dissipates the stress after a rapid and sudden deformation was applied such that the result of the relaxation test could be useful in estimating the susceptibility of materials to damage. The evaluation of mechanical properties in grains is particularly difficult because these parameters may be seriously affected by the percentage of strain, temperature and moisture content [6].

3. Mechanical properties on intact kernel

3.1. Hardness

Hardness has been considered the principal quality parameter of wheat kernels. Hardness delimits the nature of the product that will be elaborated [28]; however, other mechanical properties are essential to design the equipment used in postharvest and processing operations [3]. Marzec et al. [29] defined hardness as the mechanical resistance of the caryopsides to be modified in their nature or as a property that is determined by the measure of their behavior during focal deformation. Grain hardness is governed by genetic factors, and the *locus hardness* (H_a) is located on the short arm of the 5D chromosome, which allows the association of this grain's characteristics and environmental conditions during plant growth [30–32].

Soft, hard and durum wheat are genetically considered qualitative classes, depending on the presence and nature of puroindolin [4]. In contrast, Gazza et al. [33] reported that hereditary factors and mechanical properties are related, though there were no correlations between grain hardness and the puroindolin levels in soft and hard wheat. Osborne et al. [34] noted that the hardness of soft wheat is related to the fracture along the interphase starch/protein, whereas in hard wheat, adhesion on the interphase starch/protein is higher and is located at the cell limits.

The endosperm is the largest component of grains (>80% of weight) and is 63–72% composed of starch granules [4]. Other studies [35–37] consider that the grain hardness is determined mainly by the degree of adhesion between the starch granules and the protein matrix, whereas the endosperm density is affected by the fracture force [38].

During wheat milling, the main parameter that influences the process quality is the difference between the mechanical properties of bran and endosperm; thus, after the bran is removed, less energy is required to continue the milling [39]. Both, the endosperm and bran affect grain hardness [40]. In addition to the properties of intact wheat kernel properties, physical properties of endosperm are associated to the way of how the kernels breaks [37, 41].

Hardness is the factor that mostly affected the mechanical behavior of grains. However, there is no universally accepted definition of hardness. Some authors describe it as an individual mechanical property of a single kernel or some endosperm fragments, but it is also known as the resistance to strain or breaking. The concepts and terms used to define hardness are very extensive, as are the methods used to measure the hardness, and it is thus difficult to evaluate [1]. There is little information about the hardness of wheat kernel and its relation to other fundamental mechanical properties [15].

Since the early 1900s, studies have focused on evaluating the hardness of wheat grain. Lai et al. [42] cited Robert, who in 1910, designed equipment to quantitatively measure the fracture force of individual kernels to detect differences among hardness levels. Later, researchers developed automated equipment (CASK-HaT) that was capable of measuring the stress-strain relation between compression and the fracture force of intact kernels, thus creating the ability to distinguish between mixes of soft and hard wheat with high reliability.

Pomeranz et al. [43] developed a semiautomatic feeder instrument to measure hardness characteristics that considered the moisture content and grain size. Individual kernels were compressed between two flat surfaces, which allowed the evaluation to differentiate the degree of hardness of soft and hard wheat mixes. Nevertheless, the authors concluded that the measurements and results were strongly affected by the heterogeneity among grains (size differences and moisture content). Over time, there have been multiple equipment designs for evaluating grain hardness by applying compression loads or methods supported by cutting and puncture tests [14, 44, 45].

3.2. Elastic modulus

In addition to hardness, the elastic modulus, or Young's modulus, is considered an interesting mechanical property due to its viscoelastic nature. Shelef and Mohsenin [46] performed an experiment in which four uniaxial compression methods were evaluated in the determination of the elastic modulus on individual wheat kernels. The results concluded that the data obtained were dependent on the method that was used. Moreover, the moisture content and kernel size significantly affected those measurements.

Afkari and Minaei [15] applied a quasi-static load (at a constant velocity) on soft and hard intact wheat kernels at two moisture content levels. The values of the apparent elastic moduli were higher in hard wheat varieties with a low moisture content. In the same way, from the data obtained from the stress-strain curves, it was possible to differentiate between wheat varieties, and the effects of the geometry and shape of the kernels were detected. Elbatawi and Arafa [28] performed a test of inflection on three points, and the force distribution on the rupture of wheat kernels was determined. It was also found that forces were significantly related not to the width and height of kernels but to their thickness. The values of the apparent elastic moduli, inflection force and fracture force increased as the moisture content decreased.

Khodabakhshian and Emadi [17] described the importance of evaluating the elastic modulus by using intact kernels. They considered that this parameter allowed the measurement of firmness (a texture attribute) and that it is possible to determine the fracture stress by using the elasticity theory. The indentation as an alternative method to compression with parallel plates to assess Young's modulus. They established that when a load was applied and released at the same velocity, it was possible to evaluate the viscoelasticity (elastic and plastic work) of materials [47].

3.3. Viscoelasticity

Ponce-García et al. [6] proposed and validated an innovative method of using the compression load to evaluate the viscoelastic properties by using intact soft, hard and durum wheat kernels at different moisture levels. The viscoelastic behavior of the wheat kernels was calculated using compression experiments. The total work of compression was determined from the loading/unloading curve (**Figure 1**), represented by the Maxwell's generalized model. The area under the loading curve gives the total work done (W_t) by the loading device during compression. The reversible elastic contribution (W_e) of the total work can be deduced from the area under the unloading curve, and the energy absorbed by plastic deformation (W_p) alone is the difference between the two ($W_p = W_t - W_e$). The results showed that the yield point presented a similar fracture force in all samples, independent of the wheat class and moisture level, whereas the tendency of the force-deformation curves allowed the determination that hard wheat kernels showed a higher plastic work than durum wheat and thus had a higher elastic behavior.

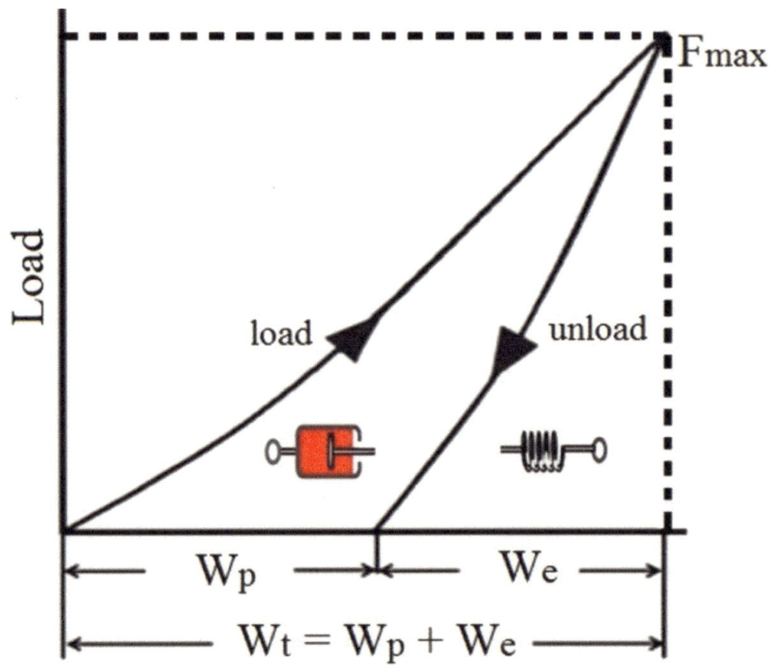

Figure 1. Compression cycle schematic. Abbreviations: F_{max}, maximum load; W_p, plastic work dissipated; W_e, elastic work performed during unloading; and W_t, total work. In Maxwell's generalized model, piston and spring represents dissipated and elastic work, respectively.

Using the latter method, Ponce-García et al. [48] evaluated the viscoelastic behavior of individual wheat kernels at two different moisture content levels through the uniaxial compression test under small strain (3%) to create experimental conditions that allowed the use of the elasticity theory to explain the viscoelasticity of wheat kernels. The results showed that this method could be a useful tool to distinguish among wheat classes, cultivars, and different moisture levels in the kernels.

The viscoelastic behavior of wheat that genotypically differs in the composition of their high molecular glutenins subunits (HMW-GS) was evaluated [49]. Those authors performed uniaxial compression tests and electrophoresis on polyacrylamide gels in presence of dodecyl sodium sulfate (SDS-PAGE) and reported an association among the viscoelastic properties of grains, the HMW-GS composition, and the viscoelastic parameters of gluten. The general tendency was that greater size and elastic work of the kernel corresponded to the best rheological quality of dough.

We are continuously looking to establish relations among protein composition, viscoelastic properties of dough, and bakery quality. In that sense, Figueroa et al. [50] conducted uniaxial compression and stress-relaxation tests on intact wheat kernels. The force-deformation curves showed at least two inflection points: the first was related to the mechanical properties of bran layers, and the second was related to the closer limit to the aleurone layer. It was concluded that bran layers showed a higher degree of elasticity than did those inside the endosperm.

The mechanical properties of wheat bran were significantly correlated with the elastic behavior of intact kernels and are related to the sedimentation volume, allelic composition of glutenin subunits, and elastic work. In the same manner, there was a directly proportional relation between the elastic modulus and kernel size. In general, kernels of hard wheat showed higher elasticity and better quality than soft wheat did [51–53].

3.4. Resistance to fracture

Several investigations have described the behavior of grains as a function of their resistance to fracture. Because of this description, the evaluation of mechanical properties of wheat kernels is strongly related to the milling processes.

Milling is a unitary operation that reduces the average volume of solid particles by dividing and/or fractioning a solid sample. The applied force depends on the magnitude, direction and velocity and allows the particles to absorb the force as a form of tension, which produces a deformation in their structure [54]. Nevertheless, when a limit is exceeded, the material is fractured, thereby provoking new surfaces. The amount of energy necessary for milling is a function of the initial and final size, the applied force and the characteristics of the raw material. Other factors that limit the reduction in size should be considered, such as the moisture and lipid contents and the sample geometry.

Gorji et al. [3] measured the resistance to fracture in terms of force and energy. The load application was carried out by placing the conditioned (tempered) grains at three moisture levels in vertical and horizontal positions and evaluating the grains at two load speeds. The force required to initiate the kernel fracture decreased when the moisture content increased. It was also established that grains in horizontal positions were more elastic and that the mechanical maximum force was associated with low moisture contents and lower test speeds.

In a similar study, Kalkan and Kara [55] determined the effect of the moisture content on individual wheat kernels submitted to *quasi-static* compression loads in two directions: (X-X) and (Y-Y). The force of rupture values in grains decreased with the increase in the moisture levels and was higher for the X-X orientation. In general, the strain of grains obtained by the rupture values showed irregular variations, independent of the moisture content.

Marzec et al. [29] developed a method that uses mechanical tests of uniaxial compression and acoustic emission to determine the quality parameters for wheat grains. The mechanical assays showed the prevalence of significant differences in almost all mechanical properties. The standard deviation was sufficiently wide to reveal the heterogeneity of the biological material, which is caused by the interactions that occur inside the endosperm between protein components and starch granules.

Finally, it can be established that studies that have reported on applying compression loads to intact wheat kernel make the measurement of mechanical properties possible, which have to be considered in future works to improve the quality and development of nondestructive methods in the evaluation of different agricultural products. It is worth noting that one of the problems affecting destructive evaluations of grains is the dissection and cutting of the sample, which make some evaluations impossible to perform. Non-destructive analyses on intact wheat kernels diminish those drawbacks and are quick tests [35].

4. Conclusion

Stress, strain, elastic modulus, stress deformation, viscoelasticity, fracture resistance and hardness are the principal mechanical properties evaluated in intact wheat kernels. Hardness clearly dominates due to its close relation with other quality characteristics of the kernels.

The moisture content is the main factor that affects the mechanical properties of grains. In general, the mechanical resistance and deformation capacity of wheat kernels decrease and increase, respectively, when the moisture content is higher. The latter applies because the energy absorption capacity is higher in moist grains than in dry grains, which incrementally increase the mechanical resistance during compression loading.

The separation of rheological parameters during kernel evaluations, such as the elastic, viscoelastic and viscous flow properties, allows the indirect measurement of wheat characteristics associated with the chemical composition (including nongluten components) and classes.

The evaluation of mechanical properties from applying compression loads on intact kernels provides important and useful information that could be considered to optimize and improve related aspects of postharvest and processing. The determination of mechanical properties should not be considered a unique method of evaluation to establish grain quality. However, it might be considered an alternative and helpful supplementary tool and complementary of other analyses.

Wheat kernels manifest a time-dependent behavior similar to other viscoelastic materials. Nevertheless, it is important to notice, at all possible times, the differences among the equipment, methods, moisture content and geometry of the grains during the development of tests.

Finally, to take advantage of the major benefits of the potential of raw materials in their natural state (intact kernel), the creation of a database that allows the evaluation of different wheat classes and cultivars according to their specific mechanical properties is recommended.

Author details

Néstor Ponce-García[1,2], Benjamín Ramírez-Wong[2*], Anayansi Escalante-Aburto[3,2], Patricia I. Torres-Chávez[2] and Juan de Dios Figueroa-Cárdenas[4]

*Address all correspondence to: bramirez@guaymas.uson.mx

1 Faculty of Agricultural Sciences, Autonomous University of Mexico State, Toluca, Estado de Mexico, Mexico

2 Research and Graduate Studies in Food Department, University of Sonora, Calle Rosales y Blvd, Sonora, Mexico

3 University of Monterrey, Health Sciences Division, Nutrition Department, Av. Ignacio Morones Prieto Poniente, San Pedro Garza Garcia, Nuevo Leon, Mexico

4 Center of Research and Advanced Studies (CINVESTAV-Queretaro), Fracc. Real de Juriquilla, Queretaro, Mexico

References

[1] Dziki D, Laskowski J. Wheat kernel physical properties and milling process. Acta Agrophysica. 2005; 6:59–71.

[2] Shewry PR. Wheat. Darwin review. Journal of Experimental Botany. 2009;60:1537–1553. DOI: 10.1093/jxb/erp058

[3] Gorji A, Rajabipour A, Tavakoli H. Fracture resistance of wheat grain as a function of moisture content, loading rate and grain orientation. Australian Journal Crop Science. 2010;4:448–452.

[4] Topin V, Radjai F, Delene JY, Mabille F. Mechanical modeling of wheat hardness and fragmentation. Powder Technology. 2009;190:215–220. DOI: 10.1016/j.powtec.2008.04.070

[5] Espitia RE, Villaseñor MHE, Peña BRJ, Huerta EJ, Limón OA. industrial quality of 28 Mexican hard seasonal wheat. II. Genetic variability and quality selection criteria. Magazine of Mexican Phytotechnique.2004;27:41–47.

[6] Ponce-García N, Figueroa JDC, López-Huape GA, Martínez HE, Martínez-Peniche R. Study of viscoelastic properties of wheat kernels using compression load method. Cereal Chemistry. 2008;85:667–672. DOI: dx.doi.org/10.1094/CCHEM-12-12-0166-R

[7] Mohsenin NN. Physical Properties of Plant and Animal Materials. Structure, Physical Characteristics and Mechanical Properties. 2nd ed. New York: Gordon and Breach Science Publishers; 1970. 17 p.

[8] Gates FK, Dobraszczyk BJ. Mechanical properties of oats and oats products. Review article. Agricultural and Food Science. 2004;13:113–123.

[9] Barbosa-Cánovas GV, Juliano P, Peleg M. Engineering properties of foods. In: Barbosa-Cánovas GV, editor. Encyclopedia of Life Support Systems (EOLSS). Developed under the Auspices of the UNESCO. Oxford: EOLSS Publishers; 2006. pp. 50–69.

[10] Shitanda D, Nishiyama Y, Koide S. Compressive strength properties of rough rice considering variation of contact area. Journal of Food Engineering. 2002;53:53–58. DOI: 10.1016/S0260-8774(01)00139-X

[11] Singh SS, Finner MF, Rohatgi PK, Buelow FH, Schaller M. Structural and mechanicals properties of corn kernels: a hybrid composite material. Journal of Materials Science. 1991;26:274–284. DOI: 0022-2461/91

[12] Mabille F, Abecassis J. Parametric modelling of wheat grain morphology: a new perspective. Journal of Cereal Science. 2003;37:43–53. DOI: 10.1006/jcrs.2002.0474

[13] Cenkowski S, Bielewicz J, Britton MG. A single kernel creep and recovery test. Transactions of the ASAE. 1991;34:2484–2490. DOI: 0001-2351/91/3406-2484

[14] Eckhoff SR, Supak WA, Davis AB. A rapid single-kernel wheat hardness tester. Cereal Chemistry. 1988;65:503–508.

[15] Afkari SAH, Minaei S. Behavior of wheat kernels under quasi-static loading and its relation to grain hardness. Journal of Agricultural Science Technology. 2004;6:11–19.

[16] Raji AO, Favier JF. Model for the deformation in agricultural and food particulate materials under bulk compressive loading using discrete element method. I: theory, model development and validation. Journal of Food Engineering. 2004;64:359–371. DOI: 10.1016/j.jfoodeng.2003.11.004

[17] Khodabakhshian R, Emadi B. Determination of modulus of elasticity in agricultural seeds on the basis of elasticity theory. Middle-East Journal of Scientific Research. 2011;7:367–373.

[18] Bargale PC, Irudayaraj J. Mechanical strength and rheological behavior of barley kernels. International Journal of Food Science and Technology. 1995;30:609–623. DOI: 10.1111/j.1365-2621.1995.tb01409.x

[19] ASAE Standard. American Society of Agricultural Engineers. Compression test of materials of convex shape. Transactions of the ASAE. 2003;3:595–603.

[20] Delwiche SR. Wheat endosperm compressive strength properties as affected by moisture. Transactions of the ASAE. 2000;43:365–373. DOI: 10.13031/2013.2713

[21] Anderssen RS, Haraszi R. Characterizing and exploiting the rheology of wheat hardness. European Food Research and Technology. 2009;229:159–174. DOI: 10.1007/s00217-009-1037-9

[22] Timbers GE, Stanley LM, Watson EL. Determining modulus of elasticity in agricultural products by loaded plungers. Agricultural Engineering. 1965;46:274–275.

[23] Peleg M. Contact and fracture elements as components of the rheological memory of solid foods. Journal of Texture Studies. 1977;8:67–76. DOI: 10.1111/j.1745-4603.1977.tb01164.x

[24] Chen P, Fridley RB. Analytical method for determining viscoelastic constants of agricultural materials. Transactions of the ASAE. 1972;15:1103–1106. DOI: 10.13031/2013.38080

[25] Borwankar RP. Food texture and rheology: a tutorial review. Journal of Food Engineering. 1992;16:1–16. DOI: 10.1016/0260-8774(92)90016-Y

[26] Waanenen K, Okos MR. Stress-relaxation properties of yellow-dent corn kernels under uniaxial loading. Transactions of the ASAE. 1992;35:1249–1258. DOI: 10.13031/2013.28727

[27] Khazaei J, Mann DD. Effects of moisture content and number of loadings on force relaxation behaviour of chickpea kernels. International Agrophysics. 2005;19:305–313.

[28] Elbatawi IE, Arafa GK. Factors affecting breaking force distribution of wheat kernel before milling. Misr Journal of Agricultural Engineering. 2008;25:1004–1025.

[29] Marzec A, Cacak-Pietrzak G, Gondek E. Mechanical and acoustic properties of spring wheat versus its technological quality factors. Journal of Texture Studies. 2011;42:319–329. DOI: 10.1111/j.1745-4603.2011.00284.x

[30] Greffeuille V, Abecassis J, Rousset M, Oury FX, Faye A, Bar LC, Lullien-Pellerin V. Grain characterization and milling behaviour of near-isogenic lines differing by hardness. Theoretical and Applied Genetics. 2006;1141:1–12. DOI: 10.1007/s00122-006-0403-2

[31] Lillemo M, Chen F, Xia X, William M, Peña RJ, Trethowan R, He Z. Puroindoline grain hardness alleles in CIMMYT bread wheat germplasm. Journal of Cereal Science. 2006;44:86–92. DOI: 10.1016/j.jcs.2006.03.004

[32] Véha A. Correlation between the kernel structure and the quality parameters on some Hungarian winter wheat varieties. Cereal Research Communications. 2007;35:1289–1292. DOI: 10.1556/CRC.35.2007.2.281

[33] Gazza L, Taddei F, Corbellini M, Cacciatori P, Pogna NE. Genetic and environmental factors affecting grain texture in common wheat. Journal of Cereal Science. 2008;47:52–58. DOI: 10.1016/j.jcs.2007.01.004

[34] Osborne BG, Henry RJ, Southan MD. Assessment of commercial milling potential of hard wheat by measurement of the rheological properties of whole grain. Journal of Cereal Science. 2007;45:122–127. DOI: 10.1016/j.jcs.2006.07.005

[35] Nielsen JP, Pedersen DK, Munck L. Development of nondestructive screening methods for single kernel characterization of wheat. Cereal Chemistry. 2003;80:274–280. DOI: 10.1094/CCHEM.2003.80.3.274

[36] Greffeuille V, Mabille F, Rousset M, Oury FX, Abecassis J, Lullien-Pellerin V. Mechanical properties of outlayers from near-isogenic lines of common wheat differing in hardness. Journal of Cereal Science. 2007;45:227–235. DOI: 10.1016/j.jcs.2006.09.002

[37] Hrusková M, Švec I. Wheat hardness in relation with other quality factors. Czech Journal of Food Science. 2009;4:240–248.

[38] Dobraszczyk BJ, Whitworth MB, Vincent JFV, Khan AA. Single kernel wheat hardness and fracture properties in relation to density and the modelling of fracture in wheat endosperm. Journal of Cereal Science. 2002;35:245–263. DOI: 10.1006/jcrs.2001.0399

[39] Peyron S, Chaury M, Rouau X, Abecassis J. Relationship between bran mechanical properties and milling behavior of durum wheat. Influence of tissue thickness and cell wall structure. Journal of Cereal Science. 2002;36:377–386. DOI: 10.1006/jcrs.2002.0473

[40] Dziki D, Laskowski J. Influence of wheat grain mechanical properties on grindingenergy requirements. TEKA Commission of Motorization and Energetics in Agriculture. 2006;6A:45–52.

[41] Evers T, Millar S. Cereal grain structure and development: some implications for quality. Journal of Cereal Science. 2002;36:261–284. DOI: 10.1006/jcrs.2002.0435

[42] Lai FS, Rousser R, Brabec D, Pomeranz Y. Determination of hardness in wheat mixtures. II. Apparatus for automated measurement of hardness of single kernels. Cereal Chemistry. 1985;62:178–184.

[43] Pomeranz Y, Martin CR, Rousser R, Brabec D, Lai FS. Wheat hardness determined by a single kernel compression instrument with semiautomated feeder. Cereal Chemistry. 1988;65:86–94.

[44] Katz R, Cardwell AB, Collins ND, Hostetter AD. A new grain hardness tester. Cereal Chemistry. 1959;36:393–401.

[45] Glenn GM. Method for classifying wheat kernels as hard or soft. US pat N°5,056,721. 1991.

[46] Shelef L, Mohsenin NN. Evaluation of modulus of elasticity of wheat grain. Cereal Chemistry. 1967;44:392–402.

[47] Gubicza J, Juhász A, Arató P, Szommer P, Tasnádi P, Vörös G. Elastic modulus determination from depth sensing indentation testing. Journal of Materials Science Letters. 1996;15:2141–2144. DOI: 10.1007/BF00241151

[48] Ponce-García N, Ramírez-Wong B, Torres-Chávez PI, Figueroa-Cárdenas JD, Serna-Saldívar SO, Cortez-Rocha MO. Effect of moisture content on the viscoelastic properties of individual wheat kernels evaluated by the uniaxial compression test under small strain. Cereal Chemistry. 2013;90:558–563. DOI: 10.1094/CCHEM-12-12-0166-R

[49] Figueroa JDC, Maucher T, Reule W, Peña RJ. Influence of high molecular weight glutenins on viscoelastic properties of intact wheat kernel and relation to functional properties of wheat dough. Cereal Chemistry. 2009:86:139–144. DOI: 10.1094/CCHEM-86-2-0139

[50] Figueroa JDC, Hernández ZJE, Véles JJ, Rayas-Duarte P, Martínez-Flores HE, Ponce-García N. Evaluation of degree of elasticity and other mechanical properties of wheat kernels. Cereal Chemistry. 2011a;88:12–18. DOI: 10.1094/CCHEM-04-10-0065

[51] Figueroa JDC, Peña RJ, Maucher T, Rayas-Duarte P, Khan K. Kernel elastic properties and sedimentation: influence of high and low molecular weight glutenin allelic composition. Cereal Chemistry. 2011b;88:41–44. DOI: 10.1094/ CCHEM-10-09-0142

[52] Figueroa JDC, Peña RJ, Rayas-Duarte P, Khan K. Influence of low molecular weight glutenin subunits on wheat kernel elasticity and sedimentation volume. Cereal Research Communications. 2011c;39:237–245. DOI: 10.1556/CRC.39.2011.2.7

[53] Figueroa JDC, Ramírez-Wong B, Peña J, Khan K, Rayas-Duarte P. Potential use of the elastic properties of intact wheat kernels to estimate millings, rheological and bread-making quality of wheat. In: Chibbar RN, Dexter J, editors. Wheat Science Dyamics: Challenges and Opportunities. Jodhpur: Agrobios (International); 2011d. pp. 317–325.

[54] Cortazar-Figueroa LM, Meléndez-Pérez R, Oliver-Hernández DM. Energy consumption and particle size distribution on cinnamon (*Cinnamomum zeylanicum*) and black pepper (*Piper niger L.*) milling. Mexican Magazine of Chemical Engineering.2008;7:123–130.

[55] Kalkan F, Kara M. Handling, frictional and technological properties of wheat as affected by moisture content and cultivar. Powder Technology. 2011;213:116–122. DOI: 10.1016/j.powtec.2011.07.015

Chapter 3

Viscoelastic Behavior of Liquid-Crystal Polymer in Composite Systems

Andreea Irina Barzic and Silvia Ioan

Additional information is available at the end of the chapter

http://dx.doi.org/10.5772/64074

Abstract

Some basic concepts regarding the viscoelastic properties of polymers in solution phase, highlighting the effects of the some experimental variables, such as system composition, temperature or time, are reviewed. The main attention is given to experiments involving shear deformation of polymer solutions, since it provides the most significant data concerning the optimal processing of final product. The response of polymers to a mechanical perturbation involves several types of molecular motions, which are reflected in the microstructure changes. The elastic constants, which determine the motion of macromolecules, become anisotropic in case of liquid-crystal polymer (LCP). The state of art concerning the viscoelastic behavior of liquid crystal polymers in composite systems is described. The main aspects that are discussed include the effects on the viscoelasticity given by molecular orientation, system composition and interactions. Finally, some particular cases are presented to illustrate the potential applications of these principles to practical problems in the processing and use of the described materials.

Keywords: viscoelasticity, liquid crystals, applications, polymers, solutions

1. Introduction

The development of new products for current technologies demands close investigation of the properties variations in the presence of external forces. The mechanical performance arises from the viscoelastic features of the material [1]. In other words, mechanical features are time dependent, whereas perfectly elastic deformation and perfectly viscous flow are idealizations that are approximately reached in some limited conditions. Viscoelastic materials present, under mechanical stress, combined characteristics of these two behaviors, a "fading memory", partial recovery, energy dissipation, etc. This type of dependence can be linear (stress – strain dependence is a straight line with a single slope) or nonlinear (stress – strain dependence presents many slopes) [2, 3].

Among the compounds with pronounced viscoelasticity, polymers are the most important in rubber and plastic industries [4, 5]. Their viscoelastic properties influence not only the mechanical reliability of the final products of these industries, but also the efficiency of processing methods at intermediate stages of production. In most cases, the designer aims to maintain the elastic deformations under a critical limit. The peculiar viscoelastic features of polymers arise from the complex dynamics of flexible chains [6]. For instance, above the glass transition temperature (T_g), the response of these materials to mechanical perturbation forces entails several types of molecular motions. The large freedom in the spatial arrangement of macromolecular compounds allows various types of motion according to the time and spatial scales. At big scales, the polymer chains present unique dynamic features that are not found in low-molecular-weight materials. These characteristics deal with motion in scales above the monomer size and are similar for polymer chains of various structures. The movement of macromolecules is reflected in various kinds of dynamic properties, including viscoelastic (rheological), dielectric and diffusion ones [6].

On the other hand, the special response to the deformation of macromolecular compounds arises from the fact that they have time- and temperature-dependent modulus, while for metals and ceramics, this property can be assumed to be constant at room temperature. The phenomenological theory of linear viscoelasticity of polymers is semiquantitatively clarified from the point of view of the effects induced by the molecular weight, temperature, concentration and other parameters [7]. However, there are still some aspects that are not entirely understood, such as the impact of molecular weight distribution or crystalline character. While not all polymers are viscoelastic to any important practical extent, the linear viscoelastic theory gives an engineering approximation for many applications in polymer and composites engineering. Even in cases that demand more elaborate approaches, the linear viscoelastic theory is a useful starting point in optimizing the mechanical performance of products containing liquid-crystal components that are used in certain devices. Particularly, liquid-crystal polymers (LCPs) present an interesting viscoleastic behavior [8, 9], which in combination with other materials can lead to composite systems with enhanced properties as demanded in modern applications [10–12].

This chapter describes the state of art in the field of LCPs in composite systems, highlighting the importance of viscoelasticity in designing advanced materials for recent technologies. Some fundamental notions, concerning the response given by these compounds under external fields, are discussed. The main attention is focused on experiments involving shear deformation of polymer solutions; generally, this is the state from which most materials are processed. The changes in viscoelastic characteristics after reinforcement of LCPs are reviewed in regard with some parameters, such as system composition, time or temperature. The practical importance of composite systems with LCPs phase is analyzed in agreement with the current demands on various industries.

2. Fundamentals on viscoelasticity in shear experiments

2.1. Viscoelastic behavior of polymers

Viscoelastic response is often used as a probe in polymer science, since it is sensitive to the material's chemistry and microstructure [13]. The response of such compounds with a viscoelastic component is complex and depends on several aspects of both experimental and structural nature.

Regarding the experimental procedure, it is known that the strain can be directly proportional to stress at any time and, in a similar manner, the stress at any particular moment in time is directly proportional to the strain depending upon which of these are considered the stimulus and corresponding response. This is generally noticed in the case of small stresses or strains. However, nonlinearities appear in the viscoelastic behavior because at increased strains the response does not range proportionally with applied field. Linear viscoelasticity can be included in general theory of mechanics of materials, and thus, macromolecular structures with viscoelastic components can be analyzed. In the dependence of modulus on normalized temperature (T/T_g), five regimes of deformation are noticed, particularly for linear amorphous compounds [14]:

- glassy regime with a large modulus (>3GPa);
- glass-transition domain in which the modulus drops from 3 GPa to 3 MPa;
- rubbery plateau with a low modulus around 3 MPa;
- viscous region when the polymer begins to flow;
- decomposition regime in which chemical breakdown occurs.

Concerning the structural aspects, one may notice that application of a stress to a polymer can determine two different atomistic mechanisms of deformation [14]. The first one is related to the distortion of the lengths and angles of the chemical bonds that move the atoms to new positions of bigger internal energy. Such motions are small and require around 10^{-12} s. The second mechanism arises from the degree of flexibility of the polymer chains that lead to large-scale re-arrangements of atoms. Facile rotation of simple carbon bonds around the main chain can induce changes in the conformation of the molecule. For structures with high mobility, the polymer can extend itself along the applied stress, generating a reduction in conformational entropy.

On the other hand, the rearrangement of the macromolecular chains after cessation of external deformation force is reflected in different relaxation time scales. At the length scale of the repeating unit, this repositioning of chains may be very fast, involving cooperativity in the conformational transitions and even as the relaxation propagates along the backbones. This implicates a growing number of segments as the time passes. At very long times, disentanglements occur and the corresponding relaxation time is long and dependent on molecular weight and molecular architecture of the system. The disentanglements strongly influence the flow properties. Considering all these aspects, the molecular response of polymers to deformation is complex, exhibiting a wide distribution of relaxation times that extend significantly in the time or frequency domain. The elastic character is mainly viewed at short time or long frequency, while at long time or short frequency, the viscous feature is prevalent.

To obtain information relevant to actual in-use conditions of viscoelastic polymer materials, mechanical characterization is often performed. Mostly, polymers are processed from solution phase or melt. The phenomena observed in such polymer systems as a result of viscoelastic deformation are mainly examined by means of dynamical mechanical spectroscopy (DMS) [15]. This is one fundamental rheological technique used to evaluate the viscoelastic properties of complex fluids and to investigate their microstructure [16]. For polymers in these states, the investigations are modified so as to enable observation of the time dependency of the material response. Although many such "viscoelastic tensile tests" have been used, one mainly encounters the following three tests:

- *Creep*: Such analyses allow measuring the time-dependent strain resulting from the application of a steady uniaxial stress. The test entails applying a constant shear stress onto a sample and observing the resulting elastic deformation and/or viscous flow. If the material has a linear behavior, stress-strain curve will be a straight line, with a slope that increases as the chosen time is decreased. The test enables to determine the following aspects:

  – a high-quality zero-shear viscosity value—by the sustained application of a shear stress that is taken at the linear viscoelastic limit for the sample;

  – the product behavior during long-term exposure to high level of stress that is still below the yield strength of the material, for example in draining, sagging, sedimentation or leveling cases;

  Creep compliance is the ratio of strain to stress in the case of time-varying strain arising from a constant stress. The dependence of creep compliance on time is described by various domains (**Figure 1a**), namely

  – Glassy: where the compliance is that of a glassy solid at short times, the latter variable having no influence;

  – Viscoelastic: observed at intermediate times, where compliance is time dependent;

  – Rubbery and flow: occurring at very long times, where compliance is typical for a rubber like solid and is independent of time;

- *Stress relaxation*: It consists of monitoring the time-dependent stress resulting from a steady strain. When instantaneous strain is applied to an ideal elastic solid, finite and constant stress will be obtained. For a linear viscoelastic solid, the instantaneous stress will be proportional to imposed strain and will decrease with time. For amorphous linear polymers at high temperatures, the stress may decay to zero, whereas in the absence of viscous flow, the stress decays to a finite value. Stress relaxation modulus dependence on time presents various regions (**Figure 1b**):

  – Glassy: found at very short times and the relaxation modulus is independent of time;

– Viscoelastic: placed at intermediate times, where relaxation modulus is time dependent;

– Rubbery and flow: noticed at very long times, relaxation modulus is similar to a rubber like solid and is independent of time;

- *Dynamic (sinusoidal) loading*: Both previous tests are convenient for the investigation of material response at long times (minutes to days), but less accurate at shorter times (seconds and less). Dynamic tests consist in determining the stress (or strain) resulting from a sinusoidal strain (or stress). This is more adequate for filling out the "short-time" range of polymer response. Application of sinusoidally stress to viscoelastic polymers determines the occurrence of a steady state. The recorded strain is also a sinusoidal signal with similar angular frequency but phase-shifted in regard with the imposed stress. This is analogous to the delayed strain observed in creep experiments. The strain lags the stress by the phase angle and this is true even if the strain rather than the stress is the controlled variable.

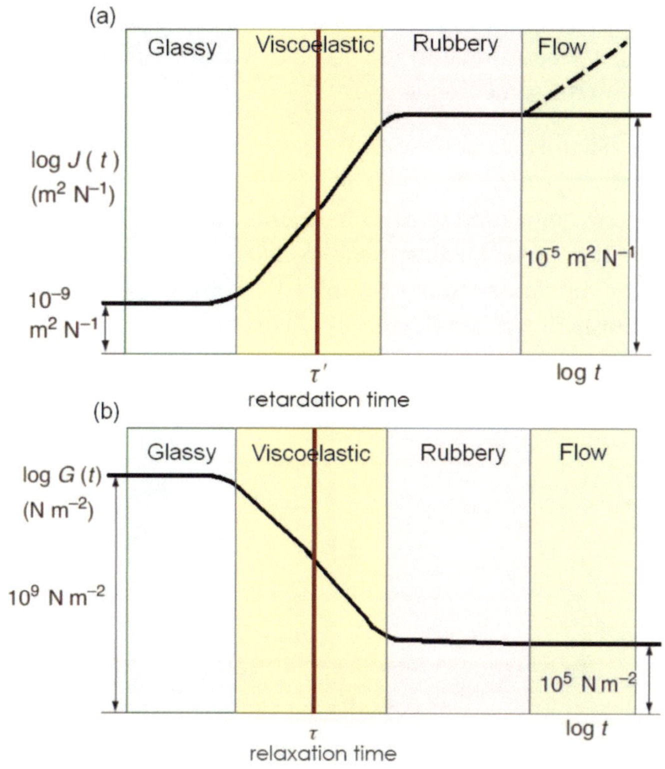

Figure 1. The dependence of creep compliance (a) and stress relaxation modulus (b) on time reflecting various behaviors of polymer systems.

In the following paragraphs, particular attention will be given to viscoelastic characterization by means of oscillatory rheology testing. The sinusoidal shear deformation, $\gamma(t)$, applied to polymer materials during such experiments, is described by the following relation (1):

$$\gamma(t)=\gamma_0 \sin(\omega t) \tag{1}$$

where γ_0 is the amplitude of applied strain and ω is the frequency of oscillation.

The shear rate, namely rate at which a progressive shearing deformation is applied to a material, is given by the relation (2):

$$\dot{\gamma}(t)=\gamma_0 \omega \cos(\omega t) \tag{2}$$

The material stress response is measured and, by the utilization of adequate theoretical models, it is correlated with the features of the polymer system under analysis. There are two different types of rheology tests, depending on the amplitude of the applied deformation [16–18]:

- Small amplitude oscillatory shear (SAOS): It is a robust, widely used rheology technique, where the amplitude of applied strain is small enough to induce a linear response of the sample. The stress is proportional to the oscillatory deformation and is represented by a sinusoidal wave with the same frequency;
- Large amplitude oscillatory shear (LAOS): It is a recent method involving large deformation to give a nonlinear response. The stress is still periodic, but not sinusoidal and can be deconvoluted in terms of a fundamental harmonic and its odd multiples. Stress is often examined in the Fourier space leading to the so-called Fourier transform rheology. Nonlinear experiments are, obviously, more complex than linear ones. However, they provide supplementary information about the morphology of microstructured polymer systems.

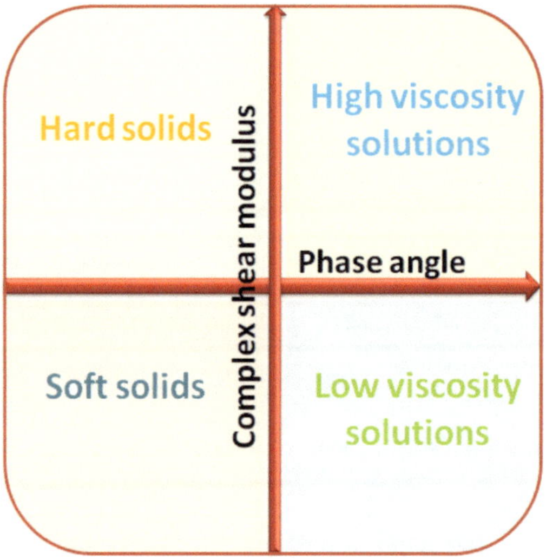

Figure 2. The viscoelastic map constructed based on phase angle and complex modulus.

Oscillatory shear testing is useful to analyze and quantify the rigidity and integrity of a polymer's internal structure as a result of a variety of phenomena, for example, flocculation and interaction of dispersed particles or droplets, or compatibility, crosslinking and entanglement of dissolved macromolecules. In most cases, oscillation rheology experiments are performed at very low applied stresses and strains, often significantly below the yield point of the polymer material. Typically, measured parameters include complex modulus (G^*), elastic (or storage) modulus (G') and viscous (or loss) modulus (G''), phase angle (δ) and tangent of the phase angle ($\tan \delta$). The viscoelastic map can be constructed based on phase angle and complex modulus (see **Figure 2**) thus allows one to distinguish between elastic solids and viscous liquids (left to right) and high to low rigidity or viscosity (top to bottom).

2.1.1. *Linear viscoelastic behavior*

Linear properties of viscoelastic polymer solutions have been investigated along with the mechanical behavior in this regime. Fundamentals on these aspects are documented in several textbooks [19–21]. SAOS measurements have been employed to characterize a very large number of complex fluids from polymer melts [22] to liquids crystalline [23] and from gels [24] to polymer blends [25, 26].

The linear domain of viscoelastic behavior can be easily determined from oscillation stress sweeps and strain sweep tests (**Figure 3**). Oscillation stress and strain sweeps offer easy-to-interpret information about the soft-solid rigidity and yield stress [4, 6]. These tests provide a deep insight into the rigidity and strength of the soft-solid polymer structures that render basic quality attributes in many manufactured products. The oscillation stress or strain sweep measurements consist in exposing the sample to small amplitude oscillatory (i.e., clockwise then counter clockwise) shear. In the beginning stages of the test, stress is sufficiently low to maintain structure. However, as the test continues the incrementing applied stress causes the ultimate disruption of structure—the yield process. The resulted structural change is manifested as a reduction of elasticity (phase angle increases) and an accompanying decrease in rigidity (complex modulus).

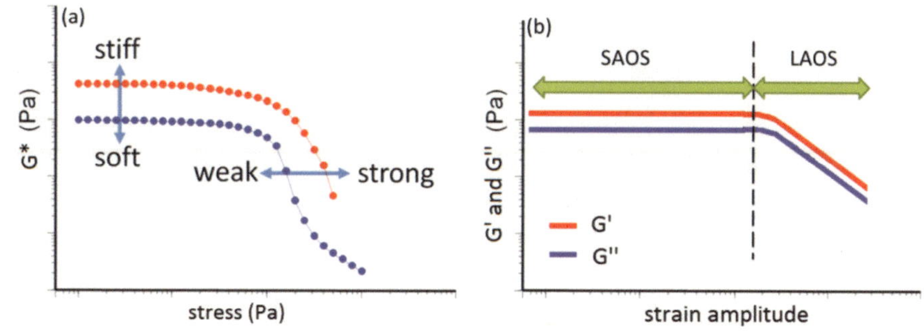

Figure 3. Schematic representation of (a) typical oscillation stress sweeps test and (b) strain sweep test at a fixed frequency, the latter determining the linear and nonlinear viscoelastic regions.

The typical shear stress response of a SAOS test is given as follows:

$$\sigma(t)=\sigma_0 \sin(\omega t+\varphi) \tag{3}$$

The frequency of oscillation (ω) of stress does not change in regard with the imposed deformation, but it is shifted by an angle $\varphi \in [0, \pi/2]$. The value of φ is equal to 0 for elastic solids, and hence, the stress is not phase shifted in relation to the strain. For viscous liquids, the value of φ is equal to $\pi/2$; therefore, in this case, the stress is in phase with the shear rate. In the linear regime, the phase angle is mainly influenced by strain frequency, while amplitude of applied strain is not reflected in the elastic or viscous components [4, 17]. Equation (3) is usually written in the following way:

$$\sigma(t)=\gamma_0\left[G'\sin(\omega t)+G''(\cos(\omega t)\right] \tag{4}$$

where:

$$G'=\frac{\sigma_0}{\gamma_0}\cos(\varphi) \tag{5}$$

$$G''=\frac{\sigma_0}{\gamma_0}\sin(\varphi) \tag{6}$$

The term G' is in phase with the strain, while the term G'' is in phase with the shear rate.

Oscillatory frequency sweep allows evaluating the nature of the structuring mechanisms present in a polymer solution or melting. The sample is subjected to small-deformation oscillations of various frequencies to estimate the structural response to shear of longer or shorter timescales. The technique can differentiate between the "relaxable" structures and those characterized by permanent elasticity, being a useful tool when trying to match textures and flow behaviors in thickened polymer systems [4]. **Figure 4** displays some typical frequency sweep tests for polymer samples.

As stated earlier, in linear regime, the amplitude of the stress is proportional to the amplitude of the imposed deformation, consequently G' and G'' do not depend on amplitude of applied strain, but on frequency only. The storage and loss modulus for typical "liquid-like" and "solid-like" fluids are shown in **Figure 4**. For the "liquid-like" sample, the elastic modulus is much lower than the viscous one and it has a frequency dependence of type: $G' \propto \omega^2$, whereas the viscous modulus is linear in frequency, namely $G'' \propto \omega^1$ [4]. In materials with "solid-like" behavior, like polymer gels, the elastic shear modulus overcomes the viscous one, and almost no dependence on angular frequency is observed. Real complex fluids often present an intermediate viscoelastic behavior.

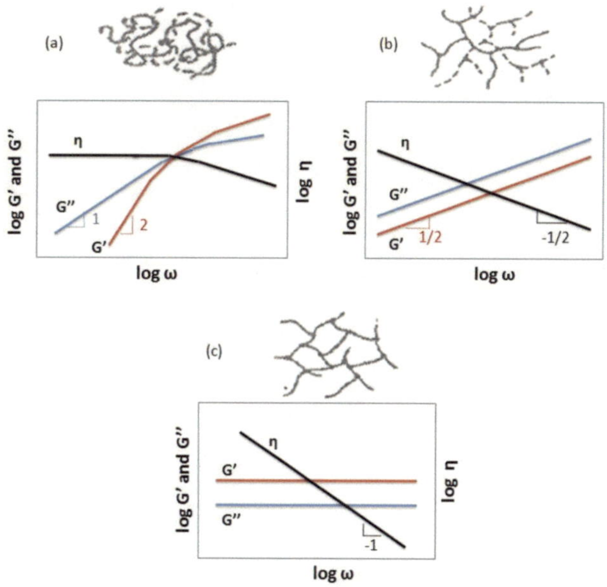

Figure 4. Typical frequency sweeps for polymer solutions (liquid-like) or melts (a), gelation or (b) and curing finished or "solid-like" materials.

2.1.2. Nonlinear viscoelastic behavior

Nonlinear viscoelastic behavior of polymer systems is intrinsically more complex than linear one, from both the experimental and the theoretical point of view. The interest for this aspect of viscoelasticity is very high because many polymeric systems are processed in the nonlinear regime. Therefore, LAOS experiments can provide essential information on the material characteristics in that specific range. Furthermore, nonlinear experiments are able to describe in detail the complex fluids microstructure. The first experimental works on nonlinear measurements were reported in the seventies [27], but only in the last two decades, they are significantly increased [28–30], owing to the significant development of experimental devices and software programs.

The shear stress response of a LAOS test is still periodic, with the same frequency ω of the imposed oscillation, but not sinusoidal [17]. It can be described by a Fourier series of odd harmonics:

$$\sigma(t)=\sum_{\substack{k=1\\k,odd}}^{\infty} a_k\cos(k\omega t)+\sum_{\substack{k=1\\k,odd}}^{\infty} b_k\sin(k\omega t) \tag{7}$$

where a_k and b_k represent the amplitude of the cosine and sine terms of the *k*-th harmonic and depend on both frequency and strain amplitude. They can be obtained as follows:

$$a_k = \frac{2}{T}\int_0^T \sigma(t)\cos(k\omega t)dt \tag{8}$$

$$b_k = \frac{2}{T}\int_0^T \sigma(t)\sin(k\omega t)dt \tag{9}$$

where T is the period of the imposed strain oscillation.

The utilization of the Euler notation leads to another form of Eq. (7):

$$\sigma(t) = \sum_{\substack{k=-\infty \\ k,odd}}^{\infty} I_{k\omega} e^{jk\omega t} \tag{10}$$

$$I_k = \frac{2}{T}\int_0^T \sigma(t) e^{-jk\omega t} dt \tag{11}$$

In expression (10), the term $Ik\omega$ is the complex coefficient of the k-th harmonic in the Fourier domain, $I_{k\omega} = a_k - jb_k$. It should be remarked that shear stress is an odd function of the strain. Therefore, only odd terms of the Fourier series are taken in consideration for in Eqs. (7) and (10).

An important aspect that one should remark is that during a LAOS test, the system is deformed with a single frequency, while the stress presents more than one frequency, so the collected signal expresses several time scales, thus providing detailed data about the analyzed polymer material. LAOS technique involves the following aspects [17]:

- a sophisticated experimental apparatus;
- an elaborated data manipulation;
- a theoretical model that allows interpretation of nonlinear results.

The first two points are widely analyzed in the literature [28, 31], while the third point is less investigated [30]. The fundamental modifications of traditional rotational rheometer to correctly implement the nonlinear DMS are reported [31]. It is essential to improve the signal-to-noise ratio of the achieved experimental signals. If appropriate practical contrivances are made, one may obtain highly sensitive detection. Since experimental data are analyzed in the Fourier domain, the technique is usually named Fourier transform rheology (FTR).

There are many advantages concerning data analysis in the Fourier rather than in the time domain. First of all, the appearance of nonlinearity is better noticed [32]. FTR can detect the nonlinear contribution at about 10 times lower strain amplitude compared to linear oscillatory measurements [31]. The resulted files of the experimental data concern only the intensity of the odd peaks, and therefore, they require a low amount of memory on the computer.

Furthermore, the Fourier coefficients are less affected by experimental noise. Generally, only the third and fifth overtones and, in particular, their absolute value are analyzed. There are not many reports concerning the separation of the real and imaginary part of the harmonics [30, 32]. The fundamental lack of FTR is the absence of theoretical models, which can correlate experimental points with the morphology of the polymers.

Regardless the viscoelastic domain in which experiments are performed, there are several inherent parameters that influence the viscoelastic behavior of polymeric systems. The primary ones are the following:

- chemical structure;
- molecular architecture;
- molecular weight and crosslinking;
- copolymers and blends;
- effect of plasticizers;
- molecular orientation;
- fillers and fibers.

2.2. Viscoelastic behavior of LCPs

The linear viscoelastic response to shear deformation was investigated for many LCPs by checking the dependence of rheological moduli on frequency. The range of strain amplitudes corresponding to linear viscoelastic domain is broad for side-chain LCPs in nematic phase, similarly to common flexible polymers. This is not maintained for main chain nematic polymers. For those in smectic phase, the range of linear viscoelasticity is narrower.

Viscoelastic behavior of LCPs depends on temperature at which the sample is subjected during oscillatory testing. For example, aqueous solutions of methylcellulose present changes in elastic modulus during the heating process from 20 to 80°C that consist in several stages, depending on concentration [33]. In the range of 20–35°C, the storage modulus increases slowly. In the following stage, G' has a more pronounced increase with temperature to about 50°C, and then, the rate of increase slows down. Above about 60°C, the polymer solutions with different concentrations show different rheological behaviors. For the samples with lower concentrations (c<4.0 g/L), elastic modulus eventually reaches a plateau value. For more concentrated solutions of 8 and 10 g/L, a third remarkable increase in G' is noticed as a result of formation of a mature gel. As for the solutions with medium concentrations, 4–6 g/L, a third appreciable increase in G' is remarked at about 66°C as a result of the occurrence of a weak gel. At higher temperatures (>65°C), the strength of the gel increases with the concentration. However, while the values of storage modulus increase monotonically with temperature, the values of this parameter do not increase monotonically as a function of concentration, as expected for normal polymer solutions. A discontinuity of change in G' at several temperatures with the concentration of cellulosic polymer in solution occurs. A similar behavior was reported for other LCP solutions in lyotropic phase [34].

The viscoelastic properties of thermotropic LCPs that have mesogenic pendant groups to flexible main chain were found to be sensitive to smectic-nematic and smectic isotropic transitions. However, they are less affected by nematic isotropic transition as time-temperature superposition applies across this transition. In contrast to main chain LCPs, the nematic side chain ones present linear viscoelastic behavior over a large range of strain amplitudes that is independent of thermal and shear histories. At very low frequencies below than the reverse of diffusion time, side chain LCPs in their nematic state have a terminal response, which is typical for viscoelastic liquids. In their smectic phase, they are still viscoelastic in the same frequency domain and approach the terminal response of a viscoelastic solid at the lowest frequencies [35].

Another report [36] concerns the comparison of viscoelastic properties of two polymers in isotropic and anisotropic phases. The poly(p-phenylene terephtalamide) (PPD-T) and HPC present similar rheological features in isotropic solutions. However, the nematic PPD-T and cholesteric HPC present different rheological row responses in that the yield stress of HCP remained almost constant with increasing concentration, whereas that of PPD-T increased significantly. At 1 rad/s both LCPs present an increase of the ratio of elastic modulus to twice loss modulus with increasing concentration. At 100 rad/s, HPC exhibits monophonic decrease in $G'/2G''$ with concentration, whereas in case of the other LCP, the ratio increase at the concentration higher than the saturated concentration. There is certain sensitivity of elastic modulus of nematic PPD-T to strain level, while for HPC, there is almost no dependence to this experimental parameter.

The evolution in time of shear moduli after flow cessation is a useful to analyze structural relaxations on LCPs. Upon flow cessation, the flow-induced orientation is lost. The variation of the moduli of a cellulose derivative with time was proved to be caused by the reformation of a chiral nematic phase that had become nematic under flow [37].

2.3. Viscoelastic behavior of LCPs in composite systems

The literature concerning the viscoelastic properties of LCPs in composite systems is not so abundant in regard with LCPs. Some reports involve oscillatory shear investigations on systems prepared from hydroxylpropyl cellulose (HPC) in lyotropic phase blended with different polymers, such as polyimide [38] and cellulose acetate phthalate [39]. Other studies concern reinforcement of LCPs with carbon nanotubes [40]. Such multiphase systems concern also blends of thermotropic LCP with commercial polymers [41].

A system consisting of HPC/polyimide presents a combination of viscoelastic behavior of each counterpart. The polyimide solutions present frequency sweep curves similar to those presented in **Figure 4(a)**. So, at the low frequencies, the loss modulus is higher than the storage modulus, with no plateau appearing in G'. At a specific frequency, this changes and rheological modulus cross each other, thus elastic modulus becoming higher [38]. The HPC in the liquid crystalline phase, the crossover frequency is higher comparatively with the case of the isotropic solutions. The dependence of the rheological moduli on frequency has slopes below unity. For HPC/polyimide composites, in the low-frequency range, the slopes are smaller comparatively to the pure components and are reduced as the lyotropic phase becomes predominant. The overlap frequency is influenced by the chemical structure of the individual polymers and by the blend composition. If the flexible counterpart is diminished, transition from the viscous to the elastic flow is delayed and the moduli crossover occurs at higher frequencies.

The HPC/cellulose acetate phthalate composites present similar behavior to those previously described. The exponents from the dependence of shear moduli versus frequency decrease on one hand, at lower values of polymer concentrations and, on the other, at lower HPC compositions in the system. In addition, the frequencies corresponding to the crossover point, which marks the transition from viscous flow to the elastic one, become higher with increasing the cellulose acetate phthalate content in composite [39].

A thermotropic liquid crystal polyester (TLCP) was reinforced with a very small quantity of modified carbon nanotube (CNT). The rheological properties of the composite are influenced by the uniform dispersion of CNT and the interactions between the CNT and TLCP, which can be enhanced by chemical modification of the filler [40]. The values of shear moduli of TLCP nanocomposites are significantly enhanced with increasing frequency and filler content. This enhancing effect was more pronounced at low-frequency region. The behavior is similar to the relaxation of typically filled polymer composite system, namely if polymer chains are fully relaxed and exhibit a characteristic homopolymer-like terminal behavior, the flow curves of polymers can be expressed by a power law with slopes like those from **Figure 4(a)**. The variations of terminal zone slopes of TLCP nanocomposites revealed the nonterminal behavior with the power law dependence of shear moduli on frequency. The decrease in the slope of for TLCP nanocomposites with the introduction of filler is based on the fact that the nanotube-nanotube or nanotube-polymer interactions produce the formation of the interconnected or network-like structures. This leads to the pseudosolid-like behavior more elasticity of TLCP nanocomposites in regard with TLCP matrix. As the shear frequency increased, the interconnected or network-like structures were disrupted by high levels of shear force and TLCP nanocomposites presented almost similar or slightly higher G' and G'' values than that of pure matrix at high-frequency region. The nanocomposites containing modified CNT have higher values of rheological moduli of TLCP, suggesting the increased interactions between modified CNT and TLCP matrix. Similar results were reported for thermotropic LCPs reinforced with a commercial polymer [41].

3. Applications

The viscoelastic features of polymer systems with self-alignment features are explored in several applications, including in tissue engineering or in display industry.

Tissues engineering represents a large subdomain of biomedicine and consists in improvement or replacement of biological functions of a damaged part of organism by utilization of a scaffold for the formation of new viable tissue. Biocompatible polymers are good candidates for cell growth substrates. In addition, the partial organization of LCPs, such as the natural derived macromolecules, is useful in preparation of cytocompatible supports with guided growth ability [42]. For example, polymer systems with chiral phase are able to direct anisotropic cell growth and play a crucial role in formation of patterns noticed in mammalian tissues [42]. The use of biologically derived polymers as LC layers for *in vitro* material production opens novel design perspectives because of the innate cellular response of these materials *in vivo*. Moreover, it was proved that the support morphology and viscoelasticity influences cell behavior [43]. If the scaffold surface has on its surface deep valleys and wide grains, then it is expected to have a greater effect on cell spreading and arrangement than alteration of the elastic modulus. Particularly, it was shown that valley depth plays a leading role in contact guidance of endothelial cells on microtextured silicone elastomer. In other words, at lower grain widths and in shallower valleys, the modulus of the substrate it appears to exhibit an essential role comparatively with rough and different nanocomposite surfaces. So, it can be stated that cell attachment, proliferation, and differentiation could be optimized by controlling the surface roughness and stiffness of the biocomposite scaffold [43].

Another important application of LCPs in composite system is the preparation of polymer dispersed liquid crystals (PDLCs) or nematic alignment layer for display purposes. The PDLC is a light-scattering material that operates on the principle of electrically modulating the refractive index of the liquid crystal in an optical isotropic, transparent solid [44]. The electro-optic material is placed between two polarizers, and its corresponding droplets, with dimensions varying between 0.1 and 10 μm, have positive dielectric anisotropy. PDLC displays demand a driving transistor array with high driving voltages, more current and less leakage. In OFF state, the polymer network follows the planar cholesteric helix and maintains its memory in previous stages by anchoring the director in suitable configuration. The helical birefringent texture reflects electromagnetic radiation from visible domain in a selective manner. In ON state, the director tends to align along the external field, but it is prevented by bulk anchoring on the polymer network. Thus, it results a scattering conical texture. Increasing the field intensity allows full alignment of director and PDLC composite becomes transparent. However, the memory regarding the initial cholesteric arrangement is permanently retained in the polymer network and on switching the field off, the optical features came back to their original state. The randomly inhomogeneous PDLC with a dilute rigid polymer network should be contrasted with liquid crystalline gel network. Cross-linking polymer chains do not phase separate from the LCP, the latter being considered as an anisotropic solvent in a homogeneous gel with a weak rubber-elastic polymer network.

On the other hand, viscoelasticity also counts when constructing polymer alignment layers for nematics. The LCP surface interaction is essential for the device reliability. The nature of this interaction is complex and is a combination of van der Waals interactions, dipolar interactions, steric factors, chemical and hydrogen bonding. Also, surface topography and mechanical features of polymer layer affect the mechanism of nematic orientation. It was assumed that anisotropic liquid crystal orientational elasticity in connection with rubbing-induced micro-grooves on polymer surface could be the key factor in designing display devices. Berreman [45] proposed that the director field adapts itself to the morphology of the polymer to reduce as much as possible the elastic strain energy arising from the distortions of the director field [45]. Introduction of soft polymers in a LCP matrix subjected to shear can determine a specific texture onto the surface of the resulted mixed alignment layer. Depending on the composite system composition, one can tune the surface texture and viscoelasticity and implicitly the ability to align uniformly nematic molecules.

4. Conclusions

This chapter describes the viscoelastic properties of polymers. The size and the large aspect ratio of these materials cause high elasticity and viscoelasticity. High elasticity is related to the ability of flexible chain macromolecules to recover from large strains, a property unique to rubber. Viscoelasticity denotes a time-varying reaction to a transient perturbation, unaccompanied by any modification in the material. Viscoelastic polymers both dissipate and store energy during deformation. These two characteristics underlie most applications of rubbery materials. The chapter provides an overview of the viscoelastic behavior during shearing of LCP solutions and also in composite systems highlighting the local and segmental motions, the chain dynamics, and also the influence of temperature, composition and time.

Acknowledgements

This work was supported by grant of the Romanian National Authority for Scientific Research and Innovation, CNCS—UEFISCDI, project PNII-ID-PCE-2011-3-0937 (no. 302/5.10.2011).

Author details

Andreea Irina Barzic and Silvia Ioan*

*Address all correspondence to: ioan_silvia@yahoo.com

Department of Physical Chemistry of Polymers, "Petru Poni" Institute of Macromolecular Chemistry, Iasi, Romania

References

[1] Christensen RM. Theory of Viscoelasticity. 2nd ed. Mineola: Dover Publications Inc; 1982.

[2] Findley WN, Lai JS, Onaran K. Creep and Relaxation of Nonlinear Viscoelastic Materials. Revised ed. New York: Dover Publications; 2011.

[3] Lakes R. Viscoelastic Materials. Cambridge: Cambridge University Press; 2009. DOI: 10.1017/CBO9780511626722

[4] Ferry JD. Viscoelastic Properties of Polymers. 3rd ed. New York: Wiley; 1980.

[5] Hyuna K, Wilhelm M, Klein CO, Cho KS, Nam JG, Ahn KH, Lee SJ, Ewoldt RH, McKinley GH. A review of nonlinear oscillatory shear tests: analysis and application of large amplitude oscillatory shear (LAOS). Prog Polym Sci. DOI: 10.1016/j.progpolymsci.2011.02.002, 36,1697–1753, 2011

[6] Watanabe H. Viscoelasticity and dynamics of entangled polymers. Prog Polym Sci. DOI: 10.1016/S0079-6700(99)00029-5, 24, 1253–1403, 1999

[7] Roland RM. Viscoelastic Behavior of Rubbery Materials. Oxford: Oxford University Press; 2011.

[8] Acierno D, Brostow W. Rheology and Processing of Liquid Crystal Polymers. London: Springer; 1996.

[9] Baek SG, Magda JJ, Cementwala S. Normal stress difference in liquid-crystalline hydroxypropylcellulose solutions. J Rheol. DOI: 10.1122/1.550404, 37, 935–945, 1993

[10] Cosutchi AI, Hulubei C, Stoica I, Ioan S. A new approach for patterning epiclon-based polyimide precursor films using a lyotropic liquid crystal template. J Polym Res. DOI: 10.1007/s10965-011-9653-z, 18, 2389–2402, 2011

[11] Dobos AM, Onofrei DM, Tudorachi N, Ioan S. Structural orientations of cellulose acetate phthalate/ethyl cellulose blends in solution. J Macromol Sci Part B. DOI: 10.1080/00222348.2015.1078625, 54, 1092–1104, 2015

[12] Dutta D, Fruitwala H, Kohli A, Weiss RA. Polymer blends containing liquid crystals: a review. Polym Eng Sci. DOI: 10.1002/pen.760301704, 30, 1005–1018, 1990

[13] Roylance D. Engineering viscoelasticity. Courses of Massachusetts Institute of Technology. http://ocw.mit.edu/courses/materials-science-and-engineering/3-11-mechanics-of-materials-fall-1999/modules/visco.pdf; 2001.

[14] Cerrada ML. Introduction to the Viscoelastic Response in Polymers. In: RA Diaz, editor. Thermal Analysis. Fundamentals and Applications to Material Characterization. Coruña: Universidade da Coruña, Servizo de Publicacións; 2005. p. 167–182.

[15] Menard KP. Dynamic Mechanical Analysis: A Practical Introduction. 2nd ed. Boca Raton: CRC Press; 2008.

[16] Shaw MT. Introduction to Polymer Rheology. New York: Wiley; 2012.

[17] Carotenuto C. Advanced mechanical spectroscopy to investigate the microstructure of complex fluids. Doctoral Thesis submitted to Napoli University, 2006.

[18] Cho KS. Viscoelasticity of Polymers Theory and Numerical Algorithms. New York: Springer; 2016.

[19] Macosko CW. Rheology: Principles, Measurements and Applications. New York: VCH; 1994.

[20] Larson RG. Constitutive Equations for Polymer Melts and Solutions. Washington: Butterworth; 1988.

[21] Collyer AA, Clegg DW. Rheological Measurements. London: Chapman and Hall; 1998.

[22] Laun HM. Description of the non-linear shear behaviour of a low density polyethylene melt by means of an experimentally determined strain dependent memory function. Rheol Acta. DOI: 10.1007/BF01567859, 17, 1–15, 1978

[23] Acierno S, Grizzuti N. Measurements of the rheological behaviour of a crystallizing polymer by an "inverse quenching" technique. J Rheol. DOI: 10.1122/1.1545080, 47, 563–576, 2003

[24] Winter HH, Chambon F. Analysis of linear viscoelasticity of a crosslinking polymer at the gel point. J Rheol. DOI: 10.1122/1.549853, 30, 367–382, 1986

[25] Tucker CL, Moldenaers P. Microstructural evolution in polymer blends. Annu Rev Fluid Mech. DOI: 10.1146/annurev.fluid.34.082301.144051, 37, 177–210, 2002

[26] Onofrei MD, Dobos AM, Stoica I, Olaru N, Olaru L, Ioan S. Lyotropic liquid crystal phases in cellulose acetate phthalate/hydroxypropyl cellulose blends. J Polym Environ. DOI: 10.1007/s10924-013-0618-7, 22, 99–111, 2014

[27] Krieger IM, Niu TF. A rheometer for oscillatory studies of nonlinear fluids. Rheol Acta. DOI: 10.1007/BF01525599, 12, 567–571, 1973

[28] Zaman AA. Techniques in Rheological Measurements. London: Chapman and Hall; 1993.

[29] Debbaut B, Burhin H. Large amplitude oscillatory shear and Fourier transform rheology for a high-density polyethylene: experiments and numerical simulation. J Rheol. DOI: 10.1122/1.1495493, 46, 1155–1176, 2002

[30] Neidhöfer T, Wilhelm M, Debbaut B. Fourier-transform rheology experiments and finite-element simulations on linear polystyrene solutions. J Rheol. DOI: 10.1122/1.1608954, 47, 1351–1371, 2003

[31] Wilhelm M, Reinheimer P, Ortseifer M, Neidhöfer T, Spiess HW. The cross-over between the linear and nonlinear mechanical behaviour in polymer solutions as detected by Fourier-transform rheology. Rheol Acta. DOI: 10.1007/s003970000084, 39, 241–246, 2000

[32] Reimers MJ, Dealy JM. Sliding plate rheometer studies of concentrated polystyrene solutions: large amplitude oscillatory shear of a very high molecular weight polymer in diethylphthalate. J Rheol, 40. DOI: 10.1122/1.550738, 40, 167–186, 1996

[33] Yin Y, Nishinari K, Zhang H, Funami T. A novel liquid-crystalline phase in dilute aqueous solutions of methylcellulose. Macromol Rapid Commun. DOI: 10.1002/marc.200600099, 27, 971–975, 2006

[34] Lee HC, Brant DA. Rheology of concentrated isotropic and anisotropic xanthan solutions. 2. A semiflexible wormlike intermediate molecular weight sample. Macromolecules. DOI: 10.1021/ma011527e, 35, 2223–2234, 2002

[35] Colby RH, Gillmor JR, Galli G, Laus M, Ober CK, Hall E. Linear viscoelasticity of side chains liquid crystal polymers. Liq. Cryst. 13(2), 233–245, 1993.

[36] Hwang SS, Kim BC, Kim KU, Hong SI. Rheological implications of mesomorphic order in lyotropic liquid crystalline polymers. Korean J Rheol, 9(1), 200–205, 1997.

[37] Appaw C. Rheology and microstructure of cellulose acetate in mixed solvent systems. Dissertation submitted at Graduate Faculty of North Carolina State University, 2004.

[38] Cosutchi AI, Hulubei C, Stoica I, Ioan S. Morphological and structural-rheological relationship in epiclon-based polyimide/hydroxypropylcellulose blend systems. J Polym Res. DOI: 10.1007/s10965-009-9342-3, 17, 541–550, 2010

[39] Dobos AM, Onofrei MD, Stoica I, Olaru N, Olaru L, Ioan S. Rheological properties and microstructures of cellulose acetate phthalate/hydroxypropyl cellulose blends. Polym Compos. DOI: 10.1002/pc.22350, 33 2072–2083, 2012

[40] Kim JY. Carbon nanotube-reinforced thermotropic liquid crystal polymer nanocomposites. Materials. DOI: 10.3390/ma2041955, 2, 1955–1974, 2009

[41] Weiss A, Wansoo HUH, Nicolais L. Novel reinforced polymers based on blends of polystyrene and a thermotropic liquid crystalline polymer. Polym Eng Sci. DOI: 10.1002/pen.760270913, 27, 684–691, 1987

[42] Kirkwood JE. Liquid crystalline collagen: matrices for oriented cell growth. Dissertation submitted to Stanfort University, 2008.

[43] Hosseini MS, Katbab AA. Effects of surface viscoelasticity on cellular responses of endothelial cells. Rep Biochem Mol Biol, 3(1), 20–28, 2014.

[44] Chang SJ, Chiao S, Lai WJ, Lin CM, Fuh AYG. Polymer dispersed liquid crystal display device for projection high definition television application. Macromol Symp. DOI: 10.1002/masy.19940840119, 84, 159–166, 1994

[45] Berreman DW. Solid surface shape and the alignment of an adjacent nematic liquid crystal. Phys Rev Lett. DOI: 10.1103/PhysRevLett.28.1683, 28, 1683–1686, 1972

Chapter 4

Thermal Properties of Hemp Fiber Reinforced Plant-Derived Polyamide Biomass Composites and their Dynamic Viscoelastic Properties in Molten State

Yosuke Nishitani, Toshiyuki Yamanaka,
Tetsuto Kajiyama and Takeshi Kitano

Additional information is available at the end of the chapter

http://dx.doi.org/10.5772/64215

Abstract

To further enhance the mechanical, thermal, and tribological properties of short natural fiber-reinforced biopolymer composites, it is very critical to understand thermal properties of these biomass composites and their dynamic viscoelastic properties in the molten state. The aim of this study is to experimentally investigate the thermal properties of hemp fiber filled plant-derived polyamide 1010 composites and their dynamic viscoelastic properties in the molten state. It was found that the addition of HF with PA1010 has a strong influence on the thermal properties such as DMA, TGA, and DSC. HF is very effective for improving the thermal and mechanical properties. The effect of alkali treatment on the dynamic viscoelastic properties of the HF/PA1010 composites in the molten state differs according to whether alkali treatment uses silane coupling agent or not. The viscoelastic properties of $NaClO_2$ are higher than those of NaOH. Silane coupling agents have a remarkable influence on rheological properties such as storage modulus, loss modulus, and complex viscosity in the low angular frequency region in the molten state, temperature dependences of rheological properties, and relationship between the phase angle and complex modulus. These rheological behaviors are also strongly influenced by the type of silane coupling agents.

Keywords: thermal properties, dynamic viscoelastic properties, plants-derived polyamide, biomass composites, hemp fiber

1. Introduction

Biopolymers and biomass polymer composites are drawing extensive interest not only as a solution for growing environmental threats but also as a solution for the depletion of petroleum in recent years [1–4]. In addition, the supply of raw materials is becoming increasingly unstable as many biopolymers are made from edible biomass such as corn. In order to solve these problems, new engineering materials made of 100% inedible plant-derived materials are strongly required. Meanwhile, the investigation of short natural (plant) fibers such as banana fiber, flax fiber, hemp fiber, ramie fiber, and sisal fiber used for reinforced biopolymer composites have attracted great interest in recent decades [4–8]. These natural fibers have some ecological advantages over inorganic fibers such as carbon and glass fibers as they are renewable, light, recyclable, and biodegradable and can be incinerated [9]. In previous studies, we conducted the development of new engineering materials such as structural materials and tribomaterials (for mechanical sliding parts such as bearing, cum, gear, and seal) based on all plant-derived materials. In particular, we investigated the rheological, mechanical, and tribological properties of natural fiber-reinforced biopolymer composites such as hemp fiber (HF)-reinforced plant-derived polyamide 1010 (PA1010) biomass composites [10–17]. PA1010 was made from sebacic acid and decamethylenediamine, which are obtained from plant-derived castor oil [18]. As castor oil is not used for food, there is no competition with human food consumption. It was found that the mechanical and tribological properties of these composites are improved when filled with hemp fibers and surface-treated using silane coupling agent. However, in order to further enhance the mechanical, thermal, and tribological properties of the short natural fiber-reinforced biopolymer composites, it is very important to understand the thermal properties of these biomass composites and their dynamic viscoelastic properties in molten state such as process ability, heat resistance, crystallinity, internal adhesion, internal microstructures, and changes and structure–property relationships [19–21]. The aim of this study is to experimentally investigate the thermal properties of hemp fiber-filled plant-derived polyamide 1010 composites and their dynamic viscoelastic properties in the molten state.

2. Thermal properties of hemp fiber-reinforced plant-derived polyamide 1010 biomass composites

2.1. Introduction

Natural fiber-reinforced plant-derived polymer biomass composites are environment friendly to a large extent and have unique performances. However, since the interfacial adhesion between the natural fiber and matrix polymer is generally poor, these biomass composites exhibit poor mechanical properties [4, 11, 22]. Interfacial adhesion can be enhanced and the mechanical properties of these composites improved by suitable surface treatment. Most fibers are pretreated before they are used as secondary phases in composite materials. The effects of the surface treatment on the mechanical properties of these biomass composites have been studied for the last two decades [3–8, 10, 11, 15–17, 22]. In particular, chemical methods such as alkali treatment (mercerization), silane treatment, and graft copolymerization and physical methods such as the corona treatment and the plasma treatment have been investigated in this field. However, only a few studies have been published on the effects of surface treatment on thermal properties of these biomass composites [14, 22–27]. The majority of natural fibers have low degradation temperatures below 200°C, which make them inadequate for processing with thermoplastics at temperatures above 200°C [27]. To further enhance the mechanical, thermal, and tribological properties of the natural fiber-reinforced biopolymer composites, it is very critical to understand the thermal properties such as heat resistance, crystallinity, internal adhesion, internal microstructures, changes, and structure–property relationships of these materials. The aim of this study is to improve the performance of all inedible plant-derived materials for new engineering materials such as structural materials and tribomaterials. Thermal properties such as dynamic mechanical analysis (DMA), thermogravimetric analysis (TGA), and differential scanning calorimetry (DSC) of hemp fiber-filled inedible plant-derived polyamide 1010 biomass composites were investigated experimentally.

2.2. Materials and methods

The materials used in this study were surface-treated hemp fiber-reinforced plant-derived polyamide 1010 biomass composites (HF/PA1010). Plant-derived polyamide 1010 (PA1010, Vestamid Terra DS16, Daicel Evonik Ltd., Japan) was used as the matrix polymer. PA1010 was made from sebacic acid and decamethylenediamine which are obtained from plant-derived castor oil [18]. Hemp fiber (HF, ϕ50–100 μm, Hemp Levo Inc., Japan) was used as a reinforcement fiber. The volume fraction of fiber V_f was fixed with 20 vol.%. HFs were precut into 5-mm-long pieces and were surface-treated by two methods: (a) alkali treatment by sodium hydroxide (NaOH) solution and (b) surface treatment using aminosilane coupling agents (S1, A-1120, 3-(2-aminoethylamino) propyltrimethoxy silane, Momentive Performance Materials Inc., USA). Alkali treatment using NaOH was employed as follows: 5% NaOH solution was placed in a stainless beaker. Then chopped HFs were added into the beaker and stirred well. This was kept at room temperature for 4 h. The fibers were then washed thoroughly with deionized water to remove excess NaOH sticking to the fibers. The alkali-treated HFs by NaOH (HF-A) were dried on air for 12 h and in a vacuum oven at 80°C for 5 h. Amino silane coupling agent was used for surface treatment. The treatment of HFs with the concentration of 1 wt.% of amino silane coupling agent was carried out in deionized water and stirred continuously for 15 min. Then, the fibers were immersed in the solution for 60 min. The surface-treated hemp fibers (HF-S1) were removed from the solution and dried in air for 12 h and in a vacuum oven at 80°C for 5 h. All the components were dried for 12 h at 80°C in a vacuum oven beforehand until the moisture level was below 0.2%, and then dry blended in a small bottle, and subsequently, the melt was mixed at 85 rpm and 220°C on a twin screw extruder (TEX-30, Japan Steel Works, Ltd., Japan). After mixing, the extruded strands of various HF/PA1010 composites were cut into 5-mm-long pieces by the pelletizer and dried again at 24 h at 80°C in a vacuum oven. Various shaped samples for various experiments were injection molded (NS20-A, Nissei Plastic Industrial, Japan). The molding conditions were as follows: cylinder temperatures of 220°C, mold (cavity) temperature of 30°C, and the injection rate of

13 cm^3/s. In addition, 2-mm-thick sheets were compression-molded under the conditions of 220°C, 3 min, and 5 MPa and cut into 5 mm × 40 mm × 1 mm-shaped specimen for measurement and dynamic mechanical analysis. To maintain the drying conditions of specimens for all measurements, they were kept in accordance with JIS K 6920–2 for at least 24 h at 23°C in desiccators after molding. Thermal properties such as dynamic mechanical analysis (DMA), thermogravimetric analysis (TGA), and differential scanning calorimetry (DSC) were evaluated. The storage modulus E', loss modulus E'', and loss tangent ($\tan\delta = E''/E'$) of the composites were measured as a function of temperature (from −100 to 200°C) using a DMA equipment (RSA3, TA instrument Japan, Inc., Japan) with a tensile fixture at a frequency of 1 Hz. Thermogravimetric analysis (TGA) was carried out in a TGA equipment (Thermo plus EVO2, Rigaku Co. Ltd., Japan). The samples used for the TGA were cut from injection-molded coupon specimens into small pieces weighing 10 mg. TGA measurement was programmed for heating from 40 to 400°C with a heating rate of 10°C/min. The differential scanning calorimetry (DSC) was measured by a DSC equipment (DSC-50, Shimadzu Co. Ltd., Japan). The samples used for the DSC were prepared like those for TGA (5 mg). The DSC measurement was scanned from −90 to 230°C with a constant heating rate of 10°C/min.

2.3. Dynamic mechanical analysis (DMA)

First, the dynamic mechanical analysis of surface-treated hemp fiber-filled polyamide 1010 biomass composites (HF/PA1010 biomass composites) is discussed. The storage modulus E' and loss tangent $\tan\delta$ (= Loss modulus E''/ Storage modulus E') are plotted as a function of temperature T for neat PA1010 (100%), untreated HF-filled PA1010 composites (HF/PA1010), HF treated by NaOH-filled systems (HF-A/PA1010), HF treated by amino silane coupling agent (S1)-filled ones (HF-S1/PA1010), and HF treated by NaOH and S1 filled ones (HF-A-S1/PA1010) in **Figure 1(a)** (E' vs. T) and **Figure 1(b)** ($\tan\delta$ vs. T), respectively. The E' of various HF/PA1010 biomass composites is higher than that of neat PA1010, indicating that HF has a strong reinforcing effect on the elastic properties of PA1010. The E' of various HF/PA1010 ones in low-temperature region less than about 50°C, which is the glass transition temperature T_g of PA1010, is the same level in spite of the kind of surface treatment. However, the E' of HF/PA1010 ones in the high-temperature region of >50°C decreases in the following sequence: HF/PA1010 (untreated) > HF-A/PA1010 > HF-A-S1/PA1010 > HF-S1/PA1010. The $\tan\delta$ curves (**Figure 1(b)**) exhibit two relaxation peaks. The first peak between at 40 and 60°C represents the glass transition temperature T_g of PA1010 and the composites, and the second one between −80 and −60°C shows the relaxation arising from the hydrogen bonds between the PA1010 chains [28]. The relaxation peak at the high-temperature region of the HF/PA1010 composites shifts slightly toward high temperature although various HF/PA1010 biomass composites are at the same level regardless of the surface treatment method. On the contrary, the relaxation peak at the lower-temperature region of neat PA1010 does not shift with the addition of HF; however, those of various surface-treated HF/PA1010 biomass composites are small in comparison with neat PA1010. In general, those of rigid fiber-filled polymer composites shift toward higher temperature due to the restriction on the movement of the polymer chains [28]. In short, flexible fibers such as the natural fiber employed in this study show the same tendency. The mechanisms of how the addition of HF affects the relaxation peak at the low-temperature region of composites needs to be studied further.

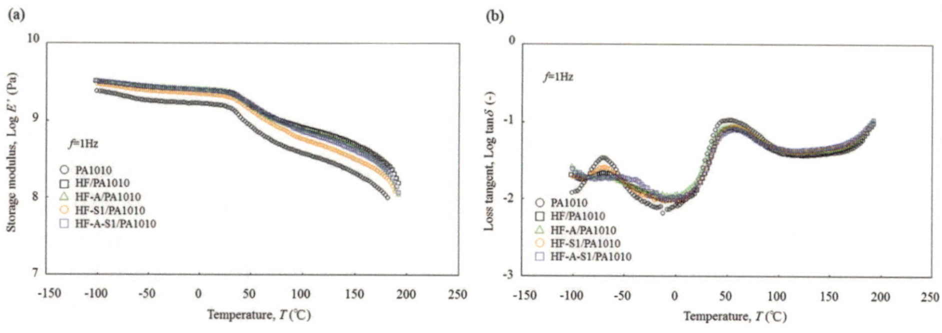

Figure 1. Dynamic mechanical properties as a function of temperature for various HF/PA1010 biomass composites. (a) Storage modulus and (b) loss tangent.

2.4. Thermogravimetric analysis (TGA)

The thermogravimetric analysis (TGA) of various surface-treated HF/PA1010 biomass composites has been discussed. **Figure 2** shows the TG curves (the weight as a function of temperature *T*) of various HF/PA1010 biomass composites. The weight of neat PA1010 is higher than that of various HF/PA1010 biomass composites over the whole temperature range. The TG curve of various HF/PA1010 biomass composites shows the evidence of two weight loss processes, while that of neat PA1010 is only one weight loss process. The first weight loss process between 80 and 200°C is attributed to the dehydration of HF as well as the thermal degradation of lignin and hemicellulose [22, 27]. The second weight loss process at about 300°C is explained in terms of the decomposition of cellulose in HF. The TG curves of various surface-

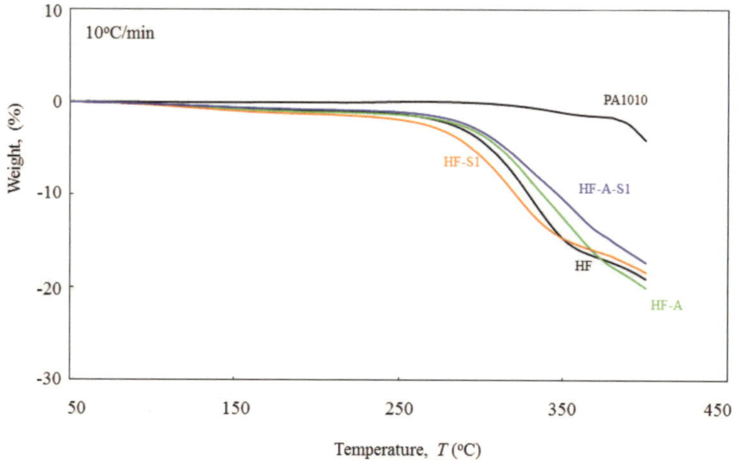

Figure 2. Thermogravimetric curves of various HF/PA1010 biomass composites.

treated HF/PA1010 biomass composites decrease in the following order: HF-A-S1/PA1010 > HF-A/PA1010 > HF/PA1010 > HF-S1/PA1010. This indicates that HF treated with the combination of NaOH and amino silane (S1) is more effective for improving the heat resistance of the HF/PA1010 biomass composites.

2.5. Differential scanning calorimetry (DSC)

The crystal form of the polymer has a strong influence on the mechanical properties of the polymer composites. Differential scanning calorimetry (DSC) was employed to evaluate the effects of the surface treatment of fiber on the crystallization behavior of HF/PA1010 biomass composites. **Figure 3** shows the DSC 1st heating (**Figure 3(a)**) and 2nd heating (**Figure 3(b)**) curves obtained at 10°C/min rate, respectively. DSC 1st heating curves in **Figure 3(a)** show a melting peak each curve, although DSC 2nd heating curves in **Figure 3(b)** have two melting peaks. This may be explained by the following mechanisms: it was pointed out recently by Li et. al. [29] that the appearance of multiple melting peaks in DSC run is probably due to rearrangement of the lamella since the polyamide crystals can be easily thickened by annealing. It is considered that the same phenomenon occurs for PA1010 used in this study. T_{m1}, which is the melting point in the lower temperature side in DSC 2nd heating curves, is attributed to the thin lamellae formed during cooling, and T_{m2}, which is in the higher ones, is attributed to the melting of the thickened crystals during the heating and annealing process. Parameters such as the melting point T_m, T_{m1}, and T_{m2} and the heat of fusion ΔH_f calculated from DSC curves for various PA1010 biomass composites are listed in **Table 1**. T_m in 1st heating curves shifts to lower temperatures when filled with the HF, and ΔH_f also decreases when HF is filled. The effect of surface treatment of fiber on the T_m and ΔH_f of HF/PA1010 biomass composites has complex behavior according to the type of the surface treatment. T_m of alkali treatment by NaOH (HF-A/PA1010 composites) shifts to higher temperature, while T_m of surface treatment by silane coupling agent (HF-S1/PA1010 composites) shifts to lower temperature. T_m of the composites surface-treated by both alkali treatment and silane coupling agent (HF-A-S1/PA1010 composites) has the immediate value between HF-A/PA1010 and HF-S1/PA1010 composites. On the contrary, ΔH_fs of HF-A/PA1010 and HF-S1/PA1010 composites are higher

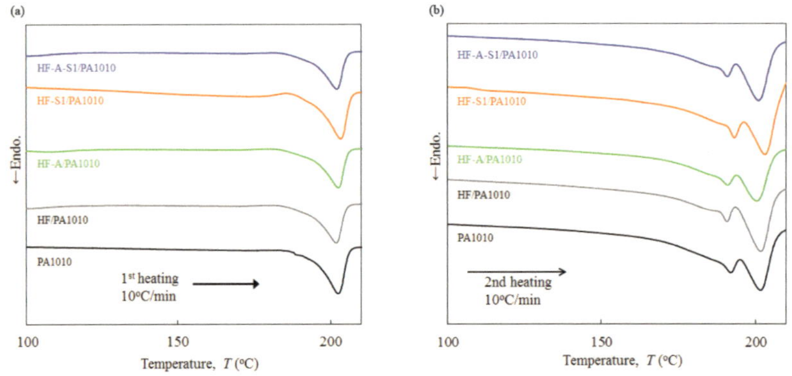

Figure 3. DSC curves of various HF/PA1010 biomass composites. (a) 1st heating and (b) 2nd heating.

than that of HF/PA1010 composites although ΔH_f of HF-A-S1/PA1010 composites is lower than that of HF/PA1010 composites. T_{m1}, T_{m2}, and ΔH_f in 2nd heating curves also have complex behavior, and these values basically decrease with filling the HF and alkali treatment, although the values slightly increase with the surface treatment by silane coupling agent. The behavior may be due to factors such as changes in crystal form with the alkali treatment and silane coupling agent treatment caused by changing in the interfacial interaction between HF and PA1010.

	1st heating		2nd heating		
	T_m (°C)	ΔH_f (J/g)	T_{m1} (°C)	T_{m2} (°C)	ΔH_f (J/g)
PA1010	203.0	18.5	192.0	201.5	38.2
HF/PA1010	201.7	14.6	190.6	201.7	34.5
HF-A/PA1010	202.4	15.3	190.9	200.4	31.4
HF-S1/PA1010	201.5	17.7	193.0	203.0	44.0
HF-A-S1/PA1010	201.9	14.5	190.9	201.0	36.6

Table 1. Parameters calculated from the DSC curves for various HF/PA1010 biomass composites.

3. Effect of alkali treatment on dynamic viscoelastic properties of hemp fiber-reinforced plant-derived polyamide 1010 biomass composites in molten state

3.1. Introduction

The interfacial adhesion between natural fiber and matrix polymer can be enhanced and the mechanical properties of natural fiber-reinforced polymer biomass composites improved by suitable surface treatment, because the interfacial adhesion between the natural fiber and matrix polymer is generally poor as mentioned earlier. The alkali treatment (mercerization) is a chemical treatment for natural fiber, which is most commonly used to reinforce thermoplastic and thermoset [3, 4, 16, 22, 24]. An important modification resulting from the alkali treatment is the disruption of hydrogen bonding in the network structure, thereby increasing surface roughness. This treatment removes a certain amount of lignin, hemicellulose, wax, and oils covering the external surface of fiber cell wall, depolymerizes cellulose, and exposes the short length crystallites. Therefore, strong effects, such as increase of the surface roughness resulting in better mechanical interlocking and increase in the number of possible reaction sites, can be expected from this treatment. Many authors have investigated the influence of alkali treatment on the various mechanical and chemical properties of natural fiber-reinforced polymer biomass composites [3, 4, 16, 22, 24]. However, although these biomass composites undergo various flows during processing by flow molding such as injection, extrusion, and compression, generally the effect of the alkali treatment on processing properties has not been studied enough [12, 30]. It is critical to understand the rheological behavior of these biomass composites in the molten state to understand the process ability, internal microstructures, changes, and structure–property relationships of these materials so as to further enhance the mechanical, thermal, and tribological properties of all plant-derived polymer-based biomass composites. Recently, we studied the effect of surface treatment, specifically alkali treatment with silane coupling agent, on the rheological properties of these biomass composites. The purpose of this study is to report the effect of alkali treatment on the dynamic viscoelastic properties of hemp fiber-reinforced plant-derived polyamide 1010 biomass composites in the molten state under oscillatory flow. Four types of surface treatments such as (a) alkali treatment by sodium hydroxide solution (NaOH), (b) alkali treatment by sodium chlorite solution ($NaClO_2$), (c) alkali treatment by NaOH solution and surface treatment by ureidosilane coupling agent (3-ureidopropyltrimethoxy silane, A-1160) (NaOH + ureidosilane), and (d) alkali treatment by $NaClO_2$ solution and surface treatment by ureidosilane ($NaClO_2$ + ureidosilane) were used for the surface treatment of hemp fiber in this study.

3.2. Materials and methods

The materials used in this study were various surface-treated hemp fiber-reinforced plant-derived polyamide 1010 biomass composites (HF/PA1010). PA1010, which is made from plant-derived castor oil, was used as the matrix polymer. Hemp fiber (HF, ϕ50–100 μm) was used as the reinforcement fiber. These materials are described in detail in the previous section 2. The volume fraction of fiber V_f was fixed with 20 vol.%. Hemp fibers were precut into 5-mm-long pieces and surface-treated by four types of surface treatments: (a) alkali treatment by sodium hydroxide solution (NaOH), (b) alkali treatment by sodium chlorite solution ($NaClO_2$), (c) alkali treatment by NaOH solution and surface treatment by ureidosilane coupling agent (S3, 3-ureidopropyltrimethoxy silane, A-1160, Momentive Performance Material Inc., USA) (NaOH + ureidosilane), and (d) alkali treatment by $NaClO_2$ solution and surface treatment by ureidosilane ($NaClO_2$ + ureidosilane). Two types of alkali treatment aqueous solutions such as sodium hydroxide solution (by NaOH) or and sodium chlorite solution ($NaClO_2$) were used as the mercerization agent. Alkali treatment by NaOH and $NaClO_2$ was employed as follows: 5% alkali agent solution was placed in a stainless beaker. Then chopped hemp fibers were added into the beaker and stirred well. This was kept at room temperature for 4 h. The fibers were then washed thoroughly with water to remove the excess of alkali agents sticking to the fibers. The alkali-treated fibers were dried in air for 12 h and in a vacuum oven at 80°C for 5 h. The surface treatment of hemp fibers with the concentration of 1 wt.% ureidosilane coupling agent (S3) was carried out in 0.5 wt.% acetic acid aqueous solution in which the pH of the solution was adjusted to 3.5 and stirred continuously for 15 min. Then, the fibers were immersed in the solution for 60 min. The surface-treated hemp fibers (HF-S) were removed from the solution and kept in air for 12 h and in a vacuum oven at 80°C for 5 h. All the components were dried for 12 h at 80°C in a vacuum oven beforehand until the moisture level was below 0.2%, and then dry blended in a small bottle, and subsequently the melt was mixed at 85 rpm and 220°C on a twin screw extruder (TEX-30, Japan Steel Works, Ltd., Japan). After mixing, the extruded strands of various HF/PA1010 biomass composites were cut in piece of about 5-mm long by the pelletizer, and dried again at 24 h at 80°C in a vacuum oven. In addition, 1-mm-thick sheets were compression-molded at the conditions of

220°C, 5 MPa, and 3 min, and cut into ϕ25-mm-disk shapes for rheological property measurements. The rheological properties in the molten state were evaluated by oscillatory flow testing using a parallel plate-type rotational rheometer (ARES, Rheometrix Scientific Co., USA). The diameter of the parallel plate was ϕ25 mm, and the gap between the two plates was fixed at 1 mm. Under such a gap condition, the test specimens were slightly compressed in the molten state. The angular frequency was varied from 10^{-1} to 10^{2} rad/s, and the strain amplitude was set as 1%. These measurements were carried out at 200–240°C. The dynamic viscoelastic properties in the molten state were evaluated by storage modulus G', loss modulus G'', loss tangent tan δ, phase angle δ (=arctan G''/G'), complex modulus $|G^*|$, and complex viscosity $|\eta^*|$. The surface of fiber and the composites fractured cryogenically in liquid nitrogen were observed using a scanning electron microscope (SEM, JSM-6360LA, JEOL Ltd., Japan). Fiber surface roughness such as arithmetic mean estimation R_a was measured using a laser microscope (VK-X200, Keyence Co., Japan). Fourier transform infrared spectroscopy (FT-IR) measurements were performed by a FT-IR spectrometer (FT/IR-6100, JASCO Co., Japan) using the attenuated total reflection (ATR) technique by a diamond prism. A total of 64 scans were taken for each sample between 400 and 4000 cm^{-1}, with a resolution of 8 cm^{-1}.

3.3. Angular frequency dependences

The dynamic viscoelastic properties of various surface-treated hemp fiber-filled plant-derived polyamide 1010 biomass composites (HF/PA1010) in molten state were evaluated by oscillatory behavior. The dynamic viscoelastic properties in the molten state are strongly dependent on the internal microstructure of the polymer composites. We shall discuss the angular frequency dependences, which is the basic variable in dynamic viscoelastic properties. **Figure 4(a)** shows the effects of alkali treatment on the relationship between storage modulus G' and angular frequency ω of various HF/PA1010 biomass composites at 220°C. G' of neat PA1010 (100%) increases with increasing ω, in agreement with the linear dynamic viscoelastic model that G' is proportional to ω^2 (log $G' \propto$ 2log ω) [31, 32]. G' of various HF/PA1010 biomass composites shows the typical storage modulus G' of highly filled systems (such as gel systems), indicating the "second rubbery plateau," i.e., the long-scale relaxation time [21, 31–35]. This

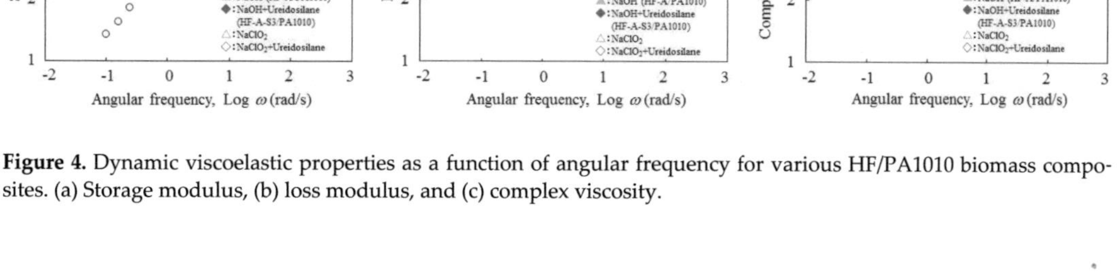

Figure 4. Dynamic viscoelastic properties as a function of angular frequency for various HF/PA1010 biomass composites. (a) Storage modulus, (b) loss modulus, and (c) complex viscosity.

behavior may be attributed to the internal structure formation such as fiber network formation, which is due to high aspect ratio of HF and the interfacial interaction between HF and PA1010. However, the slope of G' against ω of various HF/PA1010 biomass composites in low w region changes with the type of alkali treatment and decreases in the following: Untreated > NaOH > $NaClO_2$ > NaOH + ureidosilane (S3) > $NaClO_2$ + ureidosilane (S3). **Figure 4(b)** shows the effect of alkali treatment on the relationship between loss modulus G'' and ω. G'' of neat PA1010 monotonically increases with an increasing ω in the whole range of ω. This is in agreement with the linear viscoelastic model that the slope of G'' is proportional to ω ($\log G'' \propto \log \omega$). On the contrary, various HF/PA1010 biomass composites show different tendencies and the slope of G'' against w becomes small in the low w region. However, this tendency of G'' is smaller than that of G' since G' is a more sensitive viscoelastic function with respect to the structural changes of the composites compared G'' [36–38]. **Figure 4(c)** shows that the effect of alkali treatment on the relationship between the complex viscosity $|\eta^*|$ and ω. $|\eta^*|$ of neat PA1010 does not change with ω in the low ω regions, which demonstrates Newtonian behavior, and this slightly decreases with increasing ω in the high w region. On the contrary, the effect of various HF/PA1010 biomass composites rapidly decreases with increasing ω in a wide range and in magnitude orders more viscous than neat PA1010 even at high ω regions. Various HF/PA1010 biomass composites exhibit very strong shear-thinning effect, ant the curves of $|\eta^*|$ vs. ω have a slope of −45°. This behavior indicates the presence of an apparent yield stress in the low ω regions.

3.4. Effect of alkali treatment

In general, dynamic viscoelastic properties such as storage modulus G' are considered to be sensitive indicators for the quantitative analysis of morphological changes in the polymer composites as mentioned earlier. To clarify the effect of alkali treatment on the dynamic viscoelastic properties of the HF/PA1010 biomass composites, the relative storage modulus G'_r of various alkali-treated HF/PA1010 biomass composites in the low ω region (ω=0.25 rad/s) is shown in **Figure 5**. Here, the relative storage modulus is given by the values of surface-treated HF/PA1010 composites divided by that of untreated ones. The values of G'_r decrease in the following order: NaOH > $NaClO_2$ > NaOH + ureidosilane (S3) > $NaClO_2$ + ureidosilane (S3). The important findings in this results are (1) the elastic properties such as the storage modulus decreases with surface treatment by alkali and silane coupling agent, (2) the reducing ratio of G'_r of $NaClO_2$ is higher than that of NaOH, and (3) the combination of $NaClO_2$ + ureidosilane (S3) has the strongest effect on G'_r among all surface treatments in this study. The possible reasons of these phenomena are onset of self-lubrication of HF, three-dimensionally oriented fiber structure formation which may reduce the storage modulus of the surface-treated HF/PA1010 biomass composites to less than that of the untreated systems, and different roles played by the alkali treatment and silane coupling agent during the processing. It is, however, difficult to establish the exact reasons in the present stage.

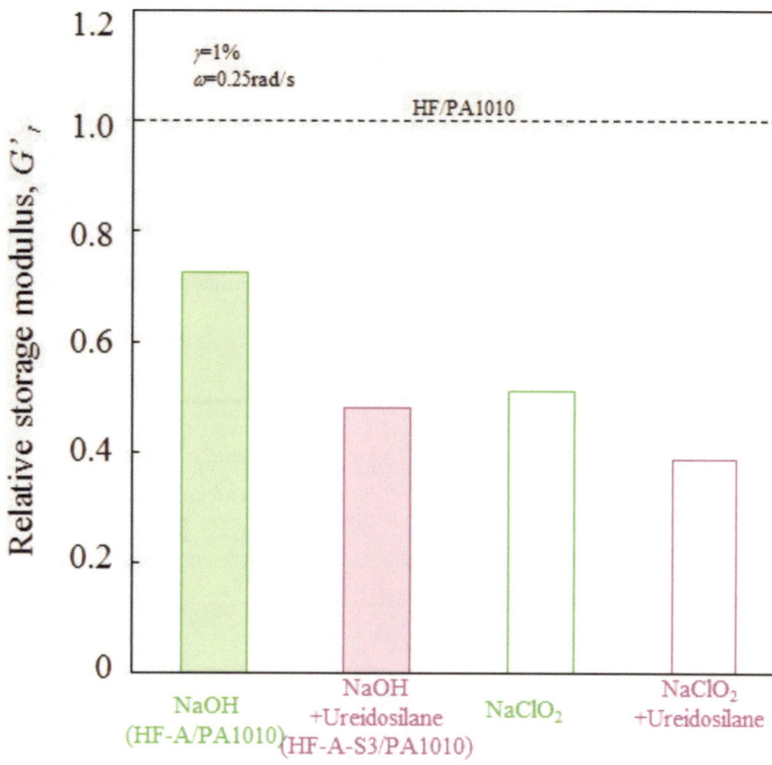

Figure 5. The relative storage modulus for various HF/PA1010 biomass composites.

3.5. Morphology and chemical analysis

To further clarify the relationship between the dynamic viscoelastic properties and internal structure formation of HF/PA1010 composites, we discuss the morphologies of these composites, which are internal structure formation such as fiber network formation and the interfacial interaction between HF and PA1010. **Figure 6** shows the SEM observation results of the HF surface before/after the alkali treatment. The surface roughness of HF increases with alkali treatment. The results of measuring the average surface roughness (arithmetic average roughness R_a) using a laser microscope are R_a = 3.2 mm (untreated), R_a= 3.4 mm (NaOH), and R_a =5.1 mm ($NaclO_2$), respectively. In particular, $NaClO_2$ treatment increases bumps and surface roughness significantly. These measurement results indicate that the attackability of HF differs according to the type of alkaline coupling agent. It is well known that alkali treatment promotes the disruption of hydrogen bonding in the network structure of natural fiber and removes the lignin and hemicellulose [4, 22]. Therefore, this treatment increases not only the surface roughness resulting in better unlocking but also the amount of cellulose exposed on the fiber surface, thereby increasing the number of possible reaction sites. Consequently, the interfacial interaction between fiber (HF) and matrix polymer (PA1010) (or silane coupling agent) also increases.

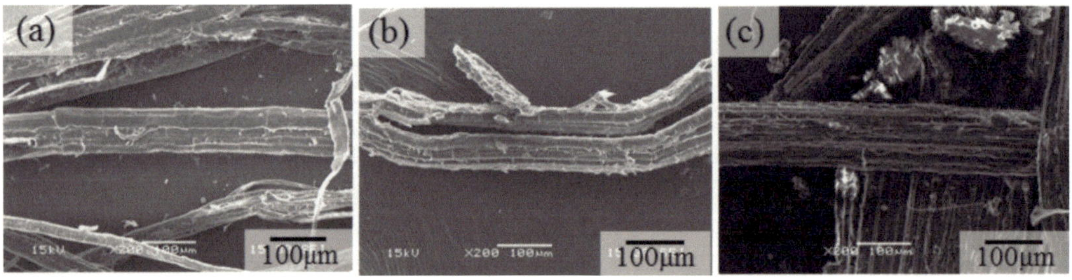

Figure 6. SEM photographs of untreated and alkali-treated hemp fibers. (a) Untreated: R_a= 3.2 mm, (b) NaOH (HF-A): R_a= 3.4 mm, and (c) $NaClO_2$: R_a= 5.1 mm.

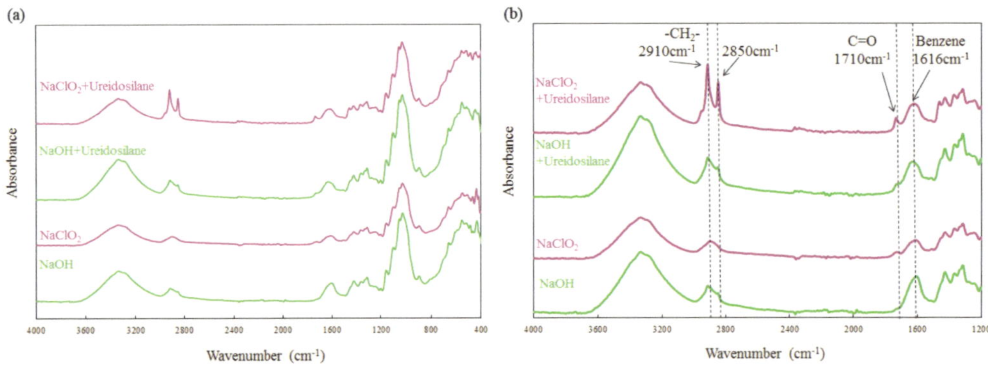

Figure 7. FT-IR spectra of various surface treated hemp fibers. (a) 400–4000 cm^{-1} and (b) 1200–4000 cm^{-1}.

It is essential to investigate chemical analyses such as Fourier transform infrared spectroscopy (FT-IR) in order to determine the chemical composition of the fiber surface. There have been various investigations on the effect of various surface treatments on the characterization of natural fiber surface [22, 26, 39, 40]. We shall discuss the observation of the characterization of the fiber surface using FT-IR. **Figure 7** shows the FT-IR spectra of various alkali-treated HFs: 400–4000 cm^{-1} (**Figure 7(a)**) and 1200–4000 cm^{-1} (**Figure 7(b)**). In the case of different alkali treatments such as NaOH and $NaClO_2$, the reduction of some vibrations, which are 1400–1500 cm^{-1} region associated with CH_2 bending of pectin, lignin, and hemicellulose, 1616 cm^{-1} related with benzene ring stretching of lignin, 2850–2910 cm^{-1} region associated with CH_2 stretching of Wax and C-H stretching of polysaccharides, and 3200–3600 cm^{-1} region corresponded with OH stretching of polysaccharides, is observed [22, 26, 39]. In particular, the peak at 1710 cm^{-1} is attributed to the C=O stretching of the acetyl groups of hemicellulose [26, 39]. The removal of hemicellulose from the fiber surfaces causes the peak to disappear [22, 26]. On the contrary, the FT-IR spectra of surface treatment by ureidosilane (S3) present some clear peaks, which are 1710 cm^{-1} corresponding with C=O stretching, 2850 and 2910 cm^{-1} related with CH_2 stretching, and 3200–3600 cm^{-1} region associated with OH stretching. These findings indicate the following: alkali treatment by $NaClO_2$ has more attackability on HF than that by NaOH.

Accordingly, the former is able to completely remove lignin, wax, and hemicellulose from hemp fiber bundles and replaces more OH groups the hemp fiber surfaces. Meanwhile, the peaks at 1710, 2850, and 2910 cm^{-1} show the presence of silane in the surface treatment by ureidosilane (S3), although the same peak is not present in the only alkali treatment. This may be attributed to the evidence of chemical bond between fiber and the silane coupling agent. Incidentally, Khan [41] and Sgriccia [39] have reported the presence of silane in fiber, which is the peak at 766 and 847 cm^{-1}. However, the concentration of silane on the fiber surfaces is too small to be detected by FT-IR in this study.

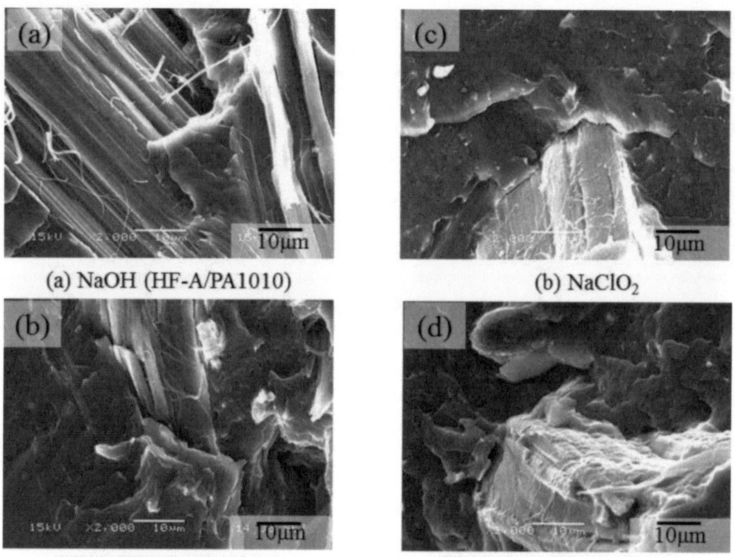

(a) NaOH (HF-A/PA1010)
(b) $NaClO_2$
(c) NaOH+Ureidosilane (HF-A-S3/PA1010)
(d) $NaClO_2$+Ureidosilane

Figure 8. SEM photographs of cryogenically fractured surface of various HF/PA1010 biomass composites.

Furthermore, to clarify the interfacial interaction between fiber (HF) and matrix polymer (PA1010), we performed SEM observation of the fractured surface cryogenically in liquid nitrogen. **Figure 8** shows SEM photographs of cryogenically fractured surface of various HF/PA1010 biomass composites: (a) NaOH, (b) $NaClO_2$, (c) NaOH + ureidosilane (S3), and (d) $NaClO_2$ + ureidosilane (S3). The results of the comparison of the results in **Figure 8** suggest that the alkali treatment by $NaClO_2$ shows larger physical contact area between HF and PA1010 than that of NaOH. The morphologies of the composites surface-treated by NaOH (**Figure 8(a)**) and $NaClO_2$ (**Figure 8(b)**) show poor interaction between HF and PA1010. This indicates poor chemical contact between fiber and matrix polymer. On the contrary, the morphologies of the composites surface-treated by both alkali treatment and silane coupling agent (**Figure 8(c)** and **(d)**) show good interaction between fiber and matrix polymer, and fiber does not leave any voids on the fracture surface. This is attributed to the chemical reaction between the ureido group in the silane coupling agent and the possible reaction site on the fiber by $NaClO_2$ alkali treatment. These results are attributed to the good combination between $NaClO_2$ and ureidosilane coupling agent (S3) for improving the rheological properties of HF/PA1010 biomass composites. These SEM image observations are in agreement with the mechanical properties such as tensile and bending results in previous report [16].

3.6. Temperature dependence

The effects of temperature on the viscoelastic properties of various surface-treated HF/PA1010 biomass composites are discussed here. The complex viscosity $|\eta^*|$ of various HF/PA1010 biomass composites is plotted against the reciprocal of the absolute temperature $1/T$ at the angular frequency ω of 25 rad/s in **Figure 9**. $|\eta^*|$ of all surface-treated HF/PA1010 biomass composites except for the NaOH systems is lower than that of the untreated ones. The effect of alkali treatment on $|\eta^*|$ of HF/PA1010 biomass composites differs for each surface treatment. In particular, NaOH + ureidosilane (S3) is most sensitive to temperature, while the untreated composites are the least. From the slope of $|\eta^*|$ vs. $1/T$ plots, the apparent activation energy E_α for flow can be calculated from the following Andrade's equation:

$$\eta = A\exp\left(\frac{E_a}{RT}\right) \quad (1)$$

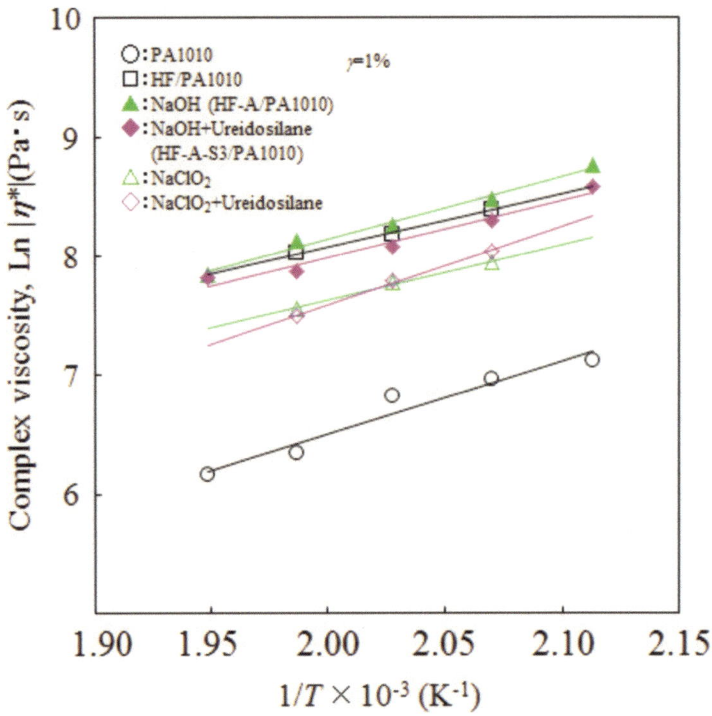

Figure 9. Temperature dependence of complex viscosity for various HF/PA1010 biomass composites.

where A is the constant value, R is the gas constant, and η was replaced by $|\eta^*|$ [12, 33]. The apparent activation energy E_a of various HF/PA1010 biomass composites is listed in **Table 2**. It was found that E_a increases in the following order: untreated (HF) < $NaClO_2$ < NaOH + ureidosilane (S3) < NaOH < pure PA1010 < $NaClO_2$ + ureidosilane (S3). It can be said that the fluidity of the materials increases with increasing Ea. Briefly, $NaClO_2$ + ureidosilane (S3) with the highest value of E_a indicates high sensitivity to temperature changes.

Code	PA1010	HF/PA1010	NaOH (HF-A/PA1010)	NaOH+ Ureidosilane (HF-A-S3/PA1010)	$NaClO_2$	$NaClO_2$+Ureidosilane
E_a (kJ/mol)	50.5	37.4	44.2	39.9	38.3	55.1

Table 2. The apparent activation energy of various HF/PA1010 biomass composites.

4. Influence of type of silane coupling agent on dynamic viscoelastic properties of hemp fiber-reinforced plant-derived polyamide 1010 biomass composites in the molten state

4.1. Introduction

It is important to analyze the flow/deformation behavior of high-performance natural fiber-reinforced polymer biomass composites during the polymer processing and to investigate the flow mechanisms and changes in the internal structure of these biomass composites. In particular, although these biomass composites undergo various flow processes, the effect of surface treatment on the rheological properties has not been studied enough [12, 21, 34, 42–44]. Therefore, there is a need for proper rheological studies on the effect of surface treatment taking into account various factors such as type of fiber, size and size distribution, degree of agglomeration, and type of surface treatment. However, the determination of the effect of surface treatment on the interface or interphase adhesion between fiber and matrix polymer is thought to be a complicated task since these factors are interrelated. We reported the effect of alkali treatment on the dynamic viscoelastic properties of hemp fiber-reinforced plant-derived polyamide 1010 biomass composites in the molten state in the preceding section. According to our survey on the previous results, the influence of the type of silane coupling agent used on the rheological properties of these biomass composites is still not well known [12]. It is therefore necessary to systematically investigate it for the further understanding of this problem. This section reports the effects of the type of silane coupling agent used on the rheological properties, which are dynamic viscoelastic properties in the molten state, investigated experimentally, for hemp fiber-reinforced plant-derived polyamide 1010 biomass composites as mentioned earlier. It discusses the dynamic viscoelastic properties in terms of various factors: angular frequency, volume fraction, various kinds of silane coupling agents, and temperature.

4.2. Materials and methods

The materials used in this study were various surface-treated hemp fiber-filled polyamide 1010 biomass composites. Since the materials, processing, and experimental methods are similar to those mentioned in Section 3.2 other than the surface treatment by silane coupling agents, details are omitted here. Hemp fibers were previously cut into 5-mm-long pieces and surface-treated by two types of surface treatment: a) alkali treatment by sodium hydroxide (NaOH) solution and b) surface treatment by silane coupling agents. Alkali treatment by NaOH was employed as follows: 5% NaOH solution was placed in a stainless beaker. Then chopped hemp fibers were then added in the beaker and stirred well. This was kept at room temperature for 4 h. The fibers were then washed thoroughly with water to remove the excess NaOH sticking to the fibers. The alkali-treated fibers (HF-A) were dried in air for 12 h and in a vacuum oven at 80 °C for 5 h. Three types of silane coupling agents such as aminosilane (S1, 3-(2-aminoethylamino) propyltrimethoxy silane, A-1120, Momentive Performance Materials Inc., USA), epoxysilane (S2, 3-glycidoxypropyltrimethoxy silane, A-187), and ureidosilane (S3, 3-ureidopropyltrimethoxy silane, A-1160) were used as the surface treatment agents. The treatment of hemp fibers with the concentration of 1 wt.% of the chosen silane coupling agent was carried out in deionized water (for S1) or 5 wt.% acetic acid aqueous solution (for S2 and S3, where pH of the solution was adjusted to 3.5) and stirred continuously for 15 min. Then, the fibers were immersed in the solution for 60 min. After treatment, fibers were removed from the solution and dried in air for 12 h and in vacuum oven at 80°C for 5 h. The volume fraction of fiber V_f was varied with 10 and 20 vol.%/ The materials used in this section are listed in **Table 3**.

Code	Alkali treatment	Silane coupling agent
HF-A	NaOH	–
HF-A-S1	NaOH	Aminosilane (S1)
HF-A-S2	NaOH	Epoxysilane (S2)
HF-A-S3	NaOH	Ureidosilane (S3)

Table 3. Code, alkali treatment and silane coupling agent used in this study.

4.3. Angular frequency dependences

The rheological properties of various surface-treated hemp fiber-filled polyamide 1010 biomass composites (HF/PA1010) in molten state were evaluated by oscillatory flow behavior. The dynamic viscoelastic properties of various surface-treated HF/PA1010 biomass composites (HF content is 20 vol.%) are plotted as a function of angular frequency ω in **Figure 10(a)** (storage modulus G') and **Figure 10(b)** (loss modulus G''), respectively. The G' of various surface-treated HF/PA1010 biomass composites shows the typical storage modulus G' of highly filled systems, indicating the "second rubbery plateau" [21, 31–35]; however, the value of G' changes with the types of silane coupling agents used. The G' of aminosilane (HF-A-S1/PA1010) and epoxysilane (HF-A-S2/PA1010) is lower than that of untreated (HF/PA1010) in wide ω regions. On the contrary, G' of ureidosilane (HF-A-S3/PA1010)-treated composites is at the same level of that of the untreated one in the low ω regions, although this is lower than that of untreated in the high ω regions. The slope of G' in the low ω region decreases in the following order: NaOH (HF-A) < ureidosilane (HF-A-S3) < epoxysilane (HF-A-S2) < aminosilane (HF-A-S1). This may be attributed to the change of interfacial interaction between HF and PA1010 according to the type of silane coupling agent used. These behaviors differ from the dynamic viscoelastic properties of the polymer composites such as glass fiber-reinforced polypropylene composites (GF/PP), in which G' of the surface-treated GF/PP by silane coupling agent is higher than that of untreated [21]. In general, the influence of surface treatment by coupling agents on the dynamic viscoelastic properties of the polymer composites is complex. This is because the surface treatment by coupling agents plays multiple roles simultaneously, coupling agent which improves the adhesion between the fiber and the polymer, lubricant which reduces friction, plasticizer which helps soften fiber and polymer, wetting agent which reduces the agglomeration of fibers, and additive for deforming fiber assembly easily and lowering the viscoelastic properties of the matrix polymer [21]. It is thought that the mechanisms mentioned earlier do not occur separately, and therefore it is very difficult to distinguish the factors of the coupling agent influencing the viscoelastic properties. On the contrary, loss modulus G'' of various surface-treated HF/PA1010 biomass composites is higher than that of pure PA1010; however, the effect of surface treatment by silane coupling agent on G'' of HF/PA1010 biomass composites does not differ according to the type of silane coupling agent used. This is because G' is a more sensitive rheological function to the structural changes of the internal structure of polymer composites compared to G''. In other words, G' is considered to be a sensitive indicator for the quantitative analysis of morphological changes in the composites as mentioned earlier. Next, the complex viscosity $|\eta^*|$ of various surface-treated HF/PA1010 biomass composites (HF content is 20 vol.%) is plotted as a function of angular frequency ω in **Figure 10(c)**. The $|\eta^*|$ of neat PA1010 does not change with ω in low ω region, which shows Newtonian behavior, and this slightly decreases with increasing ω in the high ω regions. On the contrary, the $|\eta^*|$of various HF/PA1010 biomass composites rapidly decreases with increasing ω in a wide range. The order of magnitude is more viscous than neat PA1010 even in the high ω regions. In addition, it is

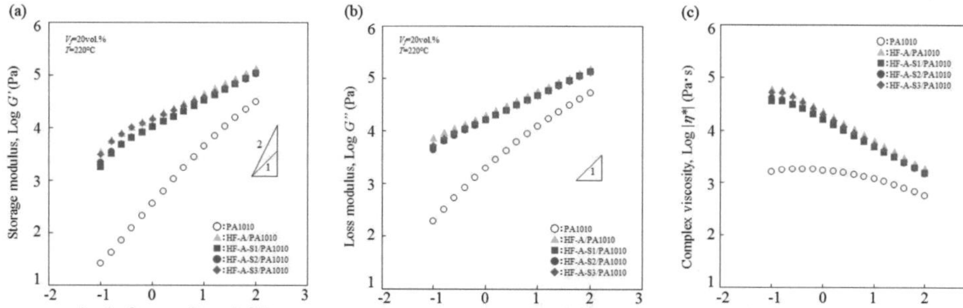

Figure 10. Dynamic viscoelastic properties as a function of angular frequency for various HF/PA1010 biomass composites. (a) Storage modulus, (b) loss modulus, and (c) complex viscosity.

interesting to note that the $|\eta^*|$ of all surface-treated HF/PA1010 composites by silane coupling agent shows a smaller value than NaOH-treated composites (HF-A/PA1010 composites).

4.4. Influence of volume fraction

The influence of volume fraction of fiber V_f on the rheological properties of various surface-treated HF/PA1010 biomass composites is discussed in this section. The relative complex viscosity $|\eta^*|_{r0}$ is plotted against volume fraction of fiber in **Figure 11**. Here, this relative value is given by the value of various surface-treated HF/PA1010 biomass composites divided by that of neat PA1010. The $|\eta^*|_{r0}$ of various surface-treated HF/PA1010 biomass composites increases with an increasing volume fraction of fiber V_f. This tendency changes with the type of silane coupling agent used. However, it is remarkably influenced by the volume fraction of fiber V_f rather than its types. These results may be attributed to the internal microstructure formed by the interaction between fibers rather than the interaction between fiber and matrix polymer. In addition, it was also found that aminosilane (HF-A-S1) is the least viscous of all silane coupling agents in this study. This behavior will be discussed in the next section.

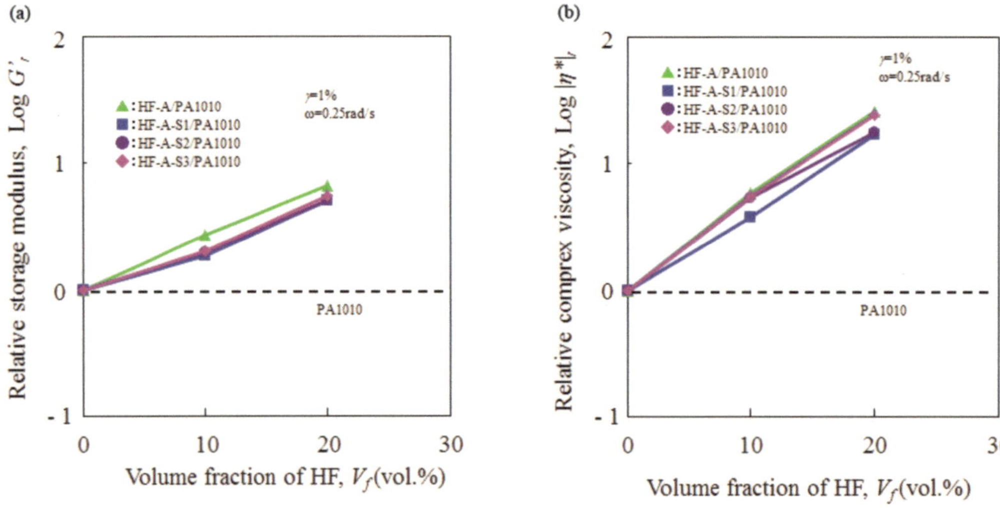

Figure 11. Influence of volume fraction of fiber on the dynamic viscoelastic properties of various HF/PA1010 biomass composites. (a) Relative storage viscosity and (b) relative complex viscosity.

4.5. Influence of type of silane coupling agent

To further clarify the effects of surface treatment by silane coupling agent on the rheological properties of HF/PA1010 biomass composites, the relative storage modulus G'_r of various surface-treated HF/PA1010 biomass composites is shown in **Figure 12(a)** (ω=0.25 rad/s) and **Figure 12(b)** (ω=25 rad/s), respectively. Here, the relative storage modulus is given by the values of surface-treated HF/PA1010 composites divided by those of NaOH-treated ones (HF-A/PA1010 composites). The values of G'_r in the low ω regions (ω = 0.25 rad/s) differ according to the type, and G'_r of aminosilane (HF-A-S1) and epoxysilane (HF-A-S2) remarkably decreases although that of ureidosilane (HF-A-S3) is almost the same as that of the NaOH-treated composites. Contrarily, the G'_r of all HF/PA1010 composites in the high ω regions (ω = 25 rad/s) is lower than that of the NaOH-treated ones and slightly decreases in the following order: ureidosilane (HF-A-S3) > epoxysilane (HF-A-S2) > aminosilane (HF-A-S1). Thus, the effect of the type of silane coupling agent used on the storage modulus of HF/PA1010 biomass composites changes with the difference in the frequency region. It is well known that the storage modulus in the low ω region is influenced by the strong interaction of fiber–matrix polymer and/or fiber–fiber and the microstructure formation by the fibers. On the contrary, this value in the high ω region is generally dominated by the matrix polymer. From the results shown above, it can be concluded that the various silane coupling agents decrease the storage modulus, and silane coupling agents used in this study play the role of the lubricant, the wetting agent, or the plasticizer, supported by the facts in **Figure 12(a)** and **(b)**. However, only ureidosilane (HF-A-S3) may play the vital role of the coupling agent which improves the interaction between fiber and matrix polymer as the G' values of ureidosilane (HF-A-S3) are higher than those of the other systems. This tendency was also observed for the mechanical and tribological properties of these biomass composites [10, 11, 16].The relative complex viscosity $|\eta^*|_r$ of various surface-treated HF/PA1010 biomass composites is shown in **Figure 12(c)** (ω=0.25 rad/s) and **Figure 12(d)** (ω=25 rad/s), respectively. Here, the relative complex viscosity is given by the values of surface-treated HF/PA1010 composites divided by that of the NaOH-treated ones (HF-A/PA1010 composites). $|\eta^*|_r$ has the same tendency as that for G'_r. Briefly, the viscoelastic properties of HF/PA1010 biomass composites decrease with surface treatments using the silane coupling agent and are highly influenced by angular frequency.

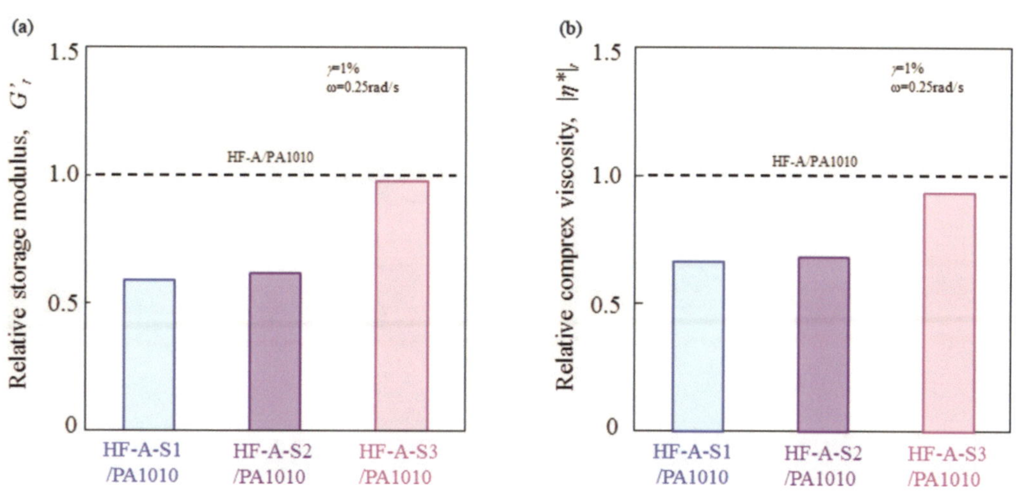

Figure 12. The relative values for various surface-treated HF/PA1010 biomass composites. (a) Relative storage modulus and (b) relative complex viscosity.

4.6. Temperature dependences

The influence of temperature on the viscoelastic properties of surface-treated HF/PA1010 biomass composites is discussed here. The complex viscosity $|\eta^*|$ of various surface-treated HF/PA1010 composites is plotted against the reciprocal value of the absolute temperature $1/T$ at the angular frequency ω of 25 rad/s in **Figure 13**. The $|\eta^*|$ of all surface-treated HF/PA1010 composites using silane coupling agent is lower than that of the NaOH-treated ones (HF-A/PA1010 composites), and the difference of the value increases with increasing temperature. The influence of temperature on $|\eta^*|$ of HF/PA1010 composites differs for each surface treatment using silane coupling agent. It is found that epoxysilane (HF-A-S2) is the most temperature sensitive, and ureidosilane (HF-A-S3) is the least. From the slope of $|\eta^*|$ vs. $1/T$ plots, the apparent activation energy E_a for flow can be calculated from the previous Andrade's equation (1). **Figure 14** shows the apparent activation energy E_a of various HF/PA1010 biomass composites. It was found that E_a increases in the following order: ureidosilane (HF-A-S3) < NaOH (HF-A) < aminosilane (HF-A-S1) < neat PA1010 < epoxysilane (HF-A-S2). It can be said that the fluidity of the materials increases with increasing E_a. Briefly, epoxysilane (HF-A-S2) with the highest value of E_a demonstrates high sensitivity to temperature changes. This means that ureidosilane (HF-A-S3) with the lowest value of E_a has stable flow processability in a wide temperature region for the HF/PA1010 biomass composites studied here.

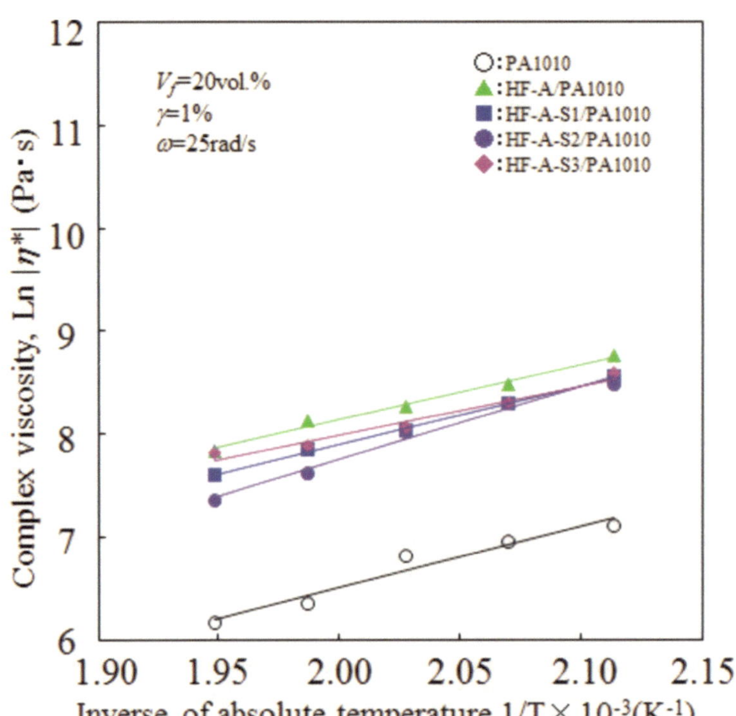

Figure 13. Temperature dependence of complex viscosity for various surface-treated HF/PA1010 biomass composites.

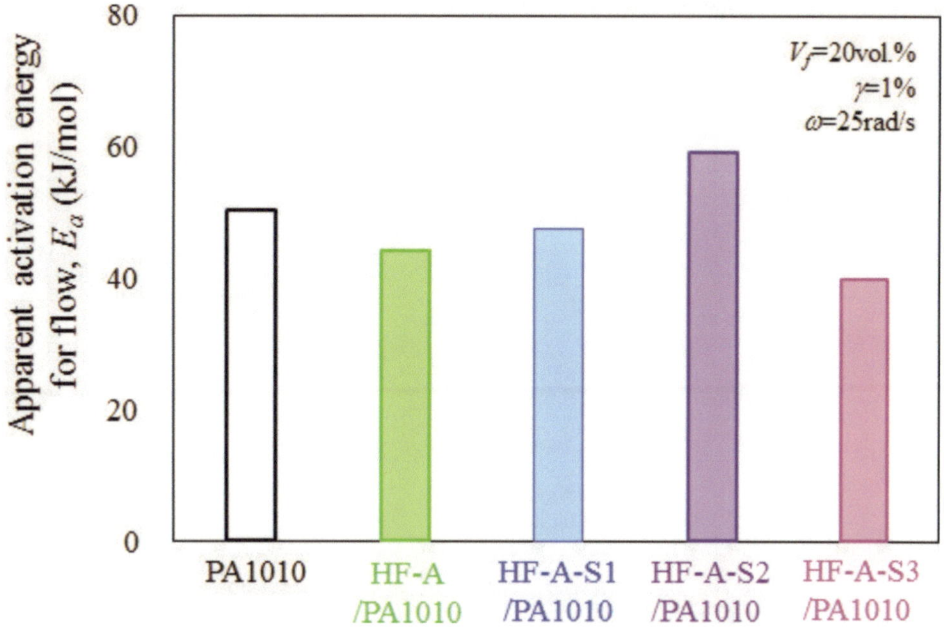

Figure 14. The apparent activation energy E_a of various HF/PA1010 biomass composites.

4.7. Complex modulus dependences

The complex modulus dependence of various surface-treated HF/PA1010 biomass composites in the molten state is discussed in this section. **Figure 15** shows the phase angle δ (=arctan G''/G') is plotted against the absolute value of the complex modulus $|G^*|$ of various surface-treated HF/PA1010 biomass composites. The van Gurp–Palmen plot [45] is drawing attention recently as another method for representing internal microstructures and their changes. This plot is considered to be a sensitive indicator for the time-temperature superposition, the presence of long chain branch of the polymer, gelation behavior, rheological percolation of polymer nanocomposites, etc. [45–47]. This is because the changes in the rheological properties, from which it is hard to understand the angular frequency dependence (G'- ω, G''- ω, tan δ – ω curves), are emphasized in this δ-$|G^*|$ plot. As seen in **Figure 15**, the δ-$|G^*|$ curve of neat PA1010 approaches 90° at the low complex modulus $|G^*|$, indicating the flow behavior of viscous fluid. On the contrary, the δ-$|G^*|$ curves of various surface-treated HF/PA1010 biomass composites show the complex behavior. However, these are of minimal value. In particular, ureidosilane (HF-A-S3), which is considered to play a strong role as a coupling agent for improving the interfacial interaction between the fiber and polymer, clearly shows the minimum value. It is also the lowest value compared to other surface treatment systems. On the contrary, the other composites which play a weak role as a coupling agent, do not show the minimum values. Thus, this δ-$|G^*|$ plot may be able to serve as an indicator for the effect of surface treatment of polymer composites.

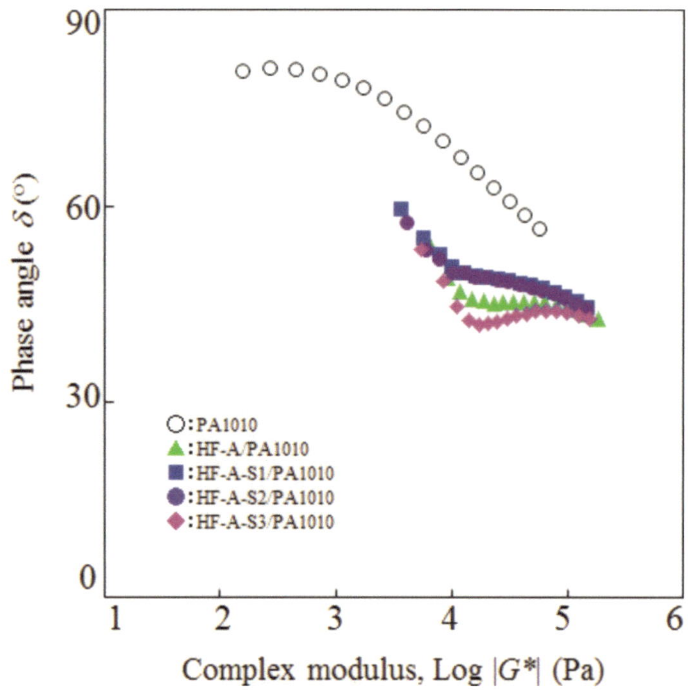

Figure 15. The phase angle is plotted against the absolute value of the complex modulus of various HF/PA1010 biomass composites.

5. Conclusion

The aim of this study is to investigate the thermal properties of hemp fiber-filled plant-derived polyamide 1010 biomass composites and their dynamic viscoelastic properties of these composites in the molten state experimentally. It was found that the addition of HF with PA1010 has a strong influence on the thermal properties such as DMA, TGA, and DSC. In particular, HF is effective for improving thermal and mechanical properties. The effect of alkali treatment on the dynamic viscoelastic properties of HF/PA1010 composites in the molten state differs according to whether the alkali treatment uses the silane coupling agent or not. In particular, the viscoelastic properties (both storage and loss moduli) of $NaClO_2$ are higher than those of NaOH. The silane coupling agents have a remarkable influence on (1) rheological properties such as storage modulus, loss modulus, loss tangent, and complex viscosity in low angular frequency regions in the molten state, (2) temperature dependence of rheological properties, and (3) relationship between phase angle and complex modulus (van Gurp–Palmen plots). These rheological behaviors were also strongly influenced by the type of silane coupling agents used. The viscoelastic properties (both storage and loss moduli) of aminosilane and epoxysilane treated composites were lower, which those of ureidosilane-treated ones were higher than the moduli of only alkali-treated composites. Ureidosilane-treated composites were the least temperature sensitive in the surface treated composites investigated here.

Acknowledgements

This work was supported by JSPS KAKENHI Grant Number 16K06750 and 25420735. We would like to thank the Functional Microstructured Surfaces Research Center (FMS, MEXT, Japan) of Kogakuin University, the Project Research of Research Institute for Science and Technology of Kogakuin University and the Ogasawara Foundation for the Promotion of Science & Engineering for funding this study, and partial support by the national budget of Czech Republic within the framework of the Centre of Polymer Systems project (reg. number: CZ. 1.05/2.1.00/ 03.0111). The authors thank Mr. J. Mukaida, Ms. M. Hasumi, Mr. K. Nakamura and many master course and undergraduate course students at the Polymeric Materials Laboratory of Kogakuin University for their assistance with the experiments.

Author details

Yosuke Nishitani[1*], Toshiyuki Yamanaka[2], Tetsuto Kajiyama[3] and Takeshi Kitano[4]

*Address all correspondence to: at13152@ns.kogakuin.ac.jp

1 Department of Mechanical Engineering, Faculty of Engineering, Kogakuin University, Tokyo, Japan

2 Tokyo Metropolitan Industrial Technology Research Institute, Tokyo, Japan

3 Jonan Branch, Tokyo Metropolitan Industrial Technology Research Institute, Tokyo, Japan

4 Polymer Centre, Faculty of Technology, Tomas Bata University in Zlin, Zlin, Czech Republic

References

[1] Wibowo AC., Mohanty AC., Misra M., Lawrence T., Drzal LT. Chopped industrial hemp fiber reinforced cellulosic plastic biocomposites: Thermomechanical and morphological properties. Thermomechanical and morphological properties. Ind. Eng. Chem. Res. 2004;43:4883–4888. DOI: 10.1021/ie030873c

[2] Petersson L., Oksman K., Biopolymer based nanocomposites. Comparing layered silicates and microcrystalline cellulose as nanoreinforcement. Sci. Technol. 2006;66:2187–2196. DOI: 10.1016/j.compscitech.2005.12.010

[3] Bledzki AK., Gassan J., Composites reinforced with cellulose based fibres. Prog. Polym. Sci. 1999;24:221–274. DOI: 10.1016/S0079-6700(98)00018-5

[4] Omar F., Bledzki AK., Fink H-P., Sain M. Biocomposites reinforced with natural fibers: 2000–2010. Prog. Polym. Sci. 2012;37:1552–1596. DOI: 10.1016/j.progpolymsci.2012.04.003

[5] John MJ., Thomas S. Biofibres and biocomposites. Carbohydrate Polymers. 2008;71:343–364. DOI: 10.1016/j.carbpol.2007.05.040

[6] Li X., Tabil L.G., Panigrahi S. Chemical treatments of natural fiber for use in natural fiber-reinforced composites: A review. J. Polym. Environ. 2007;15:25–33. DOI: 10.1007/s10924-006-0042-3

[7] Kalia S, Kaith B.S., Kaur I. Pretreatments of natural fibers and their application as reinforcing material in polymer composites – a review. Polym. Eng. Sci. 2009;49:1253–1272. DOI: 10.1002/pen.21328

[8] Xie Y., Hill CAS., Xiao Z., Militz H., Mai C. Silane coupling agents used for natural fiber/polymer composites: A review. Composites Part A. 2010;41:806–819. DOI: 10.1016/j.compositesa.2010.03.005

[9] Oksman K., Skrifvars M., J.-F. Selin. Natural fibres as reinforcement in polylactic acid (PLA) composites. Compos. Sci. Technol. 2003;63:1317–1324. DOI: 10.1016/S0266-3538(03)00103-9

[10] Hasumi M., Nishitani Y., Kitano T. Mechanical properties of hemp fiber filled polyamide 1010 composites. In: The Proceedings of Seikei-kako Symposia'11; 14–17 October 2011; Akita (in Japanese). 2011. p. 491–492.

[11] Hasumi M., Nishitani Y., Kitano T. Effect of surface treatment on the mechanical properties of hemp fiber reinforced polyamide 1010 composites. In: The Proceedings of the Polymer Processing Society 28th Annual Meeting (PPS-28); 11–15 December 2012; Pattaya. 2012. p. 07-324.

[12] Nishitani Y., Hasumi M., Kitano T. Influence of silane coupling agents on the rheological behavior of hemp fiber filled polyamide 1010 biomass composites in molten state. In: AIP Conference Proceedings; "Cleveland, Ohio, USA" 2015; 1664: 060007. Cleveland, Ohio, USA, http://dx.doi.org/10.1063/1.4918425

[13] Nishitani Y., Mukaida J., Yamanaka T., Kajiyama T., Kitano T.. Effect of processing sequence on the dynamic viscoelastic properties of ternary biomass composites (hemp fiber/ PA1010/ PA11E) in the Molten State. In: The Proceedings of Regional Conference of Polymer Processing Society (PPS-2105); 21–24 September 2015; Graz. 2015. Nat-81.

[14] Nishitani Y., Mukaida J., Yamanaka T., Kajiyama T., Kitano T. Thermal properties of hemp fiber filled polyamide 1010 biomass composites and the blend of these composites and polyamide 11 elastomer. In: The Proceedings of the 31st International Conference of the Polymer Processing Society (PPS-31); 7–11 June 2015; Jeju. 2015. p. 1116-1120.

[15] Mukaida J., Nishitani Y., Kitano T. Effect of addition of plants-derived polyamide 11 elastomer on the mechanical and tribological properties of hemp fiber reinforced polyamide 1010 composites, AIP Conference Proceedings "Cleveland, Ohio, USA" 2015; 1664: 060008. Cleveland, Ohio, USA, http://dx.doi.org/10.1063/1.4918426

[16] Mukaida J., Nishitani Y., Kajiyama T., Yamanaka T., Kitano T. Influence of types of alkali treatment on the mechanical properties of hemp fiber reinforced polyamide 1010 composites. In: The Proceedings of Regional Conference of Polymer Processing Society (PPS-2105); 21–24 September 201; Graz. 2015. Nat-104.

[17] Mukaida J., Nishitani Y., Kajiyama T., Yamanaka T., Kitano T. Effect of type of bio-TPE on the mechanical and tribological properties of plants-derived ternary composites (HF/PA1010/bio-TPE). J. Mater. Test. Res. Ass. Japan, 2016; 2016. p. 3–11.

[18] Naganuma M. Plants-derived polyamide "VESTAMID Terra" "VESTAMID HT Plus". JETI. 59; 2011. p. 88–90.

[19] Hausnerova B., Honkova N., Lengalova A., Kitano T., Saha P. Rheology and fiber degradation during shear flow of carbon-fiber-reinforced polypropylenes, Polym. Sci. Series A; 2006. p. 951–960. DOI: 10.1134/S0965545X06090100

[20] Hausnerova, B., Honkova, N., Lengalova, A., Kitano, T., Saha, P. Rheological behavior of fiber-filled polymer melts at low shear rate Part. I. Modeling of rheological properties. "Polimery/Polymers". (ISSN: 0032–2725). 2008;53:507–512.

[21] Nishitani Y., Ishii C., Kitano T. Rheological Properties of Surface Treated Glass Fiber Reinforced Polypropylenes in Molten State, Polypropylene, Dr. Fatih Dogan (Ed.), 2012. ISBN: 978-953-51-0636-4, InTech, Available from: http://www.intechopen.com/books/polypropylene/rheological-properties-ofsurface-treated-glass-fiber-reinforced-polypropylenes-in-molten-state

[22] Chand N, Fahim M. Tribology of Natural Fiber Polymer Composites. Woodhead Publishing Ltd; "Cambridge, England" 2008.

[23] Alawar A., Hamed AM., Al-Kaabi K. Characterization of treated date palm tree fiber as composite reinforcement. Compos. Part B Eng. 2009;40:601–606. DOI: 10.1016/j.compositesb.2009.04.018

[24] Aziz SH., Ansell MP. The effect of alkalization and fibre alignment on the mechanical and thermal properties of kenaf and hemp bast fibre composites: Part 1 – polyester resin matrix. Compos. Sci. Technol. 2004;64:1219–1230. DOI: 10.1016/j.compscitech.2003.10.001

[25] Abdelmouleh M., Boufi S., Belgacem M.N., Dufresne A. Short natural-fibre reinforced polyethylene and natural rubber composites: Effect of silane coupling agents and fibres loading. Compos. Sci. Technol. 2007;67:1627–1639. DOI: 10.1016/j.compscitech.2006.07.003

[26] Troedec ML., Sedan D., Peyratout C., Bonnet JP., Smith A., Guinebretiere R., Gloaguen V., Krausz P. Influence of various chemical treatments on the composition and structure of hemp fibres. Compos. Part A App. Sci. Manuf. 2008;39:514–522. DOI: 10.1016/j.compositesa.2007.12.001

[27] Araujo JR., Waldman WR., De Paoli MA. Thermal properties of high density polyethylene composites with natural fibres: Coupling agent effect. Polym. Degrad. Stab. 2008;93:1770–1775. DOI: 10.1016/j.polymdegradstab.2008.07.021

[28] Zeng H., Gao C., Wang Y., Watts PCP., Kong H., Cui X., Yan D. In situ polymerization approach to multiwalled carbon nanotube-reinforced nylon 1010 composites: Mechanical properties and crystallization behavior. Polymer. 2006;47:113–122. DOI: 10.1016/j.polymer.2005.11.009

[29] Li L., Li CY., Ni C., Rong L., Hsiao B. Structure and crystallization behavior of Nylon 66/multi-walled carbon nanotube nanocomposites at low carbon nanotube contents. Polymer. 2007;48:3542–3460. DOI: 10.1016/j.polymer.2007.04.030

[30] Choi JS., Lim ST., Choi HJ., Mohanty AK., Drzal LT., Misra M., Wibowo A. Preparation and characterization of plasticized cellulose acetate biocomposite with natural fiber. J. Mater. Sci. 2004;39:6631–6633. DOI: 10.1023/B:JMSC.0000044909.60878.97

[31] Ferry, JD. Viscoelastic Properties of Polymers. John Wiley & Sons, Incorporated; "New York, USA" 1980.

[32] Nishitani, Y., Yamada, Y., Ishii, C., Sekiguchi, I., Kitano, T. Effects of addition of functionalized sebs on rheological, mechanical, and tribological properties of polyamide 6 nanocomposites. Polym. Eng. Sci. 2010;50:100–112. DOI: 10.1002/pen.21516

[33] Nishitani, Y., Sekiguchi, I., Hausnerova, B., Nagatsuka, Y., Kitano, T. Dynamic viscoelastic properties of long organic fibre-reinforced polypropylene in molten state. Polym. Polym. Composi. 2001;9:199–211.

[34] Nishitani, Y., Sekiguchi, I., Hausnerova, B., Zdrazilova, N., Kitano, T. Rheological properties of aminosilane surface treated short glass fibre reinforced polypropylenes. Part 1: Steady shear and oscillatory flow properties in molten state. Polym. Polym. Composi. 2007;15:111–120.

[35] Nishitani, Y., Ohashi, K., Sekiguchi, I., Ishii, C., Kitano, T. Influence of addition of styrene-ethylene/butylene-styrene copolymer on rheological, mechanical, and tribological properties of polyamide nanocomposites. Polym. Composi. 2010;31:68–76. DOI: 10.1002/pc.20767

[36] Wan T., Clifford MJ., Gao F., Bailey AS., Gregory DH., Somsunan R. Strain amplitude response and the microstructure of PA/clay nanocomposites. Polymer. 2005;46:6429–6436. DOI: 10.1016/j.polymer.2005.04.105

[37] Durmus A., Kasgoz A., Macosko CW. Linear low density polyethylene (LLDPE)/clay nanocomposites. Part I: Structural characterization and quantifying clay dispersion by melt rheology. Polymer. 2007;48:4492–4502. DOI: 10.1016/j.polymer.2007.05.074

[38] Nishitani, Y., Sekiguchi, I., Ishii, C., Kitano, T. Dynamic viscoelastic properties of carbon nanofiber polyamide 66 composites in Molten State. Mater. Technol. 2010;28:135–145.

[39] Sgriccia N., Hawley MC., Misra M. Characterization of natural fiber surfaces and natural fiber composites. Compos. Part A. 2008;39:1632–1637. DOI: 10.1016/j.compositesa.2008.07.007

[40] Merlini C., Soldi V., Barra GMO. Influence of fiber surface treatment and length on physico-chemical properties of short random banana fiber-reinforced castor oil polyurethane composites. Polymer Test. 2011;30:833–840. DOI: 10.1016/j.polymertesting.2011.08.008

[41] Khan MA., Drzal LT. Characterization of 2-hydroxyethyl methacrylate (HEMA)-treated jute surface cured by UV radiation. J. Adhes. Sci. Technol. 2004;18:381–393. DOI: 10.1163/156856104773635481

[42] Shenoy AV. Rheology of Filled Polymer Systems. Kluwer Academic Publishers, Dordrecht, The Netherlands; 1999.

[43] Han CD., Van den Weghe T., Shete P., Haw JR. Effects of coupling agents on the rheological properties, processability, and mechanical properties of filled polypropylene. Polym. Eng. Sci. 1981;21:196–204.

[44] Khan SA., Prud'homme RK. Melt rheology of filled thermoplastics. Rev. Chem. Eng. 1987;4:205–272.

[45] van Gurp M., Palmen J. Time temperature superposition for polymeric blends. Rheol Bull. 1998;67:5–8.

[46] Trinkle S., Friedrich C. Van Gurp–Palmen-plot: A way to characterize polydispersity of linear polymers. Rheol. Acta. 2001;40:322–328. DOI: 10.1007/s003970000137

[47] Pötschke P., Abdel-Goad M., Alig I., Dudkin S., Lellinger D. Rheological and dielectrical characterization of melt mixed polycarbonate-multiwalled carbon nanotube composites. Polymer. 2004;45:8863–8870. DOI: 10.1016/j.polymer.2004.10.040

Chapter 5

Plasticizers for Protein-Based Materials

N. Henrik Ullsten, Mikael Gällstedt and
Mikael S. Hedenqvist

Additional information is available at the end of the chapter

http://dx.doi.org/10.5772/64073

Abstract

This study presents a practical approach to select plasticizers for proteins. It is a case study on thermoformed wheat gluten, considered here as a model protein, and it involved 30 plasticizer candidates. The approach consisted of selecting plasticisers (30 wt%) based on visual examination, rheological and molding behavior of the dough, and finally tensile data. There was no unique relationship between the torque behavior of the dough and the mechanical properties of the films. Nevertheless, the extensibility and dough analysis indicated that the most promising plasticizers were as follows: glycerol, linear glycols, ethanol amines, diols, and trimethylolpropane. Further, considering also the stiffness, it was concluded that the most efficient plasticisers were those that contained three hydroxyl groups and the linear glycols of intermediate size. Out of those, glycerol stood out as having the highest extensibility and lowest stiffness and strength. In an attempt to predict the mechanical properties of the films based on several physical data of the compounds, it was observed that there was a weak nonlinear relationship between the stiffness/strength and the size (molecular weight/molar volume), polarity and molar refractivity of the compound. The stiffness/strength decreased with an increase in these physical parameters.

Keywords: proteins, wheat gluten, plasticisers, compression molding, mixing

1. Introduction

Protein-based films have generally low oxygen permeability in dry conditions due to the high amount of hydrogen bonds. However, the hydrogen bonds also make the films brittle in dry conditions, and a plasticizer is needed for the film to have desirable ductility [1–4]. Several factors must be taken into account when choosing between different possible plasticizers, and the perfect plasticizer is probably not to be found. Plasticizer and protein materials must have similar polarity to be compatible. Insufficient dispersion of the plasticizer in the protein matrix results in a material with properties depending on the internal variation in plasticizer concentration. The molecular weight and chemistry/polarity of the plasticizer affects its diffusion properties and therefore the migration from the protein matrix and the long-term properties of the protein material. There are numerous reports on plasticized protein-based films, including matrices from plants such as soy, pea, sunflower, and wheat proteins and zein [5]. Animal-based protein matrices include sodium caseinate, keratin, gelatin, collagen, and whey and myofibrillar proteins. Plasticisers have also been an important factor to consider in biocomposites [6].

The most commonly used plasticizer for protein films is glycerol, which is miscible in most proteins, but several other plasticizers have also been studied [7–12]. Examples are polyfunctional alcohols such as sorbitol, propylene glycol, and di- and triethanolamine [1, 4, 7, 8, 10, 11, 13–20]. Often, more than one plasticizer has been used, for example water and glycerol [21], glycerol and trehalose [22], and glycerol and dendritic polyglycol [23]. An interesting direction toward new types of plasticisers is low-molar-mass proteins (hydrolysates). However, as reported by Nuanmano et al. [24] on the plasticization of fish myofibrillar proteins, that glycerol is more effective than the hydrolysates (gelatin-based) at similar contents. Both glycerol and sorbitol are harmless as plasticizers for films in contact with foodstuff and are also frequently used as sweeteners in foodstuffs [13]. It was interesting to compare the number of papers with the search words "glycerol" and "plasticizer" with those on "phthalate" and "plasticizer". The total number in the former and latter case was 48,600 and 89,800 (Google Scholar). In the years 1980, 1990, 2000, 2010, and 2015, the papers on glycerol plasticizer were 330, 520, 1270, 3640, and 5150. The same numbers for PVC plasticizer were 800, 1310, 2450, 4490, and 6020. It shows that glycerol, as a plasticizer for bio-based materials, is investigated to almost the same extent as the phthalates, being a common plasticizer for petroleum-based plastics (mainly PVC). The interest in glycerol is also increasing with time.

The potential migration of plasticizers is an important aspect to consider when choosing plasticizer. Migration of plasticizers leads to a decrease in fracture strain [11, 12]. Even though the film becomes more ductile with increasing plasticizer content, it is important to keep the content as low as possible because of barrier property issues; the gas/vapor permeabilities, generally, increase with increasing plasticizer content [25]. In general, a compromise between permeability and extensibility has to be made [11, 26–28]. In fact, a compromise has to be made also between ductility and stiffness/strength, since they normally go in opposite directions with increasing plasticizer content. Several less hygroscopic plasticizers, such as mono-, di-, or oligosaccharides and urea, have also been studied in solution cast films. Amphipolar plasticizers such as octanoic and palmitic acids, dibutyl tartrate and phthalate, and mono, di-, and tri-glyceride esters have also been studied in solvent cast protein films such as WG and zein. Still, glycerol and triethanolamine have so far seemed to be the most appropriate plasticisers for films of proteins such as whey or wheat gluten [7, 11, 12]; however, sorbitol is also a commonly used plasticizer [29]. These plasticizers have mostly been evaluated for solution cast films [30, 31], and only a limited number of studies have been reported for thermoformed gluten-based matrices with plasticizers [32–39]. In fact, for the most common plastic processing method (injection molding), only a few studies have been performed on the effects of plasticizers [40, 41].

There is no study, to our knowledge, that compares a very large set of different types of potential plasticizers for thermoformed wheat gluten films. This study focuses on a practical evaluation of potentially interesting plasticizers for use in thermoformed wheat gluten films. The outcome should be valid not just to WG, but also to proteins of different origins like pure or modified whey [42] and blood meal [43] proteins. The number of potential plasticizers is high, and these have to be carefully selected to avoid poor film properties and issues relevant, for e.g., the barrier properties, migration, aging behavior, sealing properties [44], printing properties, and cost. Which properties that are most interesting depend on the specific application, but the thermoforming and mechanical properties are of fundamental importance for most applications. This is also why these properties have been compared and evaluated, as a function of plasticizer type, in this study.

2. Experimental

2.1. Materials

The wheat gluten powder was supplied by Reppe AB, Lidköping, Sweden. The additives added for plasticizing purposes are presented in **Table 1**. The selection was based on polarity, melting, and boiling temperatures.

2.2. Blending procedure

About 350 g wheat gluten powder and 150 g of plasticizer were blended using a food processor (WATT; DUKA AB, Sweden) at the lowest speed, "Speed 1" (about 95 rpm), for 60 s. About 40 g of the blend was then transferred to a Brabender plasticorder PL2000 with a M50EHT measuring head and kneaded at 50 rpm for 2 min. The torque and the temperature were monitored with Brabender correlation program version 2.2. The starting temperature was set to 50°C.

2.3. Water adding procedure

Gluconic acid lactone, PEG 150, PEG 200, PEG 400, sorbitol, and xylose, respectively, were also mixed with small amounts of water (20% of the plasticizer content) for improved miscibility with the gluten powder. The water used was deionized and was blended in with the plasticizer.

2.4. Visual evaluation of the mixtures

The mixtures were visually evaluated during mixing. The focus was on crack formation, visible separation of the additive and the protein, and also on "apparent" dough brittleness. The visually "most promising" additive/gluten mixtures were chosen for compression molding.

Additive	T_m[a] (°C)	T_b[a] (°C)	Purity	Supplier
1,2-propanediol	-59	187.6	98%	Fluka, The Netherlands
1,3-propanediol	-27	214–216	99.6%	Aldrich, USA
1,4-butanediol	16	230	≥99.9%	Merck KGaA, Germany
2-propanol	-89	82.4	≥99.9%	Sigma-Aldrich GmbH, Germany
Adipic acid	151–154	265	≥99.9%	Sigma-Aldrich GmbH, Germany
Diethanolamine	28	217	99%	Sigma-Aldrich
Gluconic acid lactone	117	–	≥99.9%	Sigma-Aldrich GmbH, Germany
Glycerol	18	290	99.5%	Karlshamns Tefac AB, Karlshamn, Sweden
Lactic acid	53	–	≥99.9%	Sigma-Aldrich GmbH, Germany
Myvacet™ 5-07[29]	41–46	–	–	Quest International Inc., USA
Myvacet™ 9-08[29]	-12 to -14	–	–	Quest International Inc., USA
Myvacet™ 9-45[29]	4–12	–	–	Quest International Inc., USA
Octanoic acid	16–17	237	≥99.9%	Sigma-Aldrich GmbH, Germany
Diethylene glycol	-10	245	99%	Sigma-Aldrich
Triethylene glycol	-7	278	99%	Sigma-Aldrich
Tetraethylene glycol	-6	314	99%	Sigma-Aldrich
Octaethylene glycol	-4	327	≥99%	Sigma-Aldrich GmbH, Germany
Ethylene glycol	-13	197	99.5%	Merck KGaA, Germany
SAIB™ 90EA[30]	–83°C	78°C	–	Eastman Chm. Co., USA
Sorbitol	95	296	≥98%	Sigma-Aldrich, USA
Succinic acid	185–187	235	≥99.9%	Coleman & Bell Co., USA
Trimethylolpropane	58	292–297	≥99%	Perstorp Specialty Chemicals AB, Perstorp, Sweden
Triethanolamine	21	208	≥99.9%	Riedel-de Haen GmbH, Germany
Xylose	144–145	–	≥99%	Sigma-Aldrich GmbH, Germany

[a]T_m and T_b are melting and boiling points.

Table 1. Additives that were mixed with the wheat gluten powder.

2.5. Preparation of compression-molded films

The wheat gluten powder and the additives were blended using a food processor (WATT; DUKA AB, Sweden) at the lowest speed, "Speed 1" (about 95 rpm), for 30 s and thereafter at

http://dx.doi.org/10.5772/64073

"Speed 3" (about 200 rpm) for 1 min. Five grams of the mixture was then put in a frame and pressed into 0.5 mm thick films at 120°C for 3 min at a pressure of 100 bar. The compression-molded gluten sheet was cut into a square following the frame and put on Mylar films for 1 h. The press used was a Laboratory Press Polystat 200T #7105, Servitec Maschinenservice GmbH, Wustermark, Germany.

2.6. Tensile testing

The Young's modulus, tensile stress, and fracture strain (maximum tensile strain) were measured on samples, punched from the pressed gluten/additive films. Ten specimens, with a size described by EP 04/ISO 37-3 (with a test area of 4 × 40 mm), from each sample were measured by a Zwick Z010 tensile tester using the sensor 0.5 kN, controlled by TestXpert 7.11. A preload of 0.5 N applied with a speed of 100 mm/min was used. The entire tests were also performed at 100 mm/min. The tensile test and the 24-h preconditioning were performed at 50% RH and 23°C. The average values for each sample are presented.

3. Results and discussion

The molecular structure and properties of the compounds evaluated here are given in **Tables 2** and **3**. To narrow down the number of potential plasticisers the mixing/compounding and molding properties were first evaluated before going further with mechanical characterization of the compression-molded films. Very early six different compounds (four different starch syrups, octanoic acid, and pentaerythritol) were discarded due to very poor mixing with WG.

Additive	Molecular structure	Van der Waals volume
Glycerol	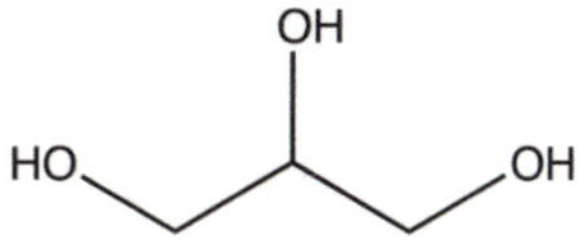	
Ethylene glycol	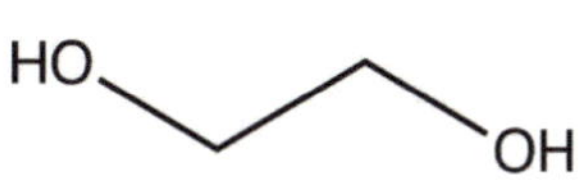	
Diethylene glycol	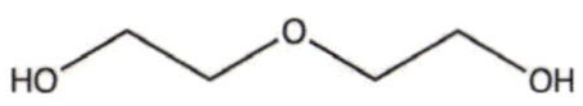	
Triethylene glycol	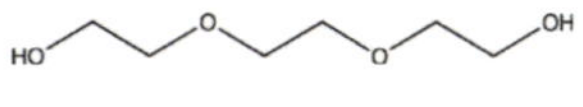	

Additive	Molecular structure	Van der Waals volume
Tetraethylene glycol	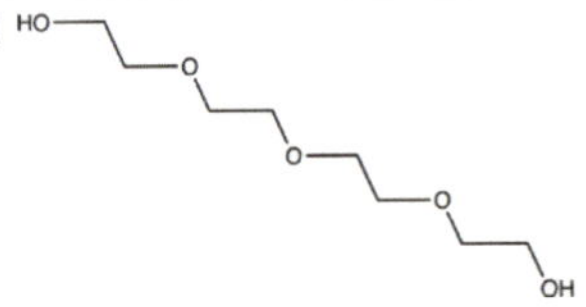	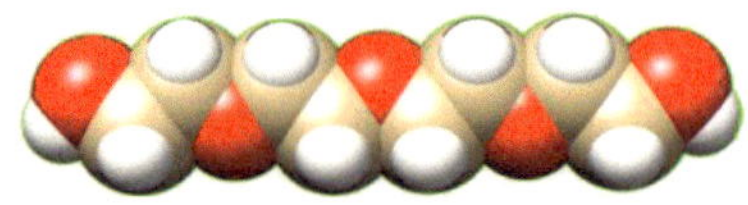
Octaethylene glycol	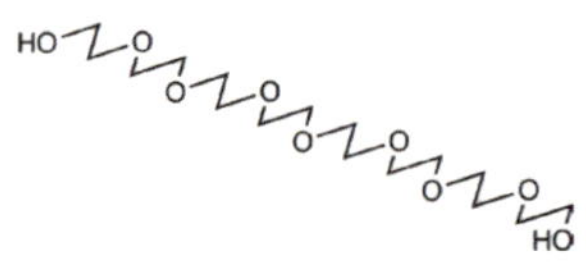	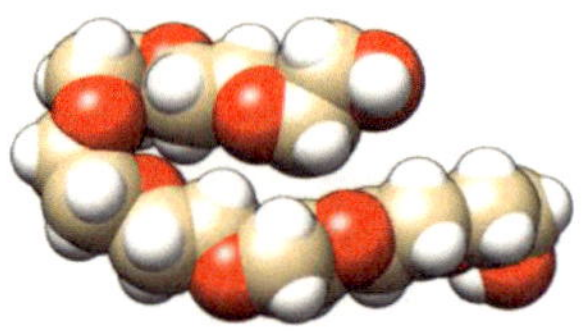
1,2-propanediol	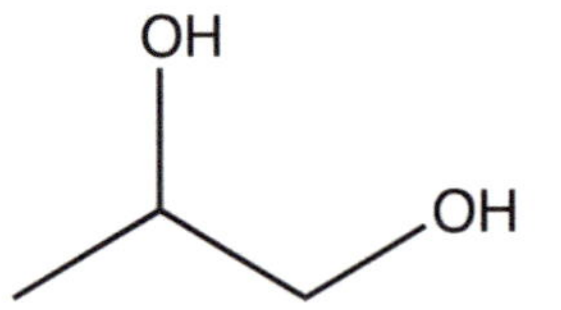	
1,3-propanediol		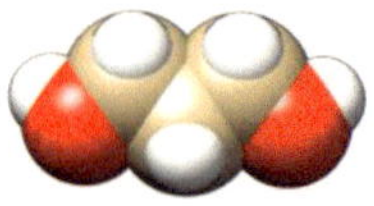
1,4-butanediol	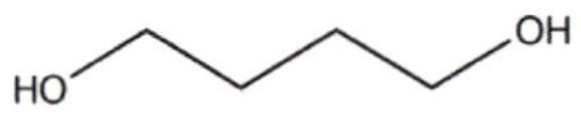	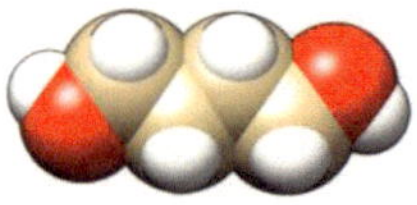
Diethanolamine	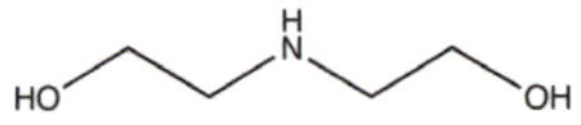	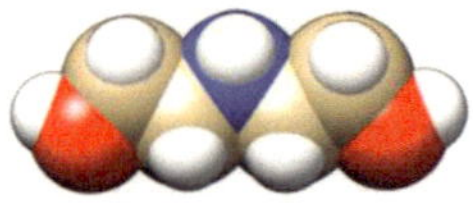
Triethanolamine	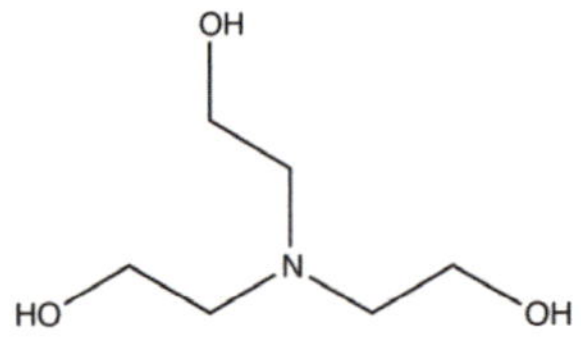	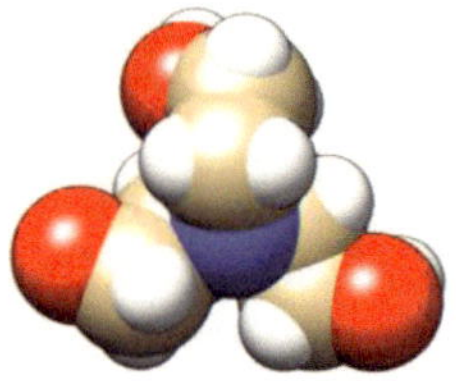
Trimethylolpropane	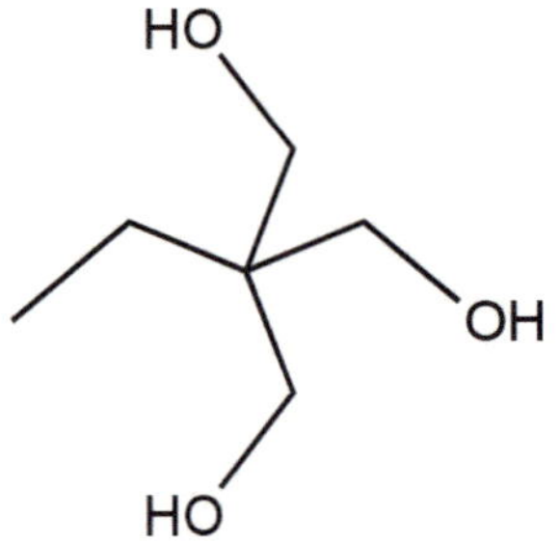	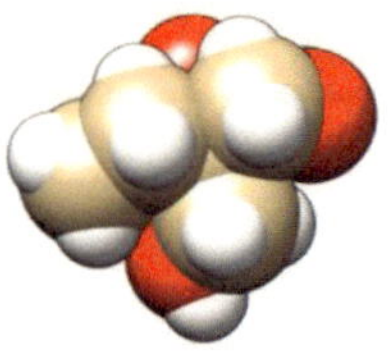

http://dx.doi.org/10.5772/64073

Additive	Molecular structure	Van der Waals volume
2-propanol	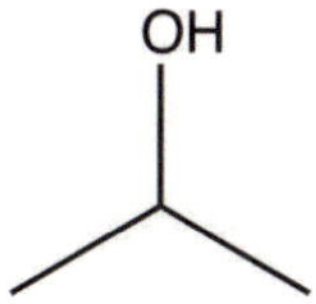	
Xylose	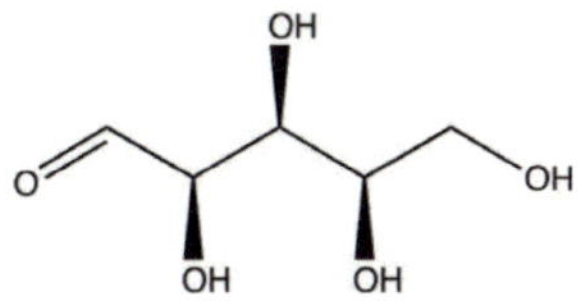	
Sorbitol	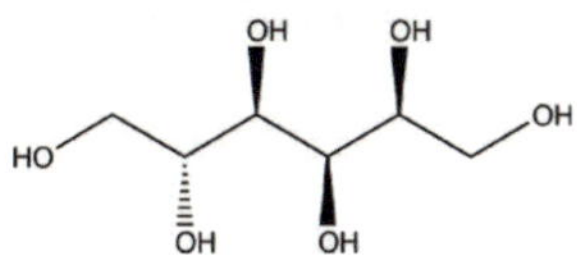	
Lactic acid	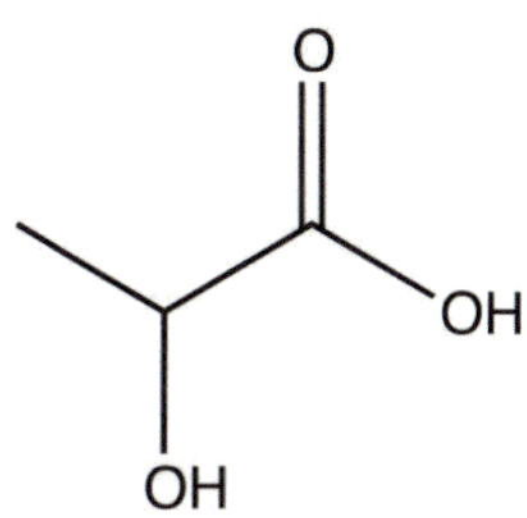	
Succinic acid	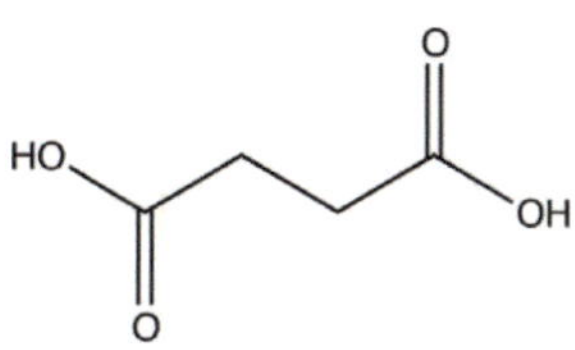	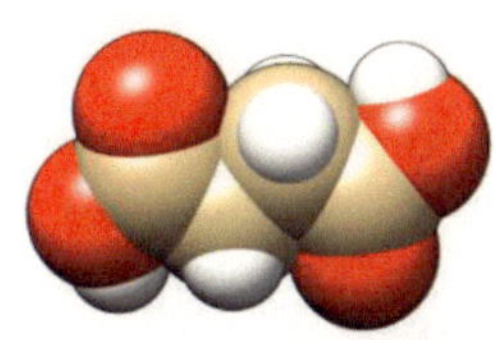
Adipic acid	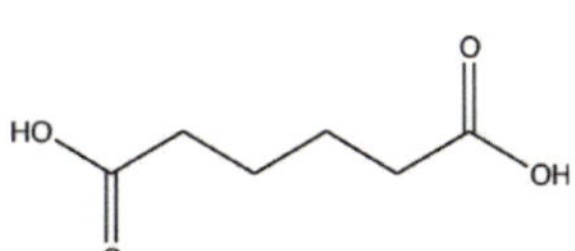	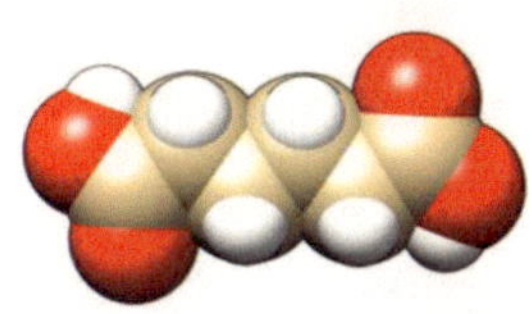
Octanoic acid	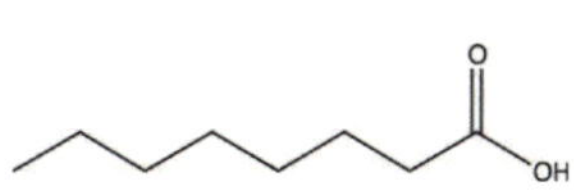	

Additive	Molecular structure	Van der Waals volume
Gluconic acid lactone	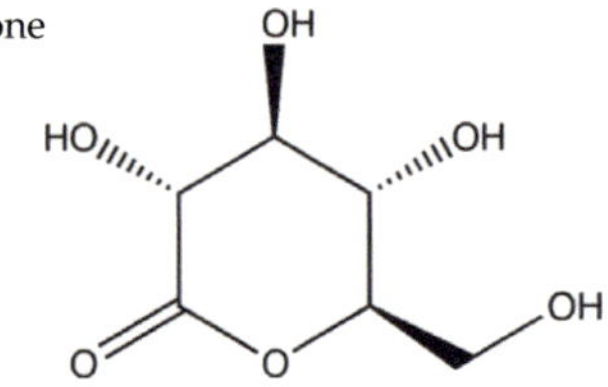	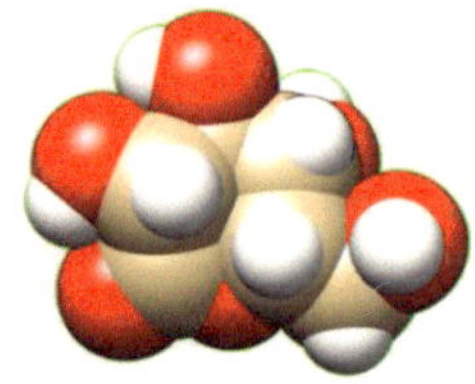
SAIB™ 90EA	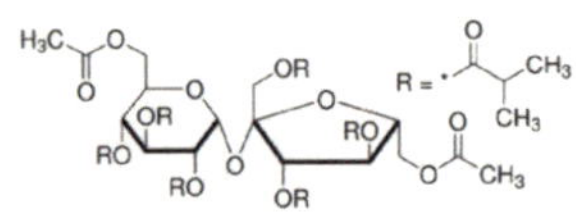	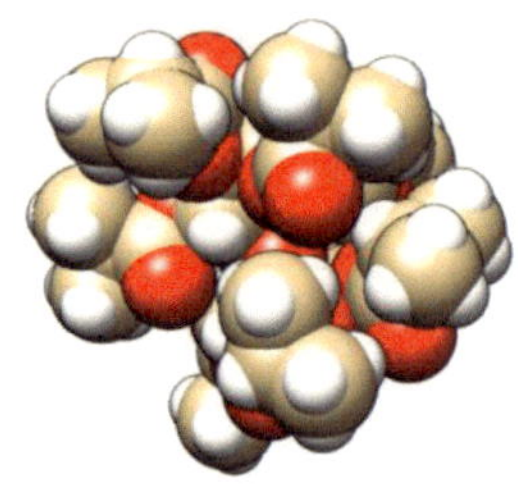
Myvacet™ 5-07	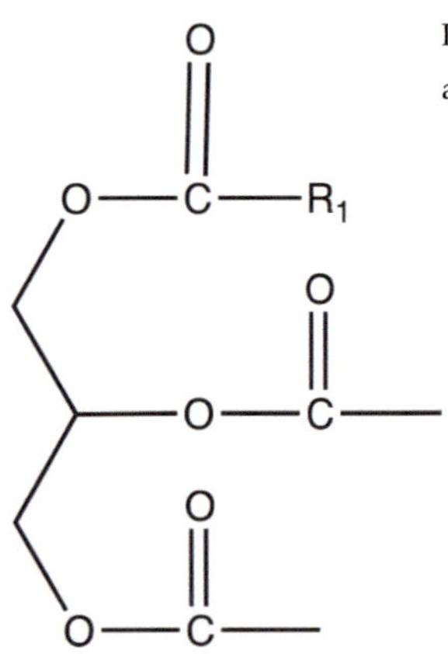	R_1: fatty acid chains based on 30% palmitic acid and 70% stearic acid
Myvacet™ 9-08	As 5-07	R_1: fatty acid chains based on 47% lauric acid, 18% myristic acid, 12% stearic acid, 10% palmitic acid, 13% others
Myvacet™ 9-45	As 5-07	R_1: fatty acid chains based on 73% oleic acid (isomers), 18% myristic acid, 13% stearic acid, 10% palmitic acid, 4% others

Table 2. Additive properties.

Additive (0.43 g additive/g WG)	Density[a] (g/cm^3)	Molecular weight (g/mol)	Crit. molar volume[b] (cm^3/mol)	Specific volume (cm^3/g)	Amount added[c] (mol/g WG)	Volume added[d] (cm^3/g WG)	Functional groups added[e] (mol/g WG)	Log P[f]	tPSA[g]	MR[h] (cm^3/mol)
Glycerol	1.260	92.1	254.5	0.794	0.0047	0.341	0.0140	-1.33	60.69	20.27
Ethylene glycol	1.200	62.1	185.5	0.825	0.0069	0.355	0.0138	-0.79	40.46	14.21
Diethylene glycol	1.120	106.1	315.5	0.893	0.0090	0.384	0.0189	-0.95	49.69	25.14

Additive (0.43 g additive/g WG)	Density[a] (g/cm³)	Molecular weight (g/mol)	Crit. molar volume[b] (cm³/mol)	Specific volume (cm³/g)	Amount added[c] (mol/g WG)	Volume added[d] (cm³/g WG)	Functional groups added[e] (mol/g WG)	Log P[f]	tPSA[g]	MR[h] (cm³/ mol)
Triethylene glycol	1.127	150.2	445.5	0.887	0.0029	0.382	0.0057	-1.1	58.92	36.06
Tetraethylene glycol	1.129	194.2	575.5	0.886	0.0022	0.381	0.0044	-1.26	68.15	46.99
Octaethylene glycol	1.130	370.4	1095.5	0.885	0.0012	0.381	0.0023	-1.88	105.07	90.68
1,2-propanediol	1.036	76.1	235.5	0.965	0.0057	0.415	0.0113	-0.47	40.46	18.9
1,3-propanediol	1.060	76.1	241.5	0.943	0.0057	0.406	0.0113	-0.68	40.46	18.81
1,4-butanediol	1.017	90.1	297.5	0.983	0.0048	0.423	0.0095	-0.23	40.46	23.41
Diethanolamine	1.097	90.1	332.5	0.912	0.0048	0.392	0.0095	-1.17	52.49	27.24
Triethanolamine	1.124	149.2	446.5	0.890	0.0070	0.383	0.0201	-1.31	63.93	38.44
Trimethylolpropane	1.150	134.2	417.5	0.870	0.0070	0.374	0.0224	-0.22	60.69	33.95
2-propanol	0.785	60.1	216.5	1.274	0.0072	0.548	0.0072	0.38	20.23	17.53
Xylose	1.525	150.1	390.5	0.656	0.0070	0.282	0.0266	-2.71	97.99	31.75
Sorbitol	1.489	182.2	461.5	0.672	0.0024	0.289	0.0142	-2.94	121.38	38.44
Lactic acid	1.200	90.1	240.5	0.833	0.0048	0.358	0.0143	-0.51	57.53	18.65
Succinic acid	1.572	118.1	307.5	0.636	0.0036	0.274	0.0073	-0.64	74.6	21.93
Adipic acid	1.360	146.1	419.5	0.735	0.0029	0.316	0.0059	0.2	74.6	31.13
Octanoic acid	0.910	144.2	507.5	1.099	0.0030	0.473	0.0030	2.43	37.3	39.7
Gluconic acid lactone	1.610	178.1	411.5	0.621	0.0024	0.267	0.0121	-2.44	107.22	35.58
SAIB™ 90EA[30]	1.110	846.9		0.901	–	0.387	–	–	–	–

[a]If not stated otherwise, the density was obtained from PubChem and at 15–25°C. The value for 1,3-propanediol was obtained from *chemicalbook.com*.
[b]Critical molar volume.
[c]Amount of additive added in moles.
[d]Amount (volume) of additive added based on the specific volume and molecular weight.
[e]Amount of functional groups added based on the molecular weight and the number of functional groups per molecule.
[f]Logarithm of the partition coefficient in n-octanol/water. The ratio of the solubility of the additive in n-octanol and water, the larger the more hydrophobic the additive is. Log P, tPSA and MR obtained from CS Chembiodraw Ultra.
[g]Total polar surface area.
[h]Molar refractivity.

Table 3. Additive properties.

3.1. Additive/WG blends that were evaluated only qualitatively (visually and by hand) during and after mixing and compression molding, due to poor miscibility/mixing and/or brittleness

Xylose and *gluconic acid lactone*, with and without water, gave both an inhomogeneous WG dough with a clear phase separation. The resulting compression-molded films were very brittle with properties similar to the molded pure wheat gluten powder. The use of *SAIB*™ 90EA [45], which is a low-viscous liquid with 90% sucrose acetate isobutyrate and 10% ethyl acetate that is used in, for example, coatings and in nail-polish lacquers, resulted in a homogeneous dough with potential for further development. However, the molded films were too brittle. Three grades of acetylated monoglyceride (AMG, *Myvacet*™ 5-07, 9-08, and 9-45 [46, 47]) were mixed with WG. The 5-07 grade (which is an AMG with 50% acetylation) has been used previously in WG formulations, although not with the primary purpose of plasticicizing the film [48]. This grade gave an inhomogeneous dough, and the resulting compression-molded film was brittle. The mechanical flexibility was too poor to proceed with further film development at this stage. *Myvacet*™ *9-08* and *Myvacet*™ *9-45* (fully acetylated using, respectively, hydrogenated coconut oil and partially hydrogenated soybean oil as fat sources) did not dissolve in the WG. The resulting molded films fell apart due to extensive brittleness. *Lactic acid* made the dough highly viscous and sticky, which prevented longer mixing in the Brabender. The pressed films were too soft and sticky to be further tested.

The use of *succinic* and *adipic acid* resulted in doughs that were quite homogeneous and viscous. The molded films were at warm conditions (when they were released from the compression molder) flexible. However, they turned out to be inhomogeneous with a significant amount of trapped air bubbles and turned brittle when cooled to room temperature. The *2-propanol* did not yield a dough when mixed with WG. The blend could be described as a wet powder rather than a dough. The compression-molded films were very brittle and could not be tensile tested. The largest glycol tested here, *octaethylene glycol* (PEG 400), yielded an inhomogeneous dough and was only very poorly mixed with the WG powder. The film had an uneven surface but parts existed which were homogeneous and smooth. However, the overall impression was, however, that the films were like pressed powder. The compression-molded films had similar appearance to molded WG powder without plasticizer and were too brittle to be tensile tested.

3.2. Additive/WG blends that were evaluated with respect to rheological and tensile properties

Glycerol serves here as a reference plasticizer due to its well-known excellent plasticizer efficiency. Its torque and temperature evolution during the 2-min mixing in the Brabender are given in **Figure 1a**. The temperature was steadily increasing from 50°C and reached the maximum temperature after 2 min (76°C). The torque was also steadily increasing reaching a maximum torque of 18.4 Nm at the end of the experiment.

The use of *ethylene glycol* (PEG 60) and *diethylene glycol* (PEG 100) resulted in doughs that seemed appropriate for, for example, extrusion. The resulting maximum and final torque values were high (similar to glycerol, **Table 4**), and the PEG 60 mixed faster with the WG powder (a faster rise in torque) then PEG 100. The reason to that is most likely the lower melting temperature of PEG 60. The maximum temperature during the 2-min mixing occurred close to or at the end of the experiment and was, for both glycols, slightly lower than for glycerol. The resulting compression-molded films were homogeneous and mechanically flexible. *Triethylene glycol* (PEG 150) yielded a dough that was sticky and that had a low viscosity/torque throughout the mixing step (**Table 4**). The resulting temperature increase was also low. An even constantly lower torque, with no temperature increase, was observed when mixing *tetraethylene glycol* (PEG 200) with WG. Still, a homogeneous dough and a flexible film were obtained.

The diols *1,2-propanediol, 1,3-propanediol,* and *1,4-butanediol* all gave homogeneous doughs and initially flexible films. However, they became less flexible after 1 day (see the mechanical section below). There were no clear correlation between molecular size and rheological behavior in this group (**Table 4**). The largest torque values and temperature increase were observed for *1,3-propanediol,* whereas the lowest torque values and temperature increase were observed for *1,4-butanediol.*

Diethanolamine and *triethanolamine* gave homogeneous doughs and flexible films. The torque remained very low throughout the mixing step and the temperature remained low. A similar torque and temperature behavior were observed for *trimethylolpropane* (TMP) (**Figure 1b**), which also resulted in a homogeneous dough and flexible film. Based on the different rheological behaviors and the apparent plasticizer efficiencies of the analyzed additive/WG combinations, it is concluded that it is not possible to predict the plasticizer efficiency based on the torque/temperature behavior during mixing. This is most clearly illustrated in **Figure 1** where two of the most efficient plasticisers gave two very different torque/temperature-time curves.

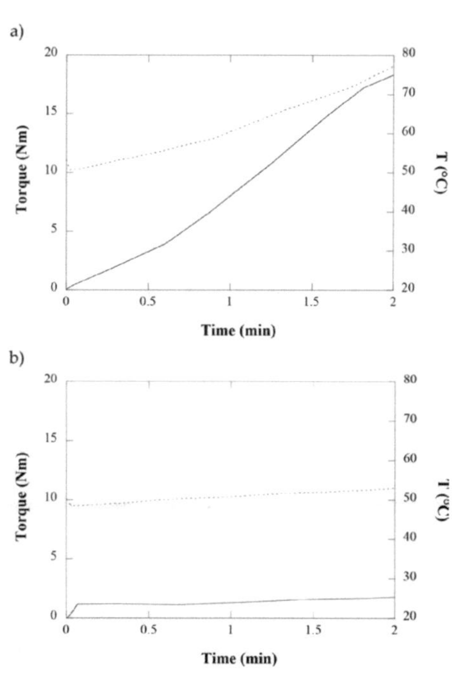

Figure 1. Torque (solid line) and temperature (broken line) as a function of mixing time in the Brabender. (a) Glycerol and (b) trimethylolpropane.

Additive	τ_{max}[a]	T at τ_{max}[b]	Time to τ_{max}[c]	τ_{Final}[d]	T_{final}[e]
Glycerol	18.4	76	2.0	18.4	76
Ethylene glycol	17.2	71	1.8	15.2	71
Diethylene glycol	15.2	70	2.0	15.2	70
Triethylene glycol	8.4	59	2.0	8.4	59
Tetraethylene glycol	0.5	50	2.0	0.5	50
1,2-propanediol	3.4	52	2.0	3.4	52
1,3-propanediol	15.1	67	2.0	15.1	67
1,4-butanediol	0.5	45	0.2	0.5	49
Diethanolamine	0.7	49	1.2	0.7	49
Triethanolamine	0.8	50	0.7	0.7	51
Trimethylolpropane	1.8	53	2.0	1.8	53

[a]Maximum torque (Nm).
[b]Temperature at maximum torque (°C).
[c]Time to maximum torque (min).
[d]Final torque (Nm).
[e]Final temperature (°C).

Table 4. The thermomixing properties of wheat gluten with plasticisers.

In **Figure 2**, the tensile curves of the compression-molded films with the most interesting additives are displayed, and in **Table 5**, their mechanical properties are given. The WG with glycerol showed the greatest fracture strain and the lowest modulus and maximum stress. The curve shape (**Figure 2a**) indicated a clear yielding (non-linearity in the curve) before fracture. The scatter in the tensile properties was also among the lowest, indicating good mixing/high miscibility (**Table 5**). TMP, not tested in wheat gluten before, showed a fracture strain above 100% and somewhat higher modulus and maximum stress as compared to glycerol. The scatter in these parameters was also higher than for glycerol. Triethanolamine and, in particular, diethanolamine were not as effective plasticisers as glycerol and TMP. However, the scatter in data (mixing efficiency) was similar to that of glycerol in the case of maximum stress and fracture strain (triethanolamine). It seemed as if the additives with three hydroxyl-terminated arms (glycerol, TMP, and triethanolamine) were better plasticizers than the linear diethanolamine.

The range in mechanical properties was large within the oligoethylene glycol family (**Figure 2b** and **Table 5**). The highest extensibility (fracture strain) was observed for triethylene glycol (**Figure 2b**), which value was close to that of TMP. In fact, its modulus was lower than that of TMP, but higher than that of glycerol. The smallest glycol (ethylene glycol) showed poor plasticizing properties with a high modulus and maximum stress and low fracture strain. The scatter in data within this family was lowest for the two most effective plasticisers (triethylene and diethylene glycols). Their scatter in modulus and maximum stress was on the same order as that of glycerol.

As for ethylene glycol, the propanediols and the 1,4-butanediol had high modulus, maximum stress, and low fracture strain (**Figure 2c** and **Table 5**). It is amazing how the plasticizing efficiency increases when a "flexible" ether linkage is inserted in the middle of the molecule (compare mechanical data of 1,4-butanediol and diethylene glycol, **Tables 2** and **4**). The effect is somewhat larger than putting an N–H group in the same place (compare 1,4-butanediol and diethyleneamine). If a third hydroxyl group is placed on the central carbon, a sizeable improvement is observed (compare 1,3-propanediol and glycerol).

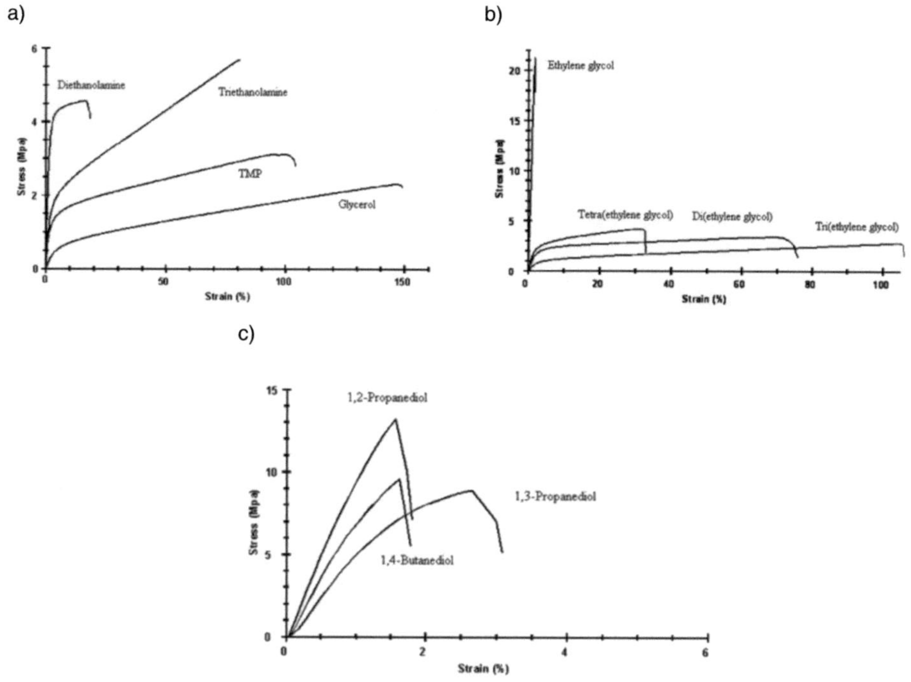

Figure 2. Tensile curves.

In general, the modulus and yield stress of tough polymers go hand in hand whereas the fracture strain is not a simple function of these [49]. Here, we show that the modulus and strength (maximum stress) are basically linearly related (**Figure 3a**). For those blends that show a clear yielding, it is also obvious that the stiffer materials also show larger yield stresses (**Figure 2**). There does, in general, not exist a clear correlation between stiffness/strength and fracture strain when tough and brittle materials of different types are investigated together. A material can be stiff and brittle or stiff and tough. Here, however, the fracture strain decreases in a nonlinear fashion with increasing stiffness (**Figure 3b**) or strength (not shown), where there is a lesser correlation at higher stiffness and strength. This shows that for the same polymer matrix (WG), the addition of additives that alter the mechanical properties, such as those with a higher or lower plasticizing ability, the four mechanical parameters (modulus, yield stress, maximum stress, and fracture strain) are clearly interrelated.

Additive	E^a	R. E^d	$\sigma_{max}{}^b$	R. σ^e	ε^c	R. $\varepsilon_b{}^f$
Glycerol	27 ± 2	7	2.4 ± 0.2	8	152 ± 10	7
Ethylene glycol	1532 ± 218	14	21.5 ± 4.2	20	2 ± 0.5	25
Diethylene glycol	133 ± 10	8	3.4 ± 0.2	6	66 ± 9	14
Triethylene glycol	77 ± 9	12	2.8 ± 0.2	7	99 ± 12	12
Tetraethylene glycol	206 ± 68	33	4.4 ± 0.9	20	34 ± 20	59
1,2-propanediol	1068 ± 103	10	13.2 ± 0.9	7	2 ± 0.2	10
1,3-propanediol	647 ± 63	10	9.0 ± 1	11	3 ± 0.9	30
1,4-butanediol	875 ± 55	6	9.9 ± 0.9	9	2 ± 0.3	15
Diethanolamine	280 ± 39	14	4.5 ± 0.3	7	16 ± 4	25
Triethanolamine	151 ± 24	16	5.7 ± 0.4	7	78 ± 6	8
Trimethylolpropane	108 ± 21	19	2.9 ± 0.6	21	104 ± 23	22

[a]Young's modulus (MPa).
[b]Maximum stress (MPa).
[c]Elongation at break (%).
[d]Relative standard deviation of modulus data.
[e]Relative standard deviation of stress data.
[f]Relative standard deviation of strain at break data.

Table 5. Mechanical properties of wheat gluten materials.

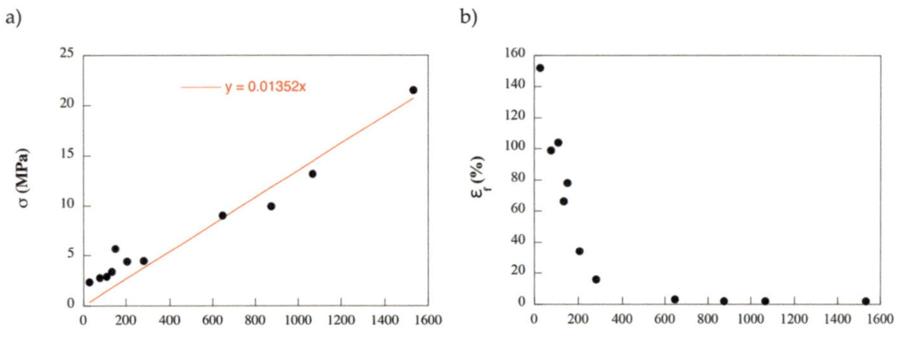

Figure 3. (a) Maximum stress versus modulus and (b) fracture strain versus modulus for the additive/WG materials in **Table 5**.

It was important to see whether it was possible to predict the plasticizer efficiency, or its trends, from any easily obtained parameter characterizing the additive. The mechanical parameters (**Table 5**) were correlated with all the physical parameters in **Table 3**. No specific correlation was observed between the mechanical properties and the density, specific volume, amount/volume of additive added (mol/g WG or cm^3/g WG), functional groups added, and log P. In fact, none of the physical parameters in **Table 3** could be used to predict the trends in the observed fracture strains. On the other hand, when plotting the size and the modulus against the critical molar volume, a trend was observed (**Figure 4a** and **b**). The stiffness decreased in a nonlinear way with increasing size of the additive, with a vanishing correlation at larger additive sizes. The modulus was correlated with the additive polarity, in terms of the tPSA (total polar surface area) (**Figure 4c**). The correlation was less than with the molecular size; however, two regions were clearly separated, a low polarity region (low tPSA) with high modulus and a high polarity region with low modulus. Finally, in **Figure 4d**, the modulus was plotted against the molar refractivity (MR), which is also a measure of the polarity (and the size) of the additive. Here, the correlation was similar as for molecular size (**Figure 4a** and **b**), with a nonlinear decrease in stiffness with increasing modulus. When plotting the same type of data for maximum stress rather than modulus, the same relationships were observed, which was not a surprise since stiffness and strength were linearly dependent (**Figure 5**). To conclude, of all the parameters listed in **Table 2**, only the molecular size and/or the polarity was affecting the mechanical properties (only stiffness and strength) in a systematic way, although the correlation was relatively poor.

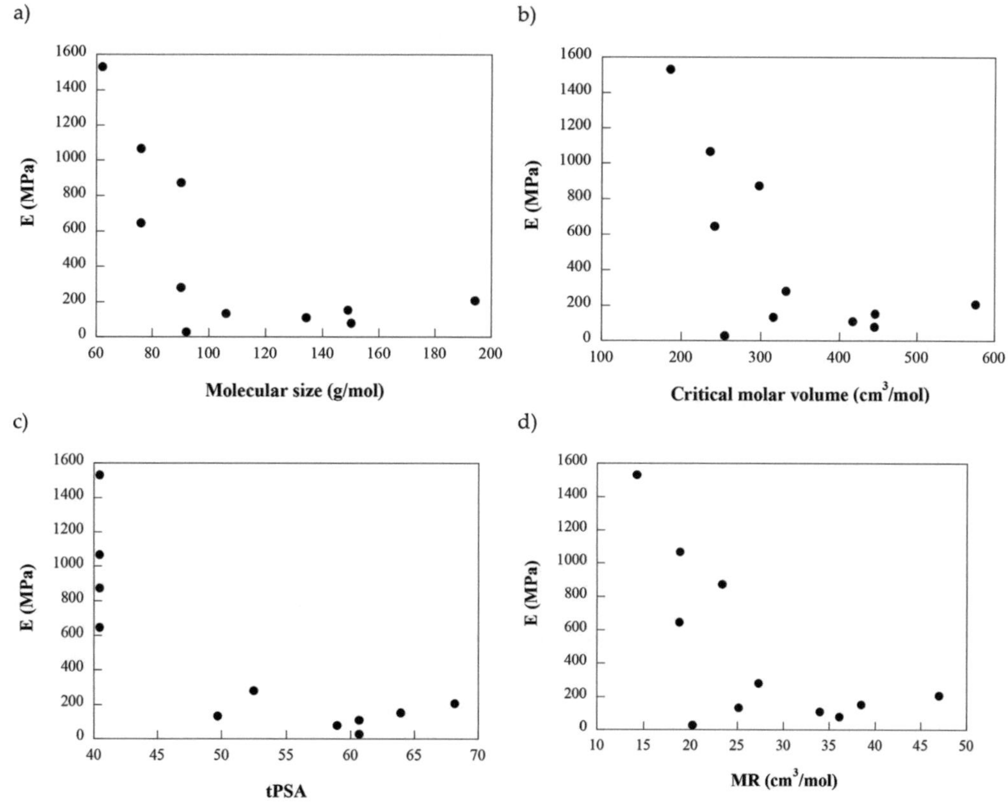

Figure 4. Modulus versus additive (a) molecular weight, (b) critical molar volume, (c) total polar surface area, and (d) molar refractivity.

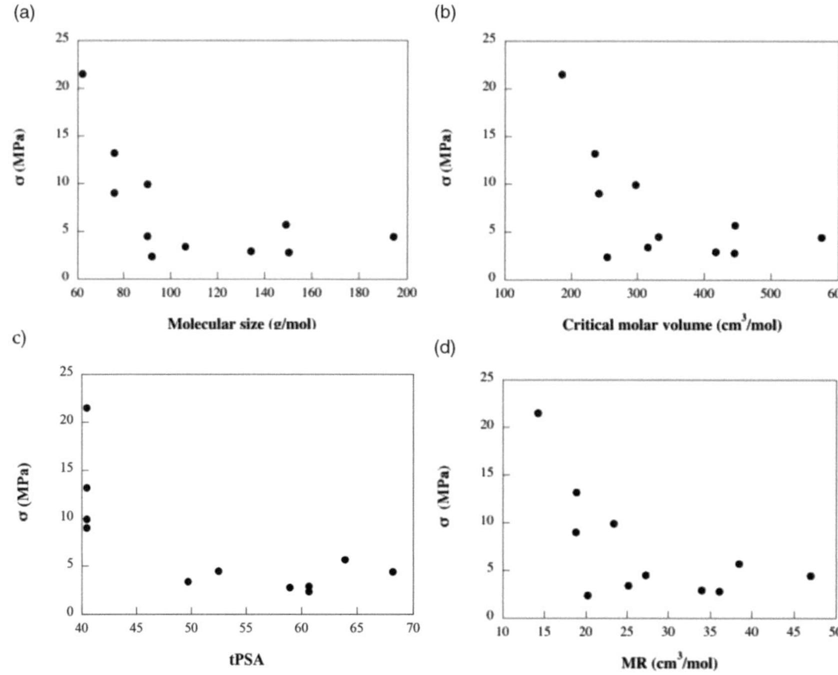

Figure 5. Maximum stress versus additive (a) molecular weight, (b) critical molar volume, (c) total polar surface area, and (d) molar refractivity.

4. Conclusions

Of the extensive number of additives/plasticizers that were tested, having different molecular weights, polarity, melting and boiling temperatures, glycerol was shown to be the most efficient plasticizer for thermoformed gluten films. The most efficient plasticisers, considering stiffness and extensibility, were those that contained three hydroxyl groups and the linear glycols of intermediate size. It should be stated, though, that only the short-term mechanical data were analyzed, and no aging and relative humidity effects were explored. All thermoformed plasticizer/gluten mixtures were studied at equal mass concentration of plasticizer, an issue that can be further elaborated; different plasticizers have different efficiency and should thus also be further studied as a function of plasticizer concentration. Still, several plasticizers in this study seemed to be fully blended with gluten after <2 min of thermomixing. Regrettably, there did not seem to be a unique relationship between the torque behavior and the final mechanical properties of the films. However, there was a weak nonlinear relationship between the stiffness/strength and the size (molecular weight, molar volume) and the polarity/polarizability (tPSA and MR) of the compound. The mixing conditions (i.e. shear forces, time, and temperature) may differ from extrusion; hence, this study should thus be considered as a first attempt to determine appropriate plasticizers for thermoformed gluten films.

Acknowledgements

Vinnova, the Swedish Governmental Agency for Innovation Systems and the "Glupack" consortium are thanked for financial support.

Author details

N. Henrik Ullsten[1,3], Mikael Gällstedt[1] and Mikael S. Hedenqvist[2*]

*Address all correspondence to: mikaelhe@kth.se

1 Innventia AB, Stockholm, Sweden

2 Department of Fibre and Polymer Technology, KTH Royal Institute of Technology, Stockholm, Sweden

3 Current address: Department of Engineering and Chemical Sciences, Karlstad University, Karlstad, Sweden

References

[1] Muzzarelli RAA, Peter MG. Chitin handbook. Grottammare: European Chitin Society; 1997. ISBN 88-86889-01-1

[2] Anker M. Edible and biodegradable whey protein films as barriers in foods and food packaging [Thesis]. Göteborg: Chalmers University of Technology; 2000

[3] Gontard N, Marchesseau S, Cuq J-L, Guilbert S. Water vapour permeability of edible bilayer films of wheat gluten and lipids. Int J Food Sci Tech. 1995;30:49–56. doi:10.1111/j.1365-2621.1995.tb01945.x

[4] Shaw NB, Monahan FJ, O'Riordan ED, O'Sullivan M. Physical properties of WPI films plasticized with glycerol, xylitol, or sorbitol. J Food Sci. 2002;67:164–167. doi:10.1111/j.1365-2621.2002.tb11377.x

[5] Gómez-Estaca J, Gavara R, Catalá R, Hernández-Muñoz P. The potential of proteins for producing food packaging materials: a review. Packag Technol Sci. 2016. doi: 10.1002/pts

[6] Jagadeesh D, Kanny K, Prashantha K. A review on research and development of green composites from plant protein-based polymers. Polym Comp. 2015;1–15. doi: 10.1002/pc.23718

[7] Pommet M, Redl A, Morel MH, Guilbert S. Study of wheat gluten plasticization with fatty acids. Polymer. 2003;44:115–122. doi:10.1016/S0032-3861(02)00746-2

[8] Pouplin M, Redl A, Gontard N. Glass transition of wheat gluten plasticized with water, glycerol, or sorbitol. J Agric Food Chem. 1999;47:538–543. doi:10.1021/jf980697w

[9] Gennadios A, Weller CL, Testin RF. Property modification of edible wheat gluten-based films. T ASABE. 1993;36:465–470. doi:10.13031/2013.28360

[10] Sánchez AC, Popineau Y, Mangavel C, Larré C, Guéguen J. Effect of different plasticizers on the mechanical and surface properties of wheat gliadin film. J Food Chem. 1998;46:4539–4544. doi:10.1021/jf980375s

[11] Hernández-Muñoz P, Hernández RJ. Effect of storage and hydrophilic plasticizers on functional properties of wheat gluten glutenin fraction films. In: Worldpak2002, 13th Proceedings of IAPRI Conference, vol 2. CRC Press LLC. 2002. pp. 481–493

[12] Irissin-Mangata J, Bauduin G, Boutevin B, Gontard N. New plasticizers for wheat gluten films. Eur Polym J. 2001;37:1533–1541. doi:10.1016/S0014-3057(01)00039-8

[13] McHugh TH, Krochta JM. Sorbitol- vs glycerol-plasti-cized whey protein edible films: integrated oxygen permeability and tensile property evaluation. J Agric Food Chem. 1994;42:841–845. doi:10.1021/jf00040a001

[14] McHugh TH, Aujard J-F, Krochta JM. Plasticized whey protein edible films: water vapor permeability properties. J Food Sci. 1994;59:416–419. doi:10.1111/j.1365-2621.1994.tb06980.x

[15] Alcantara CR, Rumsey TR, Krochta JM. Drying rate effect on the properties of whey protein films. J Food Process Eng. 1998;21:387–405. doi:10.1111/j.17454530.1998.tb00460.x

[16] McHugh TH, Krochta JM. Water vapor permeability properties of edible whey protein-lipid emulsion films. JAOCS. 1994;71:307–312. doi:10.1007/BF02638058

[17] Fairley P, Krochta JM, German JB. Interfacial interactions in edible emulsion films from whey protein isolate. Food Hydrocoll. 1997;11:245. doi:10.1016/S0268005X(97)80053-7

[18] Shellhammer TH, Krochta JM. Whey protein emulsion film performance as affected by lipid type and amount. J Food Sci. 1997;62:390. doi:10.1111/j.1365-2621.1997.tb04008.x

[19] Mahmoud R, Savello PA. Mechanical properties of and water vapor transferability through whey protein films. J Dairy Sci. 1992;75:942–946. doi:10.3168/jds.S0022-0302(92)77834-5

[20] Coltelli M-B, Wild F, Bugnicourt E, Cinelli P, Lindner M, Schmid M, Weckel V, Müller K, Rodriguez P, Staebler A, Rodríguez-Turienzo L, Lazzeri A. State of the art in the development and properties of protein-based films and coatings and their applicability to cellulose based products: an extensive review. Coatings. 2016;6:1–59. doi:10.3390/coatings6010001

[21] Duval A, Molina-Boisseau S, Chirat C, Morel M-H. Dynamic mechanical analysis of the multiple glass transitions of plasticized wheat gluten biopolymer. J Appl Polym Sci. 2016;43254(1–11). doi:10.1002/APP.43254

[22] Perez LM, Piccirilli GN, Delorenzi NJ, Verdini RA. Effect of different combinations of glycerol and/or trehalose on physical and structural properties of whey protein concentrate-based edible films. Food Hydrocoll. 2016;56:352–359. doi:10.1016/j.foodhyd.2015.12.037

[23] Bozic M, Majeric M, Denac M, Kokol V. Mechanical and barrier properties of soy protein isolate films plasticized with a mixture of glycerol and dendritic polyglycerol. J Appl Polym Sci. 2015;132(1–12):41837. doi:10.1002/app.41837

[24] Nuanmano S, Prodpran T, Benjakul S. Potential use of gelatin hydrolysate as plasticizer in fish myofibrillar protein film. Food Hydrocoll. 2015;47:61–68. doi:10.1016/j.foodhyd.2015.01.005

[25] Tong Q, Xiao Q, Lim LT. Effects of glycerol, sorbitol, xylitol and fructose plasticisers on mechanical and moisture barrier properties of pullulan–alginate–carboxymethylcellulose blend films. Int J Food Sci Techn. 2013;48:870–878. doi:10.1111/ijfs.12039

[26] Ali Y, Ghorpade VM, Hanna MA. Properties of thermally-treated wheat gluten films. Ind Crop Prod. 1997;6:177–184. doi:10.1016/S0926-6690(97)00216-1

[27] Gontard N, Ring S. Edible wheat gluten film: influence of water content on glass transition temperature. J Agric Food Chem. 1996;44:3474–3478. doi:10.1021/jf960230q

[28] Gennadios A, Brandenburg AH, Park JW, Weller CL, Testin RF, Water vapor permeability of wheat gluten and soy protein isolate films. Ind Crop Prod. 1994;2:189–195. doi:10.1016/0926-6690(94)90035-3

[29] Jost V, Stramm C. Influence of plasticizers on the mechanical and barrier properties of cast biopolymer film. J Appl Polym Sci. 2015;133(1–9):42513. doi:10.1002/app.42513

[30] Olabarrieta I, Cho S-W, Gällstedt M, Sarasua J-R, Johansson E, Hedenqvist MS. Aging properties of films of plasticized vital wheat gluten cast from acidic and basic solutions. Biomacromolecules. 2006;7:1657–1664. doi:10.1021/bm0600973

[31] Olabarrieta I, Gällstedt M, Ispizua I, Sarasua J-R, Hedenqvist MS. Properties of aged montmorillonite–wheat gluten composite films. J Agric Food Chem. 2006;54:1283–1288. doi:10.1021/jf0522614

[32] Ullsten NH, Gällstedt M, Johansson E, Gräslund A, Hedenqvist MS. Enlarged processing window of plasticized wheat gluten using salicylic acid. Biomacromolecules. 2006;7:771–776. doi:10.1021/bm050822u

[33] Ullsten NH, Gällstedt M, Cho S-W, Johansson E, Hedenqvist MS. Properties of extruded vital wheat gluten sheets with sodium hydroxide and salicylic acid. Biomacromolecules. 2009;10:479–488. doi:10.1021/bm800691h

[34] Gällstedt M, Mattozzi A, Johansson E, Hedenqvist MS. Transport and tensile properties of compression-molded wheat gluten films. Biomacromolecules. 2004;5:2020–2028. doi: 10.1021/bm040044q

[35] Redl A, Gontard N, Guilbert S. Determination of sorbic acid diffusivity in edible wheat gluten and lipid based films. J Food Sci. 1996;61:116–120. doi:10.1111/j.1365-2621.1996.tb14739.x

[36] Zárate-Ramírez LS., Martínez I, Romero A, Partal P, Guerrero A. Wheat gluten-based materials plasticised with glycerol and water by thermoplastic mixing and thermo-moulding. J Sci Food Agric. 2011;91:625–633. doi:10.1002/jsfa.4224

[37] Zárate-Ramírez LS, Romero A, Bengoechea C, Partal P, Guerrero A. Thermo-mechanical and hydrophilic properties of polysaccharide/gluten-based bioplastics. Carbohyd Polym. 2014;112:24–31. doi:10.1016/j.carbpol.2014.05.055

[38] Zárate-Ramírez LS, Romero A, Martínez I, Bengoechea C, Partal P, Guerrero A. Effect of aldehydes on thermomechanical properties of gluten-based bioplastics. Food Bioprod Proc. 2014;92:20–29. doi:10.1016/j.fbp.2013.07.007

[39] Zubeldía F, Ansorena MR, Marcovich NE. Wheat gluten films obtained by compression molding. Polym Test. 2015;43:68–77. doi:10.1016/j.polymertesting.2015.02.001

[40] Perez V, Felix M, Romero A, Guerrero A. Characterization of pea protein-based bioplastics processed by injection moulding. Food Bioprod Process. 2016;97:100–108. doi:10.1016/j.fbp.2015.12.004

[41] Cho S-W, Hedenqvist MS, Johansson E, Gällstedt M. Injection-molded nanocomposites and materials based on wheat gluten. Int J Biolog Macromol. 2010;48:146–152. doi: 10.1016/j.ijbiomac.2010.10.012

[42] Winkler H, Vorwerg W, Schmid M. Synthesis of hydrophobic whey protein isolate by acylation with fatty acids. Eur Polym J. 2015;62:10–18. doi:10.1016/j.eurpolymj.2014.10.014

[43] Bier JM, Verbeek CJR, Lay MCJ. Using synchrotron FTIR spectroscopy to determine secondary structure changes and distribution in thermoplastic protein. J Appl Polym Sci. 2013;130:359–369. doi:10.1002/APP.39134

[44] Cho S-W, Ullsten H, Gällstedt M, Hedenqvist MS. Heat-sealing properties of compression-molded wheat gluten films. J Biobased Mater Bioenergy. 2007;1:56–63. doi:10.1166/jbmb.2007.006

[45] SAIB data sheet: sucrose acetate isobutyrate. Eastman Chemical Company. USA. 1994

[46] Product sheet: Myvacet™. Quest International. The Netherlands. 2000.

[47] US patent 2010/0044924 A1. Cap DS. Candle refill kit and method of use.

[48] Gennadios A, Weller CL, Testin RF. Property modification of edible wheat gluten-based films. Trans ASAE. 1993;36:465–470. doi:10.13031/2013.28360

[49] Hedenqvist MS, Ritums JE, Condé-Brana M, Bergman G. Sorption and desorption of tetrachloroethylene in fluoropolymers: effects of the chemical structure and crystallinity. J Appl Polym Sci. 2003;87:1474. doi:10.1002/app.11645

Chapter 6

Rheological Properties of Carbon Nanofiber-Filled Polyamide Composites and Blend of these Composites and TPE

Yosuke Nishitani and Takeshi Kitano

Additional information is available at the end of the chapter

http://dx.doi.org/10.5772/64531

Abstract

For the purpose of developing new engineering materials with sufficient balance among mechanical, electrical, processability, triboloical properties, etc., in this study, we investigated the dynamic viscoelastic properties of carbon nanofiber (CNF) filled polyamide (PA) composites and the blend of these composites and thermoplastic elastomer (TPE) in the molten state, which were mainly obtained in our previous studies. It was found that vapor grown carbon fiber (vapor grown carbon fiber) has a stronger influence on the dynamic viscoelastic properties of the composites in the molten state. Rheological percolation thresholds seem to exist between 1vol.% and 5vol.% of VGCF contents. On the other hand, the effect of the addition of TPE (styrene-ethylene/butylene-styrene copolymer (SEBS) and maleic anhydride grafted SEBS (SEBS-g-MA)) on the dynamic viscoelastic properties of VGCF/PA6 composites in the molten state differed at each viscoelastic value. It was clarified that the dynamic viscoelastic properties of VGCF/PA6/SEBS-g-MA ternary composites are higher than those of VGCF/PA6/SEBS ones. Furthermore, the influence of processing sequences on the dynamic viscoelastic properties of VGCF/PA6/SEBS-g-MA composites in the molten state differed according to the mixing steps of materials. These may be attributed to the change in the internal structure caused by addition of TPE, type of SEBS and processing sequences.

Keywords: dynamic viscoelastic properties in molten state, carbon nanofiber, polyamide, composites, polymer blends

1. Introduction

Recently, nanofiller-filled thermoplastic composites are attracting considerable interest [1–5]. One of the advantages of these nanocomposites is that the volume ratio of the fillers in the systems is much lower than that of conventional thermoplastic composites, offering a distinct advantage in that less energy is required when these nanocomposites are processed for industrial applications. Graphitic carbon nanofilaments such as single-wall carbon nanotube (SWCNT) and carbon nanofiber (CNF) have been of great research interest [6, 7]. Especially, CNF-filled thermoplastic composites have been proven to be very beneficial in terms of the improved mechanical, thermal, and electrical properties [8–14]. CNF disperses well in the polymer matrix and has a diameter of about 100–150 nm, which is relatively larger than other nanofilaments. It can therefore serve as a low-cost alternative to SWCNT. In our previous works, we investigated the rheological, mechanical, and tribological properties of various carbon fibers (CFs) such as polyacrylonitrile-based carbon fiber (PAN-CF) and pitch-based carbon fiber (Pitch-CF) and two kinds of vapor grown carbon fiber (VGCF and VGCF-S) and filled thermoplastic (polyamide (PA), polybutylene terephthalate (PBT), etc.) [15–22] composites in order to obtain new polymer nanocomposite-based engineering materials for use as the structural materials and tribomaterials of mechanical sliding parts such as gears, bearings, cams, seals, etc. The mechanical and tribological properties improved when filled with various CFs. Fiber properties such as aspect ratio (length/diameter) and fiber diameter have a stronger influence on the viscoelastic properties than fiber content. We also demonstrated the optimum mechanical and tribological properties of various CF-filled polymer composites. However, to further enhance the mechanical, electrical, processability, tribological properties, etc. in these CNF-filled thermoplastic composites, it is very critical to understand the rheological behavior of these thermoplastic composites in the molten state such as processability, internal microstructure, changes, and structure-property relationships [23–25].

Moreover, to date, little interest has been paid to multicomponent polymer composites such as polymer blends based on CNF-filled thermoplastic composites for engineering materials [18, 22, 26, 27]. To develop new engineering materials with sufficient balance among mechanical, electrical, processability, tribological properties, etc., various properties need to be well balanced. Several investigations have proved that multicomponent polymer composites, which are ternary blends based on nanofiller-filled composites such as nanofiller/polymer/polymer or nanofiller/polymer/rubber, etc., are effective for realizing good balances among various properties [18, 22, 26–32]. Above all, their remarkable performance stems from their morphologies in immiscible blends. It is well known that the morphologies of immiscible polymer blends are controlled by the chemical reaction between components during melt blending [32]. In addition, the morphologies of multicomponent polymer composites such as phase structure, fiber dispersion, fiber localization, etc. are strongly influenced by the design of material composition and by the processing sequence of melt mixing using a twin extruder. The former is the material design of multicomponent materials: specifically, the type of materials, composition ratio, whether compatibilizer is used or not, functional groups, surface treatment of filler, etc. [32, 33]. On the other hand, the latter is the mixing technique of multicomponent composites [22, 34, 35]. The processing sequences during melt mixing using

http://dx.doi.org/10.5772/64531

a twin screw extruder affect the morphologies of multicomponent composites. As a result, the physical properties are also strongly influenced by the processing sequence. However, there are only a few published studies on the effects of the addition of thermoplastic elastomer (TPE) and the processing sequence on dynamic viscoelastic properties of these CNF-filled multicomponent composites [18]. For the purpose of developing new engineering materials with sufficient balance among mechanical, electrical, processability, tribological properties, etc., we investigated the dynamic viscoelastic properties of CNF-filled polyamide composites and the blend of these composites and TPE in the molten state, which were obtained mainly in our previous studies. In particular, this study discusses the effects of the filling of CNF, addition of TPE, and processing sequence on dynamic viscoelastic properties in the molten state.

2. Dynamic viscoelastic properties of carbon nanofiber-filled polyamide 66 composites in the molten state

2.1. Introduction

CNF is nanosized diameter CF and is used as fillers for polymer composites. One of the advantages of CNF-filled composites is that the volume ratio of fillers in the systems is much lower than that of conventional polymer composites. These CNF-filled polymer composites have been proven to be very beneficial in terms of improved mechanical, electrical, tribological properties, etc. [5, 8–14]. In our previous studies, we investigated the mechanical and tribological properties of various vapor grown carbon fibers (VGCF), which are a type of CNF-filled polymer composites: polyamide (PA) [20–22], polybutylene terephthalate (PBT) [15, 17–19], polyimide (PI) [16], etc. were used as matrix polymer. We found that the mechanical and tribological properties improved when filled with VGCF, and demonstrated the optimum mechanical and tribological properties for various VGCF-filled polymer composites. However, to further enhance the mechanical, electrical, processability, tribological properties, etc. in these CNF-filled thermoplastic composites, it is critical to understand the rheological behavior of these composites in the molten state such as processability, internal microstructure, changes, and structure-property relationships [23–25]. Several studies have proven that VGCF-filled polymer composites such as VGCF/polycarbonate (PC) [8, 10], VGCF/polypropylene (PP) [10, 11, 36], VGCF/polystyrene (PS) [37], VGCF/PBT [15, 19], etc. demonstrate distinctive rheological behaviors. However, there is still a lack of comprehensive understanding of the rheological properties of these VGCF-filled composites, in particular, when VGCF content is low. The purpose of this study is to experimentally investigate the rheological properties, which is the dynamic viscoelastic properties in the molten state, of VGCF/PA66 composites. Particularly, this study discusses the effect of the filling of VGCF, volume fraction, strain, angular frequency, and temperature on the dynamic viscoelastic properties in the molten state.

2.2. Materials and methods

The materials used in this study were carbon nanofiber (CNF)-filled polyamide 66 (PA66) composites in the molten state, which are used in the first step of the fabrication of these composites. PA66 (UBE Nylon 2020P, Ube Industries, Ltd., Japan) was used as the matrix polymer. Vapor grown carbon fiber (Showa Denko K. K., Japan) which is a type of CNF was used as the filler. Its fiber diameter and initial length are 150 nm and 10 μm, respectively. VGCF-filled PA66 composites (VGCF/PA66) with the VGCF content of 1, 5, and 10 vol% were prepared in this study. All the components were dried for 5 h at 100°C in a vacuum oven beforehand until the moisture level was 0.2%, then dry blended in a small bottle, and subsequently the melt was mixed at 110 rpm and 295°C in a twin screw extruder (TEX-30, Japan Steel Works, Ltd., Japan). After mixing, the extruded strands of VGCF/PA66 composites were cut into 5 mm long pieces by the pelletizer, and dried again for 5 h at 100°C in a vacuum oven. In addition, 1 mm thick sheets were compression molded at the conditions of 270°C, 5 MPa, and 3 min, and cut into ϕ25 mm disk shapes for measuring the rheological properties. The rheological properties in the molten state were evaluated by oscillatory flow testing using a parallel plate-type rotational rheometer (ARES, Rheometric Scientific Co., USA and ARES-G2, TA Instruments Japan Inc., Japan). The diameter of the parallel plate was ϕ25 mm, and the gap between the two plates was fixed at 1 mm. Under such gap conditions, the test specimens were slightly compressed in the molten state. First, the strain sweep test was performed from initial strain value 10^{-1} to a final strain value of 10^2 in percent at the angular frequency of 3.5 rad/s. Second, in frequency sweep testing, a small strain amplitude oscillatory shear was applied to the sample. The angular frequency was varied from 10^{-1} to 10^2 rad/s, and the strain amplitude was set as 1%. These measurements were carried out for all samples at 270, 280, and 290°C. The dynamic viscoelastic properties in the molten state were evaluated by storage modulus G', loss modulus G'', loss tangent tan δ, phase angle δ (= arctan G''/G'), complex modulus $|G^*|$, and complex viscosity $|\eta^*|$.

2.3. Strain dependences

Viscoelastic properties of VGCF/PA66 composites are discussed in this section. First, strain sweep testing was carried out to characterize the strain dependence of viscoelastic properties and also transition from linear to nonlinear viscoelastic ones of VGCF/PA66 composites and then to estimate the change in the internal microstructure of these composites. Since the storage modulus G' is a more sensitive rheological function to the structural changes of the nanocomposites than the loss modulus G'', only the storage modulus curves are presented in this study [38]. The strain dependence of the storage modulus G' of VGCF/PA66 composites is shown in **Figure 1**. This measurement was carried out for all samples at 280°C and the angular frequency of ω at 3.5 rad/s. The strain dependences of G' show remarkably different behavior according to the volume fraction of VGCF V_f. The G' of neat PA66 (100%) and VGCF/PA66 with V_f= 1 vol % (VGCF1) exhibits a linear behavior, in which the modulus is a constant over a wide range of strain amplitude. On the other hand, the G' of VGCF/PA66 with V_f = 5 vol% (VGCF5) and V_f = 10 vol% (VGCF10) show linear plateau only in low strain regions and rapidly decrease with increasing strain, which is an independent transition from linear to nonlinear behavior. It was observed that the value of the plateau modulus G'_p changes with increasing V_f. These results may be due to the increase in the fiber-matrix interactions, and to the increase in the surface area of fiber with increasing V_f.

http://dx.doi.org/10.5772/64531

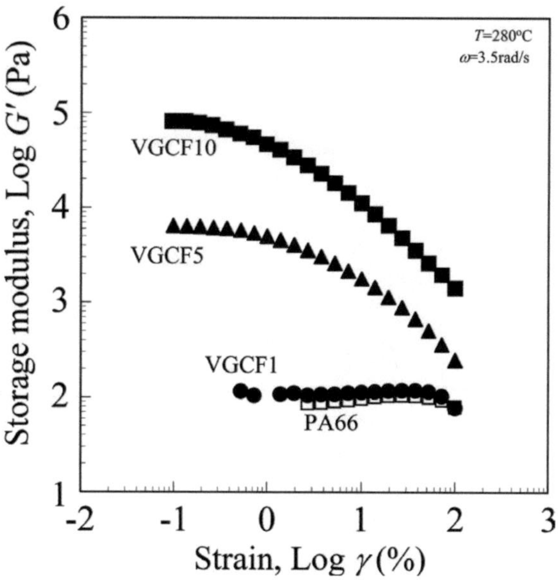

Figure 1. Storage modulus as function of strain for VGCF/PA66 composites at 3.5 rad/s and 280°C.

To clarify the transition point from linear to nonlinear behavior, the strain dependences of relative storage modulus G'_r (= G'/G'_p) of VGCF/PA66 composites are shown in **Figure 2**. Here, the relative storage modulus G'_r is given by the values of VGCF/PA66 composites divided by that of the G'_p of VGCF/PA66 composites. The regions of linear viscoelastic behavior for neat PA66 and VGCF1 are clearly observed over a whole range of strain. However, those of VGCF5 and VGCF10 decrease abruptly in the G'_r, which shows nonlinear behavior. The transition point which appeared in the deviation region from linear to nonlinear viscoelastic behavior is defined as critical strain γ_c. γ_c was generally taken as the strain value at the G' equal to 90% of the plateau modulus (0.9 G'_p) [19, 38], which is shown in the broken line in **Figure 2**. The relationship between critical strain value γ_c and volume fraction of fiber V_f of VGCF/PA66 composites at 3.5 rad/s and 280°C is shown in **Figure 3**. γ_c of VGCF/PA66 composites varies sharply from V_f, and exhibits a straight relation between γ_c and V_f on this log-log plot. That is power law of γ_c and V_f ($\gamma_c \propto V_f^{\,n}$). Here, n is the slope of power law and is found to be –2.55 from the relation in **Figure 3**. This may be attributed to the change in the internal structure of the composites such as fiber dispersion, fiber-matrix, and fiber-fiber interactions caused by increasing V_f [19, 39]. In general, it is well known that the dependence of critical strain value on the volume fraction of fiber of polymer composites and nanocomposites can be explained in terms of the change of internal microstructure formation, the alignment of the anisotropic nanofiller to the strain direction, and the breakdown of fiber-fiber interaction of the physical networks [19, 32, 38]. These behaviors have been reported for other nanofillers such as clay, silica, etc. filled polymer nanocomposites [39, 40]. In this study, it was found that the volume fraction of fiber has a strong influence on the strain dependence of viscoelastic properties.

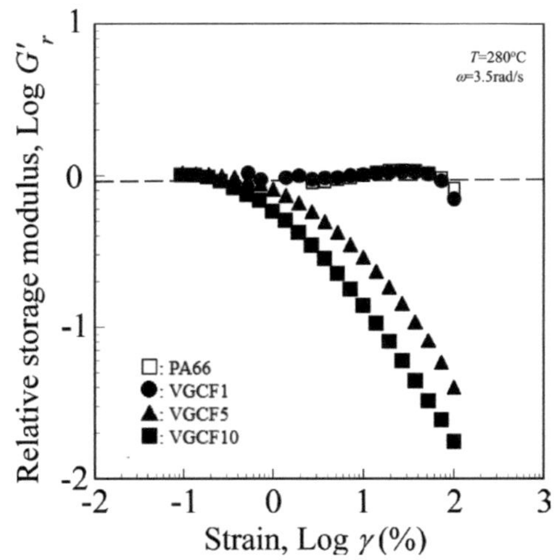

Figure 2. Relative storage modulus as function of strain for VGCF/PA66 composites at 3.5 rad/s and 280°C.

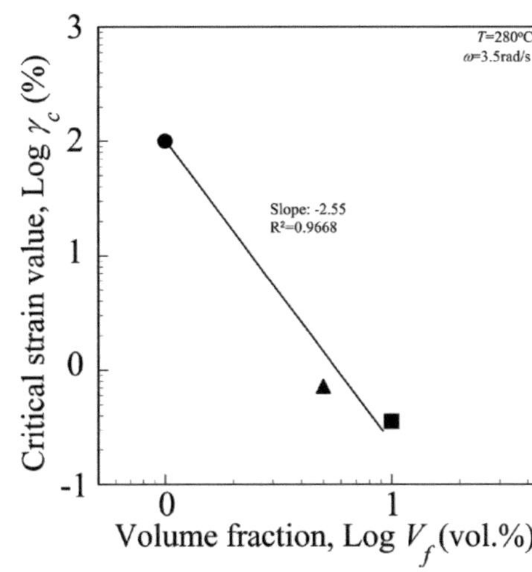

Figure 3. Dependence of critical strain on volume fraction of VGCF for VGCF/PA66 composites at 3.5 rad/s and 280°C.

2.4. Angular frequency dependences

The rheological properties of VGCF/PA66 composites were evaluated by oscillatory flow behavior, which are considered to be strongly dependent on the internal microstructure of these composites. The dynamic viscoelastic properties of these composites are plotted as a function of angular frequency ω at 280°C and strain of 1% in **Figure 4(a)** (storage modulus G'), **Figure 4(b)** (loss modulus G''), **Figure 4(c)** (loss tangent tan δ), and **Figure 4(d)** (complex viscosity $|\eta^*|$), respectively. The slope of G' and G" against ω in low and high ω regions are listed in **Table 1**. The G' of neat PA66 increases with increasing ω; however, the slope of G'- ω

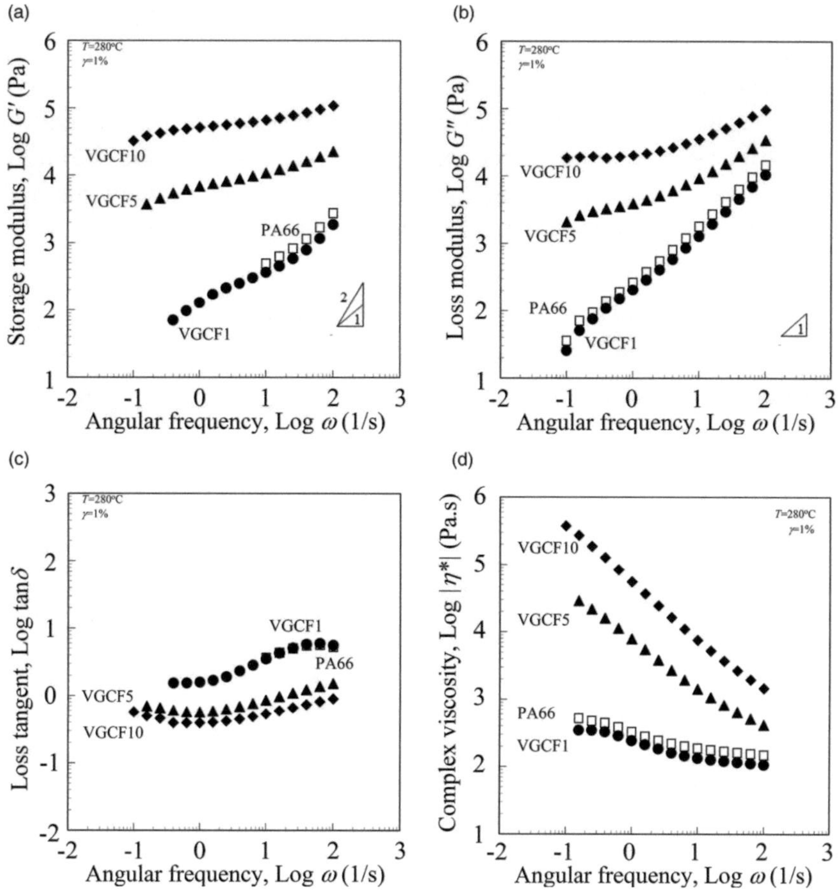

Figure 4. Dynamic viscoelastic properties as a function of angular frequency for VGCF/PA66 composites at 1% strain and 280°C. (a) Storage modulus. (b) Loss modulus. (c) Loss tangent. (d) Complex viscosity.

in high ω region is as small as 0.95. The G' of neat PA66 is generally proportional to ω^2 ($\log G' \propto \log \omega^2$) in the linear dynamic viscoelastic models [41, 42]. Thus, the slope of G' against ω is 2, which is $\log G' \propto 2\log \omega$, although that of neat PA66 used in this study has a different tendency. This may be attributed to the material properties of PA66 used such as molecular structure, molecular weight distribution, orientation, etc., although the reason is not clear because the G' of neat PA66 in low ω region is not measured because it is below the detection limit of the G' measured using this apparatus. On the other hand, the G' of VGCF/PA66 composites has a different behavior according to the volume fraction of VGCF V_f. That of VGCF1 has the same tendency as that of neat PA66; however, that of VGCF is slightly small than that of neat PA66 in the whole range of ω. The slope of G' against ω of VGCF1 in low and high ω region becomes higher than that of VGCF1 in the middle ω region. Moreover, G' of VGCF5 and VGCF10 is remarkably higher than that of PA66, and VGCF 5 and VGCF10 show the typical G' of highly filled systems [42, 43], although V_f in the composites is only 5 vol%.

The slopes of G' against ω become small, indicating the "second rubbery plateau," i.e., the long-scale relaxation time [25, 41, 42]. Hence, the viscoelastic properties of VGCF/PA66 composites change from liquid-like (viscous behavior) to solid-like (elastic behavior) according to V_f. This second rubbery plateau behavior suggests the presence of an apparent yield stress, which can be attributed to the strong fiber network formed due to the size of VGCF. This is in accordance with theoretical expectations and experimental observations for highly filled systems [43–47]. Similar phenomena have been reported by other systems [25, 31–33].

Code	Slope of G'-ω		Slope of G''-ω	
	Low ω	High ω	Low ω	High ω
PA66	-	0.95	1.05	0.90
VGCF1	0.65	0.94	1.17	0.91
VGCF5	0.41	0.38	0.39	0.61
VGCF10	0.22	0.26	0.05	0.48

Table 1. Slope of viscoelastic properties as function of angular frequency curves of VGCF/PA66 composites in low and high frequencies.

On the contrary, the loss modulus G'' in **Figure 4(b)** has the same tendency as G'. The G'' of VGCF1 is slightly smaller than that of neat PA66. The slope of G'' against ω of VGCF1 is also about 1, which agrees with the linear viscoelastic model that the slope of G'' is proportional to ω ($\log G'' \propto \log \omega$). On the other hand, the G'' of VGCF5 and VGCF10 is higher than the G'' of neat PA66 in the whole ω region, and the slope of G'' against ω becomes small in low ω regions. However, there is a tendency of G'' being smaller than that of G' since the G' is a more sensitive viscoelastic function to the structural changes of the composites than G'' [20]. From the results of loss tangent tan δ against ω in **Figure 4(c)**, tan δ of VGCF1 is similar to that of neat PA66 (100%), and have a maximum peak in high ω region. The values of tan δ for neat PA66 and VGCF1 are higher than 1 in the whole ω region ($\tan \delta > 1$). This indicates that PA66 and VGCF1 behave like liquid, which is a viscosity-dominant property. In contrast, VGCF5 and VGCF10 have different tendencies, and the values of tan δ for VGCF5 and VGCF10 are smaller than 1 ($\tan \delta < 1$) except that of VGCF5 in high ω regions. Thus, VGCF5 and VGCF10 behave like solid materials, which is an elastic-dominant property. Complex viscosity $|\eta^*|$ in **Figure 4(d)** shows the different tendencies according to the V_f. $|\eta^*|$ of neat PA66 and VGCF1 slightly decreases with increasing V_f. In addition, their smaller value than neat PA66 in the whole ω region is a very interesting behavior. On the other hand, the $|\eta^*|$ of VGCF5 and VGCF10 abruptly decreases with increasing ω, which exhibits a very strong shear thinning effect. The curves of $|\eta^*|$ versus ω have a slope of –45°. This behavior indicates the presence of an apparent yield stress in low ω regions. Furthermore, the curve of $|\eta^*|$ versus ω of VGCF1 is remarkably different from that of VGCF5 in low ω regions. This may be attributed to the changes in the fiber network formation. This tendency thus indicates the presence of the transition point of the internal structure, that is, the rheological percolation threshold, which is almost equal to the gel transition point from liquid-like to solid-like behavior.

http://dx.doi.org/10.5772/64531

2.5. Influence of volume fraction

The influence of volume fraction of VGCF V_f on the rheological properties of VGCF/PA66 composites is discussed here. The relative complex viscosity $|\eta^*|_r$ is plotted against volume fraction of VGCF V_f in **Figure 5**. Here, the relative value is given by the values of VGCF/PA66 composites with various V_f divided by that of neat PA66. Three types of angular frequencies: ω = 0.25, 2.5, and 25 rad/s were used as a parameter. The $|\eta^*|_r$ of VGCF/PA66 composites at various ω has a minimum peak at 1 vol%. It then rapidly increases with increasing V_f, and it seems to be remarkably influenced by ω. These results may be attributed to the internal microstructure due to the interaction between fiber and polymer matrix. Especially, the particular phenomena of the complex viscosity of the composites with low fiber content (V_f= 1 vol%), which are smaller than those of neat PA66, are observed in this study. Similar phenomena have been reported by other systems [15, 19, 20, 36]. These phenomena may be due to the self-lubrication of VGCF and also due to the two dimensionally oriented fiber structure formation, which has the possibility to reduce the viscoelastic properties of the composites less than that of matrix polymer. However, it is difficult to confirm these reasons in the present situation. Furthermore, VGCF/PA66 composites sharply increase with V_f between 1 and 5 vol%. This critical composition is regarded as a rheological percolation. At high fiber content, the fiber-fiber interaction is more pronounced. Thus, these results demonstrate that fiber size parameters such as diameter, length, and aspect ratio have strong influence on the viscoelastic properties of VGCF/PA66 composites.

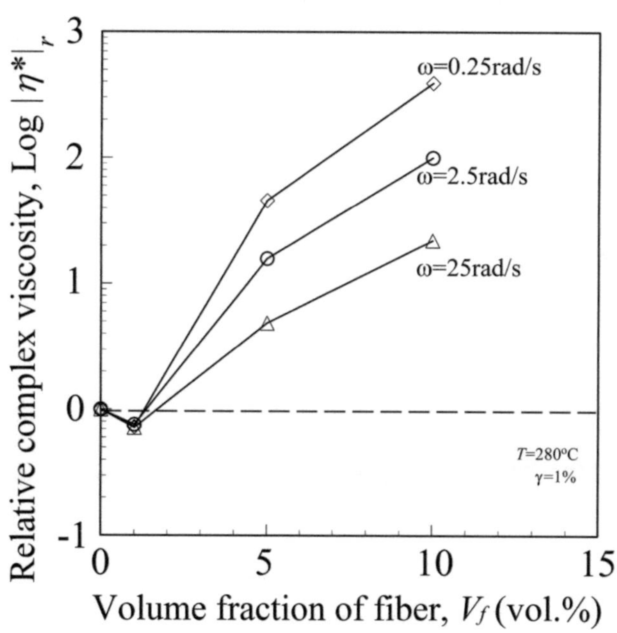

Figure 5. Influence of volume fraction of fiber on relative complex viscosity for VGCF/PA66 composites at 1% strain and 280°C.

2.6. Relationship between storage modulus and loss modulus

The relationship between storage modulus and loss modulus at 1% strain of VGCF/PA66 composites is shown in **Figure 6**. This log G' versus log G'' plots are called as "Han plot" or "modified Cole-Cole plot" [33, 46, 48], which is an analogous Cole-Cole plot used in dielectric spectroscopy. Such plot was used by Han et al. [33] to investigate temperature-induced changes in the microstructure of homopolymers, block copolymers, and polymer blends, and by Potschke et al. [46] to indicate the change of the microstructure in CNT-filled polymer composites such as CNT/PC nanocomposites. It was proposed that the shift and change in the slope of the log G' versus log G'' curves indicate that the microstructure changes significantly with addition of CNT. The curves of various VGCF/PA66 composites in this study show the different behavior related to V_f. The neat PA66 and VGCF1 draw the same curves, and G' is smaller than G'', which means the viscoelastic properties are dominated by the viscous properties. On the other hand, G' at a given G'' of VGCF5 and VGCF10 increases significantly with increasing V_f. Thus, the slope of G' versus G'' decreases with increasing V_f. In addition, the curves of a part of VGCF5 and the whole VGCF10 are located higher than the broken line in **Figure 6**, which contributes to a more dominant role for the elastic properties than the viscous ones. Similar behavior has been reported by other carbon nanotube-filled polymer composites such as CNT/PC [46], CNT/PLA [49], etc. Also, Potschke et al. [46] and Kitano et al. [44] found a similar response of first normal stress difference versus shear stress for glass-fiber-filled polyethylene melts under steady shear conditions. In this study, VGCF clearly present fiber content dependence, confirming the presence of strong interactions between fiber and polymer matrix caused by the nanoscale fiber size such as fiber diameter. Moreover, the decrease of the slope with the increase of fiber content indicates that composites become more heterogeneous due to increasing the interactions between fiber and fiber. It is found from these results that VGCF/PA66 composites change the microstructure according to the fiber content.

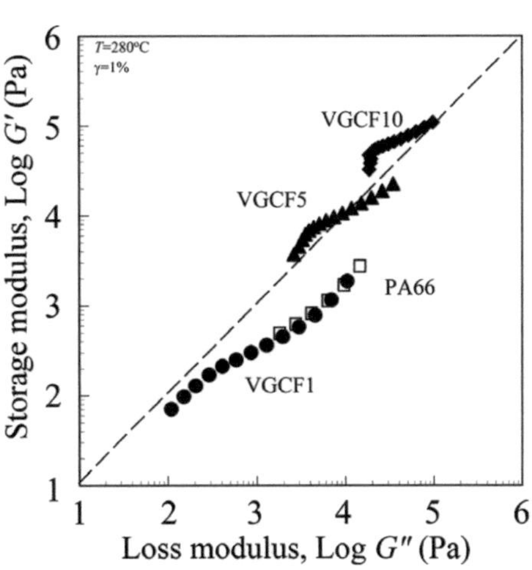

Figure 6. Relationship between storage modulus and loss modulus for VGCF/PA66 composites for VGCF/PA66 composites at 1% strain and 280°C.

http://dx.doi.org/10.5772/64531

2.7. Complex modulus dependences

The influence of complex modulus on the viscoelastic properties of VGCF/PA66 composites is discussed in this section in order to clarify the apparent yield stress and the rheological percolation behavior. A long-scale relaxation time was observed by the curves of G'–ω of VGCF/PA66 in the long-scale time region (in low ω regions) as mentioned earlier. This may be due to the fiber network formation in the composites, indicating the presence of an apparent yield stress in low ω regions. The apparent yield value of $|G^*|$ is thought to be a critical value which determines whether material can flow or not. To discuss this apparent yield stress, the complex viscosity $|\eta^*|$ is plotted against complex modulus $|G^*|$ in double logarithmic coordinates for VGCF/PA66 composites as shown in **Figure 7**. In general, the curves of $|\eta^*|$ versus $|G^*|$ show the following behavior: in the case of Newtonian fluids, although $|\eta^*|$ slightly increases with decreasing $|G^*|$, $|\eta^*|$ has a constant value behavior. On the contrary, in non-Newtonian fluids, such as highly filled polymer systems, $|\eta^*|$ does not have the constant value behavior, and $|\eta^*|$ dramatically increases with slightly decreasing $|G^*|$. In particular, the $|\eta^*|$ of high fiber content samples is almost independent of $|G^*|$. This dramatic behavior in which the $|G^*|$ of high fiber content samples decreases gradually in high ω regions may be due to the remarkable change in the structure of VGCF in high ω regions. It can be estimated from $|\eta^*|$ versus $|G^*|$ curves in the low ω region that there are apparent yield values [25]. In this study, $|\eta^*|$ of neat PA66 and VGCF1 gradually increases with decreasing $|G^*|$, which indicates that there is no apparent yield stress. However, the $|\eta^*|$ of VGCF5 and VGCF10 increases rapidly with a slight decrease in $|G^*|$, which shows that there are the yield values. The yield values $|G^*|_y$ of VGCF5 and VGCF10 are 30 and 300 kPa, respectively. Thus, $|G^*|_y$ is strongly dependent on the fiber content. This behavior was reported by other fiber-filled polymer composites and nanocomposites, such as GF/PP [25, 50], organic fiber/PP [42], CNT/UPR [51], etc. In addition, these curves of $|\eta^*|$ versus $|G^*|$ of VGCF/PA66 composites sharply increase with V_f between 1 and 5 vol%, indicating the presence of the rheological percolation threshold. From these results, the internal structure, changes, and yield stress can be estimated from the curves of $|\eta^*|$ versus $|G^*|$.

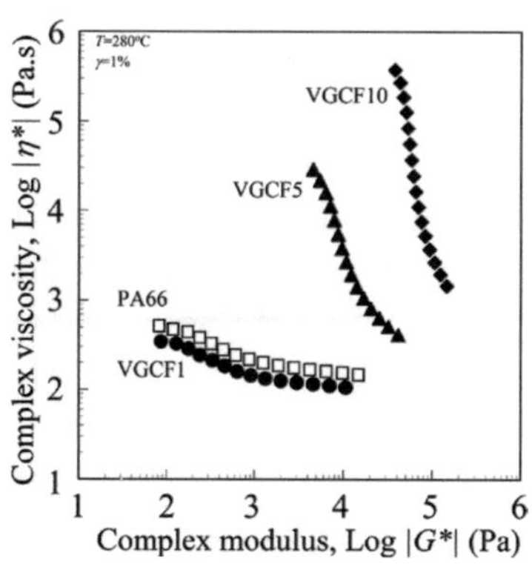

Figure 7. Relationship between complex viscosity and complex modulus for VGCF/PA66 composites at 1% strain and 280°C.

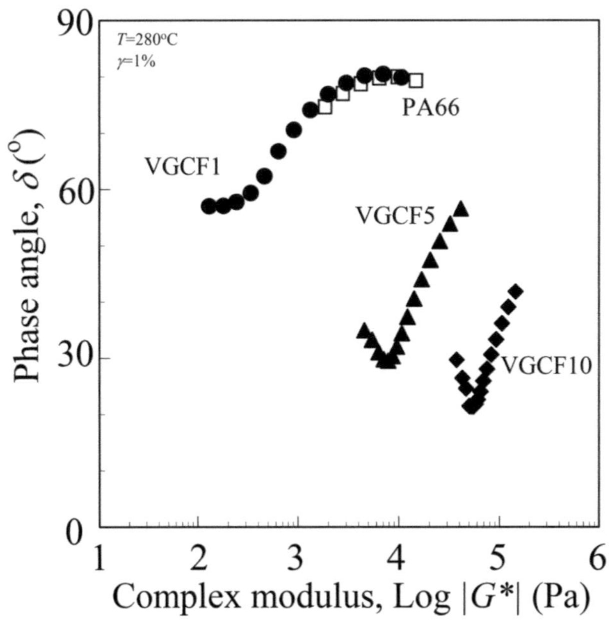

Figure 8. Phase angle versus absolute value of the complex modulus plots (van Gurp-Palmen plot) for VGCF/PA66 composites at 1% strain and 280°C.

On the other hand, the internal structure of the composites and their changes can be identified by van Gurp-Palmen plot [52]. This plot is drawing attention recently as another means of representing the internal microstructures and their change. This plot is considered to be a sensitive indicator for the time-temperature superposition, the presence of long chain branch of the polymer, the polymer entanglement, the gelation behavior, rheological percolation of the polymer nanocomposites, etc. [46, 52–55]. This is due to the emphasis on the change in rheological properties in this δ-$|G^*|$ plot, which are particularly difficult to understand in terms of angular frequency dependence ($G'-\omega$, $G''-\omega$, $\tan\delta-\omega$ curves, etc.). **Figure 8** shows that the phase angle δ (= arctan G''/G') is plotted against the absolute value of the complex modulus $|G^*|$ of VGCF/PA66 composites. The δ-$|G^*|$ curves of neat PA66 and VGCF1 show the same tendency, where δ of neat PA66 and VGCF1 is relatively higher than that of other composites although δ decreases gradually with decreasing $|G^*|$. However, δ of VGCF5 and VGCF10 decreases abruptly with a slight decrease in $|G^*|$, and the δ-$|G^*|$ curves of VGCF5 and VGCF10 demonstrate minimum peak behaviors. It is usually assumed that the change in rheological properties near the percolation threshold of a filler network embedded in a viscoelastic liquid is equivalent to the so-called "liquid-solid transition" (or gelation behavior) [47]. It is well known that the gelation point is to plot the loss tangent (tan δ) versus the ω. At gelation point, tan δ is frequency independent. On the other hand, the curve of δ-$|G^*|$ corresponds to a plateau in low complex modulus in this van Gurp-Palmen plot. Previous researches have suggested the following phenomenon in polymer nanocomposites such as CNT-filled polymer systems [46, 47, 53]: In liquid-like behavior, the δ-$|G^*|$ curve approaches 90° at the low complex modulus $|G^*|$, where in solid-like behavior, the δ-$|G^*|$ curves approaches 0° at the low complex modulus $|G^*|$. However, in the case of VGCF/PA66 composites in this study, the δ-$|G^*|$ curves of neat PA66 and VGCF1 not only demonstrate liquid-like behavior but also solid-like ones at low complex modulus. Furthermore, the δ-$|G^*|$ curves of VGCF/PA66 composites demonstrate remarkable solid-like behavior with increasing V_f. The $|G^*|_{d\text{-min}}$, which is a plateau modulus, is determined by the complex modulus of minimum value of δ in δ-$|G^*|$ curves [56]. The influence of volume fraction of VGCF on the $|G^*|_{d\text{-min}}$ of VGCF/PA66 composites is shown in **Figure 9**. As seen in **Figure 9**, $|G^*|_{d\text{-min}}$ of VGCF/PA66 composites increases sharply with increasing V_f, and exhibits a straight relations between $|G^*|_{d\text{-min}}$ and V_f on this log-log plot. The slope of $|G^*|_{d\text{-min}}$ versus V_f on log-log plot is found to be 2.61 ($|G^*|_{d\text{-min}} \propto V_f^{2.61}$) from the relation in **Figure 9**. Thus, this δ-$|G^*|$ plot may be able to serve as an indicator for the rheological percolation.

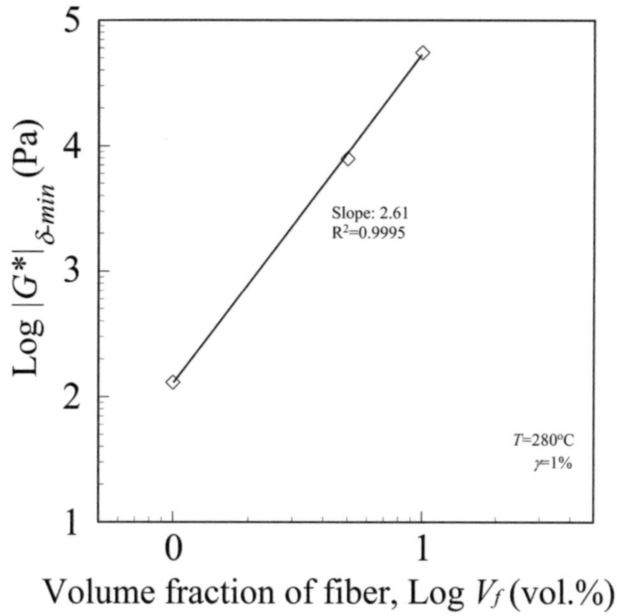

Figure 9. Influence of volume fraction of fiber on absolute value of complex modulus at minimum phase angle for VGCF/PA66 composites at 1% strain and 280°C.

2.8. Temperature dependences

The influence of temperature on the viscoelastic properties of VGCF/PA66 composites is discussed here. The complex viscosity $|\eta^*|$ of VGCF/PA66 composites is plotted against the reciprocal of the absolute temperature $1/T$ at the angular frequency ω of 100 rad/s and at temperatures 270, 280, and 290°C in **Figure 10**. The influence of temperature on $|\eta^*|$ of VGCF/PA66 composites shows the same tendency although the slope of $|\eta^*|$ versus $1/T$ plots slightly differ for each V_f. Especially, the value of VGCF1 is the smallest in this plot in the whole temperature range. From the slope of $|\eta^*|$ versus $1/T$ plots, the apparent activation energy E_a for flow can be calculated from the following Andrade's equation:

$$\eta = A\exp\left(\frac{E_a}{RT}\right) \quad (1)$$

where A is the constant value, R is the gas constant, and η was replaced by $|\eta^*|$ [42, 56]. The apparent activation energy E_a of VGCF/PA66 composites is listed in **Table 2**. E_a is not a constant value for filled systems, but dependent on the volume fraction of fiber [42]. However, it was found that E_a of VGCF/PA66 composites have a complex behavior according to V_f and E_a increases in the following order: VGCF5 < neat PA66 < VGCF10 < VGCF1. It can be said that the fluidity of the materials increases with increasing E_a. In short, VGCF1 with the highest value of E_a indicate high sensitivity to temperature change. In general, it is well known that the increase of V_f decreases the fluidity of the composites due to the formation of the fiber network such as the interaction between fiber and fiber. However, E_a in this study shows the different behavior. This may be attributed to the change in these internal structures at high angular frequency of 100 rad/s.

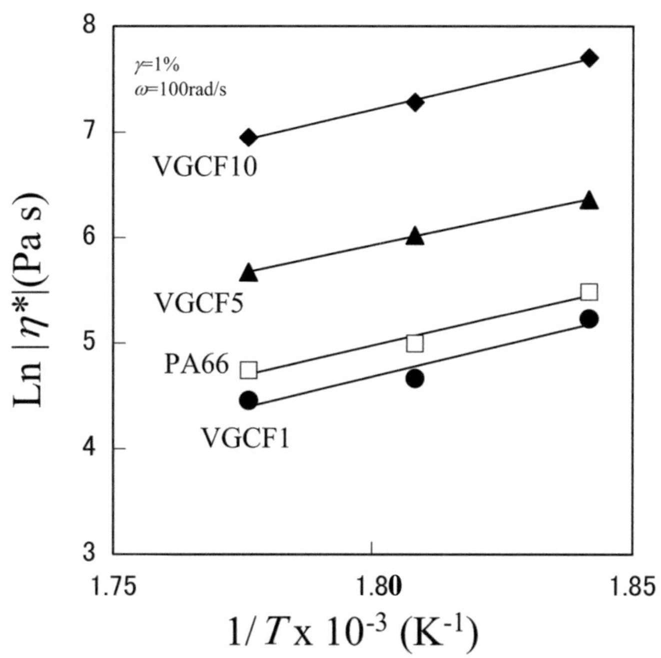

Figure 10. Temperature dependences of complex viscosity for VGCF/PA66 composites at the angular frequency of 100 rad/s and 1% strain. (a) VGCF/PA6/SEBS. (b) VGCF/PA6/SEBS-g-MA.

Code	PA66	VGCF1	VGCF5	VGCF10
E_a (kJ/mol)	94.7	98.5	87.1	95.9

Table 2. Apparent activation energy of flow for VGCF/PA66 composites.

3. Effect of addition of thermoplastic elastomer on dynamic viscoelastic properties of carbon nanofiber-filled polyamide 6 composites in molten state

3.1. Introduction

Recently, there has been considerable discussion on engineering materials containing of nanosized filler such as CNF, CNT, Clay, etc. filled polymer composites based on the multi-component polymer systems such as binary and ternary polymer blends [18, 22, 31, 32], because they can improve various physical properties with sufficient balances between mechanical and tribological properties by controlling the internal structure such as phase structure and filler dispersion. In our previous works, we considered the relationship between the structural and physical properties of filled polymer composites based on the multicomponent polymer systems required for constructing the technology of internal structures control and investigated the rheological, mechanical, and tribological properties of VGCF-filled polymer blends of PBT and TPE [17, 18], clay-filled polymer blends of PA and TPE [31, 32], etc. It was found that the rheological, mechanical, and tribological properties are improved by the addition of third components such as TPE, PP, PE, etc. However, the rheological properties of filled polymer composites based on the multicomponent polymer systems, in particular, their relationship between rheological properties and internal structure, are yet to be fully clarified. To further enhance the various physical properties, in these CNF-filled polymer composites based on the multicomponent polymer systems, it is very critical to understand the rheological behavior of these multicomponent composites in the molten state such as processability, internal microstructure, changes, and structure-property relationships. The aim of this study is to report the effects of the addition of TPE on the dynamic viscoelastic properties of VGCF/PA6 composites in the molten state. In particular, this study discusses the effects of the addition of TPE, type of TPE, volume fraction of VGCF, strain, and angular frequency on the dynamic viscoelastic properties in the molten state.

3.2. Materials and methods

The materials used in this study were ternary nanocomposites: VGCF-filled polyamide 6 (PA6) composites and the blend of these composites and styrene-ethylene/butylene-styrene copolymer (SEBS), which are called VGCF/PA6/SEBS ternary composites. VGCF (Showa Denko K.K, Japan, $d = 150$ nm, $l = 10$ μm) was used as a filler. PA6 (1013B, Ube industries, Ltd., Japan) was used as a matrix polymer. Two types of SEBS: SEBS (standard, Tuftec H1052, Asahi Kasei Chemicals Co., Japan) and SEBS-g-MA (maleic anhydride (MA) grafted SEBS, Tuftec M1943) were used as the blending materials. The content of maleic anhydride (MA) was determined at 10 mg CH_3ONa/g by the titration method. The ratio of styrene to ethylene/butylene in both block copolymer was 20/80 by wt%. Details such as code, manufacturer, etc. are listed in **Table 3**. The composition of PA6 and SEBS (or SEBS-g-MA) was fixed at 80/20 wt%, and three kinds of VGCF volume fraction were selected 0, 1, and 5 vol%. Prior to mixing, VGCF-X, PA6, and SEBS were dried in a vacuum oven at 80°C for 12 h until the moisture level was 0.2%. All the components were dry blended in a small plastic bottle, and subsequently melted and mixed at 85 rpm and 240°C in a twin screw extruder (TEX-30, Japan Steel Works, Ltd., Japan). After mixing, the extruded strands of these ternary composites were cut into 5 mm long pieces by a pelletizer and were dried again at 80°C for 12 h in the vacuum oven. In addition, 1 mm thick sheets were compression molded at the condition of 240°C, 5 MPa, and 3 min, and cut into ϕ25 mm disk shapes for rheological properties measurements.

Materials	Grade	Manufacturer	Note
PA6	1013B	Ube Industries, Ltd.	-
VGCF	VGCF®	Showa Denko K.K.	Fiber diameter = ϕ150 nm Fiber length = 10 μm
SEBS	Tuftec™ H1052	Asahi Kasei Chemicals Corp.	Styrene/Ethylene-Butylene ratio = 20/80
SEBS-g-MA	Tuftec™ M1943	Asahi Kasei Chemicals Corp.	Maleic anhydride functionalized SEBS Styrene/Ethylene-Butylene ratio =20/80

Table 3. Materials used in this study.

Since the experimental methods such as dynamic viscoelastic properties in the molten state are same as the one in Section 2.2, other than measurement temperature of 240°C and morphology observation method, details are omitted here. To clarify the internal structure of these ternary composites such as the dispersion of SEBS (or SEBS-g-MA) and VGCF in PA6 matrix polymer, the surface of samples fractured cryogenically in liquid nitrogen was observed using scanning electron microscope (SEM, EDX-WET SEM, JSM-6360LA, JEOL Ltd., Japan). The cryogenically fractured surface was etched in toluene for 24 h to remove the dispersed SEBS particles. A quantitative analysis of dispersed SEBS particle size was carried out from several SEM microphotographs using two kinds of image processing software (Adobe Photoshop, Adobe and Image J, NIH). The software used identifies each individual dispersed SEBS particle and evaluates its area A. From these findings, the apparent particle size d was calculated as follows:

$$d = \left(\frac{4A}{\pi}\right)^{\frac{1}{2}} \tag{2}$$

The measured particle size was characterized by evaluating number average diameter d_n, weight average diameter d_w, and volume average diameter d_v from more than 250 particles defined as [32, 57, 58]:

$$d_n = \frac{\sum n_i d_i}{\sum n_i} \tag{3}$$

$$d_w = \frac{\sum n_i d_i^2}{\sum n_i d_i} \tag{4}$$

$$d_v = \frac{\sum n_i d_i^4}{\sum n_i d_i^3} \tag{5}$$

where n_i is the number of dispersed particles having diameter d_i. The ratio of d_w/d_n is an indication of polydispersibility, which is the distribution of dispersed particles.

On the other hand, the fiber length of VGCF and its distribution were measured for samples which had been melted and mixed by an ordinary method: burning off the matrix polymer in a furnace at 550°C for 4 h, wetting fibers in water added with trace surfactant, and spreading them on Al stage (for scanning electron micrographs, SEM). The fiber length distribution of VGCF was observed from SEM. The length of at least 250 fibers was scanned in different regions for evaluating them accurately. A quantitative analysis of the fiber length was made from several micrographs using two kinds of image processing software as mentioned earlier. The number average fiber length l_n and weight average fiber length l_w were calculated according to the following equations [19, 23]:

$$l_n = \frac{\sum N_i l_i}{\sum N_i} \tag{6}$$

$$l_w = \frac{\sum N_i l_i^2}{\sum N_i l_i} \tag{7}$$

where N_i is the number of fibers and l_i is the fiber length. The fiber distribution l_w/l_n was calculated from above equations.

3.3. Strain dependences

The strain dependence of the dynamic viscoelastic properties of ternary composites (VGCF/PA6/SEBS) is discussed here. It was measured by strain sweep testing in order to characterize the transition from linear to nonlinear viscoelastic properties of the ternary composites and also to estimate the change in internal microstructure such as the dispersion and localization of VGCF and SEBS, etc. of these ternary composites. Only the storage modulus G' curves are presented in this study since G' is the more sensitive rheological function to the structural changes of the composites than other functions as mentioned in Section 2.3. The strain dependence of storage modulus G' of ternary composites (VGCF/PA6/SEBS) is shown in **Figure 11(a)** (VGCF/PA6/SEBS composites) and (b) (VGCF/PA6/SEBS-g-MA composites), respectively. This measurement was carried out for all samples of temperature at 240°C and the angular frequency at 3.5 rad/s. The strain dependences of G' show remarkably different behavior according to the type of material such as V_f, with or without SEBS and type of SEBS. G' of neat PA6 exhibits a linear behavior, in which the modulus is a constant over a wide range of strain amplitude γ. The G' of VGCF1/PA66 binary composites has the same tendency as neat PA6, and is slightly higher than that of neat PA6. On the other hand, G' of PA6/SEBS binary blends and VGCF1/PA6/SEBS ternary composites shows the same linear behaviors as neat PA6; however, it is higher than that of VGCF1/PA6 composites. On the contrary, the G' of VGCF5/PA6 composites and VGCF5/PA6/SEBS ternary composites shows the linear plateau only in low strain regions, and rapidly decreases with increasing γ, which are independent transition from linear to nonlinear behavior. The G' of VGCF5/PA6/SEBS ternary composites is higher than that of VGCF5/PA6 composites, and the magnitude of the increase is almost 1 digit (decade) on the log scale. On the other hand, the G' of SEBS-g-MA systems, which are PA6/SEBS-g-MA and VGCF/PA6/SEBS-g-MA, shows a different behavior from the SEBS systems. G' of PA6/SEBS-g-MA shows a linear behavior in wide strain region until several 10%, and is higher than that of PA6/SEBS blends. The G' of VGCF1/PA6/SEBS-g-MA and VGCF5/PA6/SEBS-g-MA has the same tendency as that of PA6/SEBS, and their values, which are the plateau modulus G'_p, increase with increasing V_f. It was found from these results that G'-γ curves are changed by material composition factors such as the addition of VGCF and SEBS, volume fraction of fiber V_f, and type of SEBS, which is SEBS and SEBS-g-MA. In addition, these phenomena are distinguished as the following mechanisms: the addition and volume fraction of VGCF emphasize the nonlinear behavior and increase the plateau modulus. The addition of SEBS increases the magnitude value of the plateau modulus. Especially, the type of SEBS changes the magnitude value of the plateau modulus and the critical strain value, which is the transition point from linear to nonlinear viscoelastic behavior. These may be attributed to the changes in the internal microstructure formation of the composites, which is the dispersion of VGCF and SEBS, interaction between fiber and matrix polymer, localization, size of dispersed SEBS particles and VGCF, agglomerations of VGCF, interaction between fiber and fiber, etc.

To further clarify the transition point from linear to nonlinear behavior, the relationship between critical strain value γ_c and volume fraction of fiber V_f of VGCF/PA6, VGCF/PA6/SEBS, and VGCF/PA6/SEBS-g-MA composites at 3.5 rad/s and 280°C is shown in **Figure 12** in the semilogarithmic scale. Here, the critical strain value γ_c can be calculated from the curves of the relative storage modulus G'_r (= G'/G'_p) – strain γ of these composites using the method in Section 2.3. γ_c of various VGCF/PA6 composites abruptly decreases with increasing V_f and exhibits a straight relations between γ_c and V_f in the semilogarithmic plot. The slope of the γ_c - V_f plot gradually increases in the following order: VGCF/PA6 (–9.95) < VGCF/PA6/SEBS (–6.50) < VGCF/PA6/SEBS-g-MA (–3.24). It was found from these results that material composition factors such as the addition of VGCF and SEBS, volume fraction of fiber V_f, and type of SEBS, which is SEBS and SEBS-g-MA, have strong influences on the strain dependence of viscoelastic properties. In particular, the strain dependences of ternary composites have the different behavior because of these factors. These behaviors may be considered to be due to the change in the internal structure of the composites as mentioned earlier.

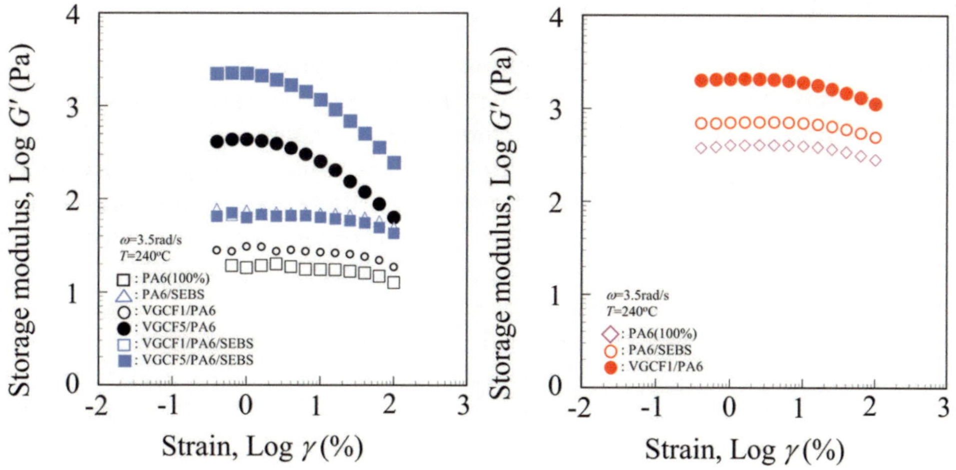

Figure 11. Storage modulus as a function of strain for ternary composites (VGCF/PA6/SEBS) at 3.5rad/s and 240°C. (a) VGCF/PA6/SEBS. (b) VGCF/PA6/SEBS-g-MA.

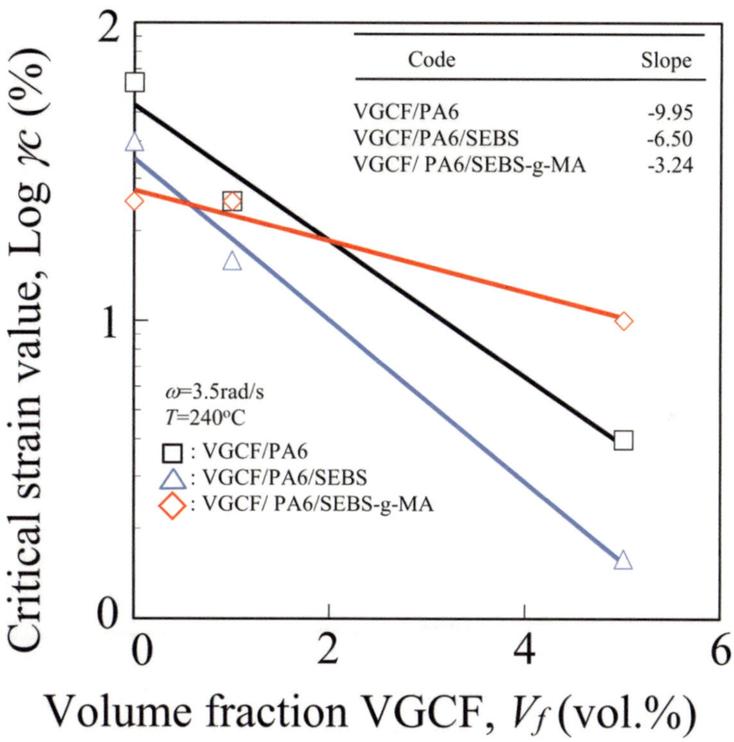

Figure 12. The dependence of critical strain on volume fraction of VGCF for ternary composites (VGCF/PA6/SEBS) at 3.5rad/s and 240°C.

3.4. Angular frequency dependences

The influence of these materials composition factors on the dynamic viscoelastic properties of ternary composites (VGCF/PA6/SEBS) in the molten state is discussed in this section. These properties are strongly dependent on the internal microstructure formation of the polymer composites. We shall discuss the angular frequency dependence, which is the basic variable in these properties. The dynamic viscoelastic properties of these ternary composites are plotted as a function of ω at temperature 240°C and strain 1% in **Figure 13(a)** (storage modulus G', V_f= 1 vol%), **(b)** (loss modulus G'', V_f= 1 vol%), **(c)** (storage modulus G', V_f= 5 vol%), and **(d)** (loss modulus G'', V_f= 5 vol%), respectively. In **Figure 13(a)**, the slopes of G' and G'' against ω in the low and high ω regions are listed in **Table 4**. The G' of neat PA6 increases with increasing ω, disagreeing with the linear viscoelastic model [41, 42]. The slope of G'- ω in high ω regions of neat PA6 is 1.57. The G' of VGCF1 shows the same tendency as that of neat PA6, and is slightly higher than that of neat PA6. PA6/SEBS blends show the typical G' of the multicomponent systems [31, 32]. The slope of G' against ω, in particular, in low ω regions becomes small with the addition of SEBS, indicating the "second rubbery plateau," i.e., the long scale relaxation time [25, 41, 42]. This may be attributed to the heterogeneous structure such as sea-island structures in PA6/SEBS blends. G' of VGCF1/PA6/SEBS composites is higher than that of PA6/SEBS blends, which shifts toward high elasticity (solid-like) with the addition of VGCF caused by the changes in the internal structures such as interaction between fibers or agglomerations. Furthermore, the G' of PA6/SEBS-g-MA blends and VGCF1/PA6/SEBS-g-MA composites is much higher than that of PA6/SEBS blends and VGCF1/PA6/SEBS composites, and the slopes of G' against ω of these SEBS-g-MA systems are visible in low ω regions, indicating the formation of small dispersed phase and the new structure in these polymer blends and composites. In **Figure 13(b)**, loss modulus G'' of these ternary composites shows a slightly different behavior from the G'. The values of G'' of these materials in low ω regions increase in the following orders: PA6 < VGCF/PA6 < PA6/SEBS < VGCF1/PA6/SEBS < PA6/SEBS-g-MA < VGCF1/PA6/SEBS-g-MA composites, although those of these materials in high ω regions have different behaviors: PA6 < PA6/SEBS < VGCF/PA6 < VGCF1/PA6/SEBS < PA6/SEBS-g-MA < VGCF1/PA6/SEBS-g-MA composites. In general, storage and loss moduli in low ω regions are dominated by the internal structure, and those in high ω regions dominated by the polymer matrix. However, the little reverse phenomenon can be explained in the present stage. On the other hand, in **Figure 13(c)** and **(d)** with VGCF 5 vol% loading systems, the G' of various composites shows remarkably the typical storage and loss moduli of highly filled systems, indicating the "second rubbery plateau." However, another behavior, which is crossover of VGCF5/PA6/SEBS and VGCF5/PA6/SEBS-g-MA composites, is clearly seen at ω = 1–10 rad/s. In particular, in low ω regions, the G' and G'' of VGCF5/PA6/SEBS composites are higher than those of VGCF5/PA6/SEBS-g-MA composites. In short, VGCF5/PA6/SEBS composites demonstrate solid-like behavior. These behaviors may be explained by the changes in the internal structure caused by the dispersed phases, the interaction between fiber and polymer matrix, the interaction between fibers, agglomerations, etc. Therefore, it is necessary to understand these behaviors to observe the morphology of ternary composites (VGCF/PA6/SEBS). From the results of the angular frequency dependences of storage and loss moduli of ternary composites mentioned above, these dependences show different behaviors according to the

http://dx.doi.org/10.5772/64531

addition of VGCF and SEBS, type of SEBS, and volume fraction of fiber. The addition of SEBS was changed by the tendency of G' and $G'' - \omega$ curves, which shows remarkably the "second rubbery plateau" in low ω regions. In particular, these dependences appear to be conspicuously high for the composites with SEBS-g-MA, except for only VGCF5/PA6/SEBS composites in low ω regions. On the other hand, the addition of VGCF and volume fraction of fiber slightly enhanced these dependences.

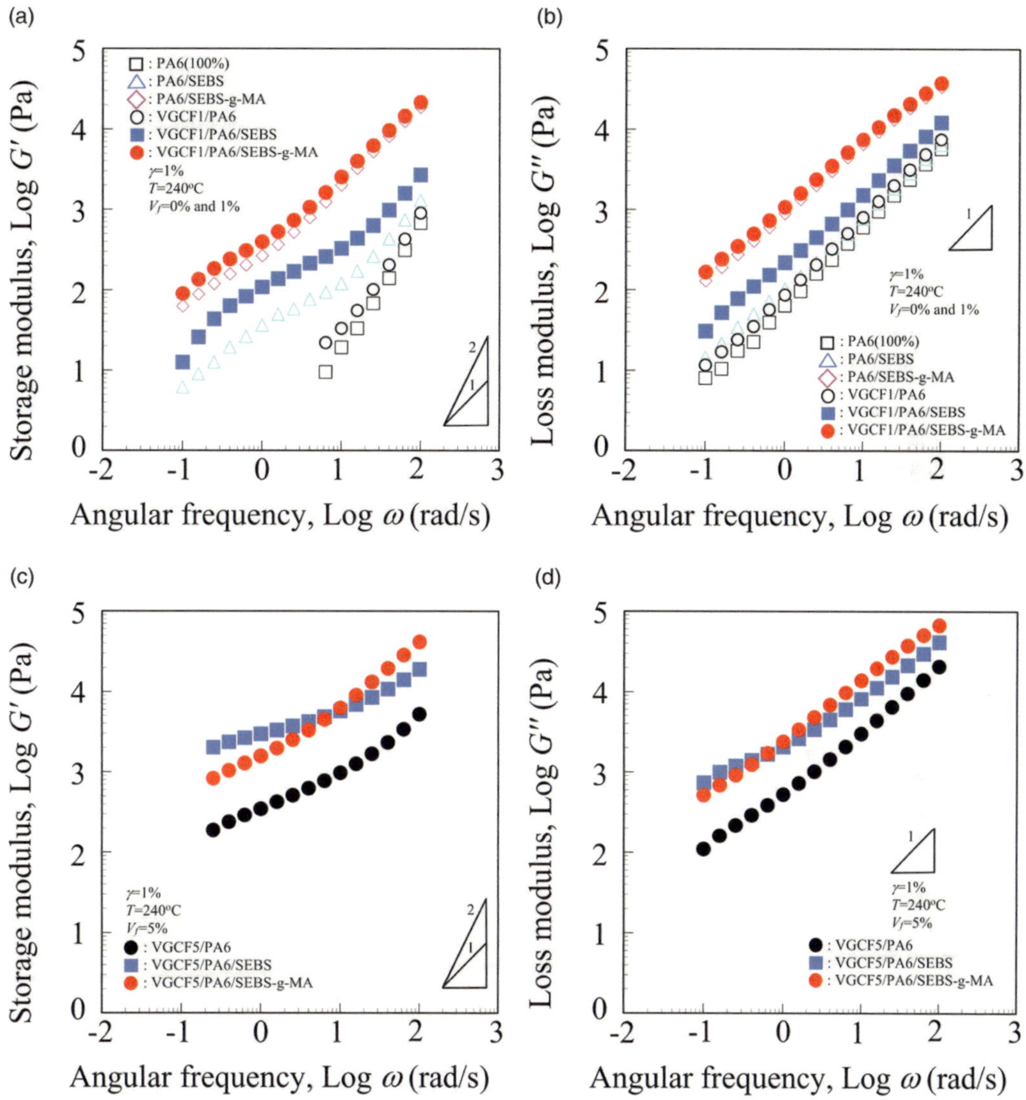

Figure 13. Dynamic viscoelastic properties as a function of angular frequency for ternary composites (VGCF/PA6/SEBS) composites at 1% strain and 240°C. (a) Storage modulus (V_f=0% and 1%). (b) Loss modulus (V_f=0% and 1%). (c) Storage modulus (V_f=5%). (d) Loss modulus (V_f=5%)

Code	Slope of G'-ω		Slope of G''-ω	
	Low ω	High ω	Low ω	High ω
PA6 (100%)	-	1.57	0.91	0.98
PA6/SEBS	0.78	1.04	0.86	0.92
PA6/SEBS-g-MA	0.62	0.97	0.84	0.71
VGCF1/PA6	-	1.45	0.87	0.97
VGCF5/PA6	0.58	0.72	0.67	0.84
VGCF1/PA6/SEBS	0.91	0.92	0.84	0.91
VGCF5/PA6/SEBS	0.44	0.52	0.42	0.71
VGCF1/PA6/SEBS-g-MA	0.63	0.93	0.80	0.70
VGCF5/PA6/SEBS-g-MA	0.56	0.82	0.66	0.69

Table 4. Slope of viscoelastic properties as function of angular frequency curves of ternary composites (VGCF/PA6/SEBS) in low and high frequencies.

3.5. Influence of volume fraction and type of SEBS

The influence of volume fraction of fiber and type of SEBS on the rheological properties of ternary composites (VGCF/PA6/SEBS) is discussed in this section. The relative storage modulus G'_r is plotted against the volume fraction of fiber V_f in **Figure 14(a)** (in low ω regions, ω = 0.25 rad/s) and (b) (in high ω region, ω = 25 rad/s), respectively. Here, this value is given by the value of various ternary composites (VGCF/PA6/SEBS) divided by that of neat PA6. G'_r of all ternary composites (VGCF/PA6/SEBS) increases with increasing V_f. This tendency changes with the type of SEBS and the angular frequency regions. The increase ratios of G'_r

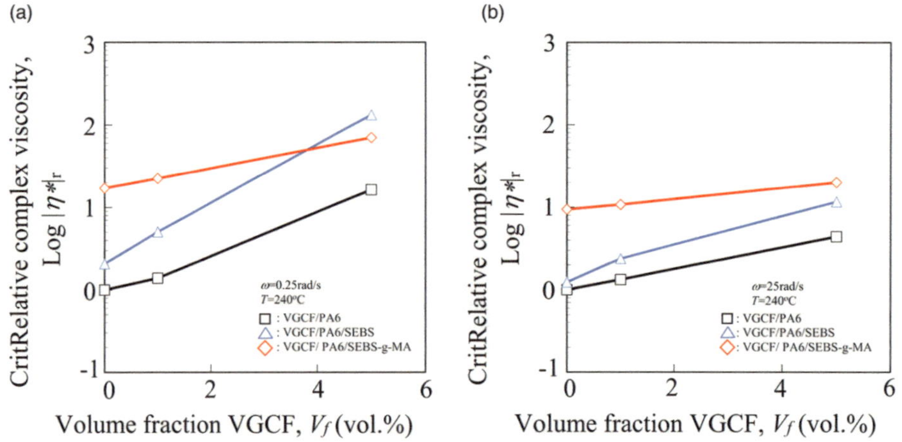

Figure 14. Influence of volume fraction of fiber on relative complex viscosity for ternary composites (VGCF/PA6/SEBS) at 1% strain and 240°C. (a) Relative complex viscosity. (b) Relative complex viscosity.

against V_f in low ω regions (**Figure 14(a)**, ω = 0.25 rad/s) are larger than those in high ω regions (**Figure 14(b)**, ω = 25 rad/s). It should be noted that G'_r of VGCF5/PA6/SEBS composites is higher than that of VGCF5/PA6/SEBS ones. In general, storage modulus in low ω regions is dominated by the internal structure as mentioned earlier. Thus, it is necessary for understanding this behavior to observe the internal structure of these ternary composites, and the morphologies of these composites will be discussed in the next section.

3.6. Morphology

To further clarify the relationship between the dynamic viscoelastic properties and internal structure of ternary composites (VGCF/PA6/SEBS), we discuss the morphologies of these composites, which are the internal structure such as fiber network formation and dispersed SEBS particles. We observed the cryogenically fractured surfaces of various ternary composites, which were etched by toluene in order to remove the dispersed SEBS particles using a scanning electron microscope. **Figure 15** shows the SEM photographs of the etched fracture surfaces of PA6/SEBS blends (**Figure 15(a)**), VGCF1/PA6/SEBS composites (**Figure 15(b)**), VGCF5/PA6/SEBS composites (**Figure 15(c)**), PA6/SEBS-g-MA blends (**Figure 15(d)**), VGCF1/PA6/SEBS-g-MA composites (**Figure 15(e)**), and VGCF5/PA6/SEBS-g-MA composites (**Figure 15(f)**), respectively. Here, **Figure 15(a)-(c)** was observed at the magnification of 5000, and **Figure 15(d)-(f)** was observed at 20,000. Each polymer blends and ternary composite exhibited typical separate spherical phases (dispersed SEBS particles) in PA6 continuous matrix domains. **Table 5** summarizes the various data of the dispersed SEBS particles for each ternary composite, calculated by image processing from SEM photographs such as number average diameter d_n, weight average diameter d_w, volume average diameter d_v, and polydispersibity d_w/d_n. The sizes of dispersed SEBS particles change with the type of SEBS and volume fraction of VGCF, and these decrease in the following orders: PA6/SEBS > VGCF1/PA6/SEBS > VGCF5/PA6/SEBS >> PA6/SEBS-g-MA > VGCF1/PA6/SEBS-g-MA > VGCF5/PA6/SEBS-g-MA composites. This means that the size of dispersed SEBS particles decreases with the SEBS with functional groups of maleic anhydride (MA), and these SEBS particles decrease with increasing V_f. These morphological changes are the consequence of graft reaction between the PA6 matrix and MA in SEBS, which separates spherical particles in continuous matrix domains. MA groups are expected to induce strong interaction and reaction due to high reactive functional groups such as amine and carboxyl end groups and acid amide in the main chain of PA [32]. These observations indicate that dynamic viscoelastic properties shown earlier correlate closely with the size of dispersed SEBS particles. On the other hand, we have to consider the presence of VGCF, which may complicate these behaviors. Each ternary composite shows good dispersion of VGCF, and the location of VGCF is in PA6 continuous matrix domains. For comparison, it was not observed by VGCF in toluene solution for etching in order to remove the SEBS particles. The number average fiber length l_n, weight average fiber length l_w, fiber distribution l_w/l_n, and number average aspect ratio $(a_r)_N$ are listed in **Table 6**. The number average aspect ratio $(a_r)_N$ is defined as l_n/d, where d is the average fiber diameter of VGCF ($d = \phi 150$ nm). The fiber lengths values of l_n, l_w, and $(a_r)_N$ of VGCF/PA6 and VGCF/PA6/SEBS-g-MA decrease with increasing V_f, although those of VGCF/PA6/SEBS increase with V_f. Fiber distribution l_w/l_n has complex behavior; however, VGCF/PA6/SEBS-g-MA has the highest values in this study. What should be noted is that the fiber length value of VGCF5/PA6/SEBS is much longer than those of other composites except for VGCF1/PA6 composites. These results coincide with the dynamic viscoelastic values earlier, that is, the storage modulus of VGCF5/PA6/SEBS is crossover with that of VGCF5/PA6/SEBS in low ω regions. This may be attributed to the change in internal structure by the interactions of fiber-fiber and fiber-matrix, and agglomerations caused by the increase in fiber length values. Therefore, the internal structure of ternary composites was found to correlate closely with the dynamic viscoelastic properties of ones. In general, it is well known that the fiber length and the aspect ratio of these compounds in the twin screw extruder depend on concentration, characteristics of these compounds, and the addition of third component, which suggests fiber degradation may be due to the interactions between fiber and matrix, and between fibers. However, it is difficult in the present stage to understand the exactly reason why the fiber length of VGCF5/PA6/SEBS are longer than those of other composites in this study.

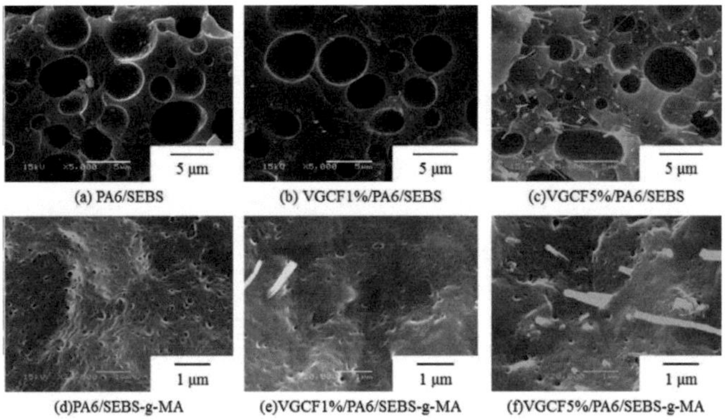

Figure 15. SEM micrographs of etched fracture surface for the ternary composites (VGCF/PA6/SEBS) (SEBS content is 20 wt% and etched with toluene for 24 h to remove the dispersed phases): (a) PA6/SEBS, (b) VGCF 1%/PA6/SEBS, (c) VGCF 5%/PA6/SEBS, (d) PA6/SEBS-g-MA, (e) VGCF 1%/PA6/SEBS-g-MA, and (f) VGCF 5%/PA6/SEBS-g-MA. (a) VGCF fiber length. (b) SEBS particle size.

Material	d_n (nm)	d_w (nm)	d_v (nm)	d_w/d_n
PA6/SEBS	2930	3750	4880	1.28
VGCF1/PA6/SEBS	2810	3920	4940	1.39
VGCF5/PA6/SEBS	2260	3060	4250	1.36
PA6/SEBS-g-MA	88	97	121	1.11
VGCF1/PA6/SEBS-g-MA	81	88	104	1.08
VGCF5/PA6/SEBS-g-MA	51	59	95	1.17

Table 5. Dispersed SEBS particle size of ternary composites (VGCF/PA6/SEBS).

http://dx.doi.org/10.5772/64531

Material	l_n (μm)	l_w (μm)	l_w/l_n	$(a_r)_N$
VGCF1/PA6	3.28	3.64	1.11	21.9
VGCF5/PA6	2.91	3.23	1.11	19.4
VGCF1/PA6/SEBS	2.90	3.12	1.08	19.3
VGCF5/PA6/SEBS	3.04	3.38	1.11	20.3
VGCF1/PA6/SEBS-g-MA	2.28	3.12	1.37	15.2
VGCF5/PA6/SEBS-g-MA	2.22	2.96	1.33	14.8

Table 6. Fiber length, its distribution, and aspect ratio of ternary composites (VGCF/PA6/SEBS).

Furthermore, to clarify the relationships between the morphologies of ternary composites and the dynamic viscoelastic properties of ones, it is necessary to investigate the effect of morphologies of ternary composites on the dynamic viscoelastic properties in detail. G' is replotted as a function of average weight fiber length of VGCF l_w in **Figure 16(a)**, and average weight diameter of dispersed SEBS particles d_w in **Figure 16(b)**, respectively. The fiber length l_w dependence of G' has a complex behavior. It is fundamentally influenced by the volume fraction of fiber and the addition of SEBS rather than the fiber length. In VGCF/PA6/SEBS systems, the longer the fiber length is, the more the increase in G'. On the other hand, the size of diameter of dispersed SEBS particles d_w dependence is not clearly recognized, and G' is strongly influenced by the type of SEBS and the volume fraction of VGCF.

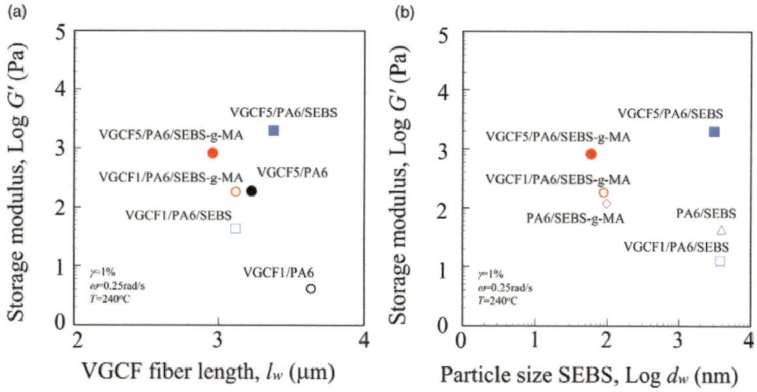

Figure 16. Influence of internal structure parameters such as SEBS particles size and fiber length on storage modulus of ternary composites (VGCF/PA6/SEBS). (a) VGCF fiber length. (b) SEBS particle size.

4. Effect of processing sequence on the dynamic viscoelastic properties of ternary composites (VGCF/PA6/SEBS-g-MA) in molten state

4.1. Introduction

We discussed the dynamic viscoelastic properties of ternary composites (VGCF/PA6/SEBS) in the molten state in the previous section and concluded that the effect of the addition of TPE on the dynamic viscoelastic properties of VGCF/PA6 composites in the molten state differs according to the viscoelastic value. It was clarified that the dynamic viscoelastic properties of VGCF/PA6/SEBS-g-MA ternary composites are higher than those of VGCF/PA6/SEBS ones. This may be attributed to the change in the internal structure caused by addition of TPE. Thus, it is important to further investigate the relationship between rheological properties and internal structure of these ternary composites. It is well known that the morphologies of these ternary composites are influenced by the processing sequences [34, 35]. Several investigations have been conducted on the effect of processing sequences at melt mixing by twin extruder on the relationship between the morphology and the physical properties of ternary composites such as PA/Clay/SEBS [35, 59], VGCF/PBT/TPE [60], VGCF/PA6/SEBS [22], etc. However, there is not enough information on reliable relations between the internal structure and rheological properties of these ternary composites. The purpose of this study is to report the effect of processing sequences on the dynamic viscoelastic properties of VGCF-filled polymer blends of PA6 and SEBS-g-MA (VGCF/PA6/SEBS-g-MA ternary composites).

4.2. Materials and methods

The materials used in this study were ternary composites (VGCF/PA6/SEBS-g-MA). Since the materials, composition, processing, and experimental methods are the same as in Section 3.2 other than the processing sequence using the twin extruder, they are omitted here. The composition of PA6 and SEBS-g-MA was fixed as 80/20 by weight fraction, and three kinds of VGCF volume fraction were selected as 0, 1, and 5 vol%. All the components were dried for 12 h at 80°C in a vacuum oven beforehand until the moisture level was below 0.2%. Four different processing sequences were carried out: (1) VGCF, PA6 and SEBS-g-MA were mixed simultaneously (process A), (2) VGCF was mixed with PA6 (VGCF/PA6 composites) and then these composites were blended with SEBS-g-MA (process B), (3) SEBS-g-MA was blended with PA6 (PA6/SEBS-g-MA blends) and then these blends were mixed with VGCF (process C), and (4) VGCF were mixed with SEBS-G-MA (VGCF/SEBS-g-MA composites) and blended with PA6 (process D) and then attempted to prepare the ternary composites (VGCF/PA6/SEBS-g-MA). **Figure 17** shows the schematic diagram of four different processing sequences for ternary composites (VGCF/PA6/SEBS-g-MA). The melt of these materials was mixed at 85 rpm and 240°C in a twin screw extruder (TEX-30, Japan Steel Works, Ltd.). After mixing, the extruded strands of these ternary composites were cut in piece of about 5 mm long by a pelletizer and were dried again at 80°C for 12 h in the vacuum oven. In addition, 1 mm thick sheets were compression molded at the condition of 240°C, 5 MPa, and 3 min, and cut into ϕ25 mm disk shapes for rheological properties measurements.

4.3. Angular frequency dependences

The effect of processing sequences on the dynamic viscoelastic properties of ternary composites (VGCF/PA6/SEBS-g-MA) is discussed here. The dynamic viscoelastic properties of these ternary composites with 5 vol% VGCF prepared by various processing sequences are plotted as a function of angular frequency ω in **Figure 18(a)** (storage modulus G') and **(b)** (loss modulus G''), respectively. Each storage and loss moduli increase with increasing ω; however, the magnitude value of G' and G'', in particular, in low ω regions, changes according to the processing sequences in the following orders: Process A > Process C > Process B = Process D. The decreases of G' and G'' are closely related to the number of mixing of VGCF, and the twice mixing methods of VGCF which remix VGCF such as processes B and C are more effective than unimixing methods such as processes A and C. These may be attributed to the change in the internal structure of these ternary composites by different processing sequences. Another important point is the difference in the order of decrease in G' according to the angular frequency region. In short, the G'- ω curve of process B is crossover to that of processes C and D. This is suggested by the change in the internal structure according to ω. To clarify the effect of processing sequences on G' of these ternary composites, the relative storage modulus G'_r in low ω regions (ω = 0.25 rad/s) is shown in **Figure 19(a)** (V_f= 1 vol%) and **(b)** (V_f = 5 vol%), respectively. Here, this relative value is given by the value of various ternary composites prepared by different processing sequences divided by that of process A. G'_r change according

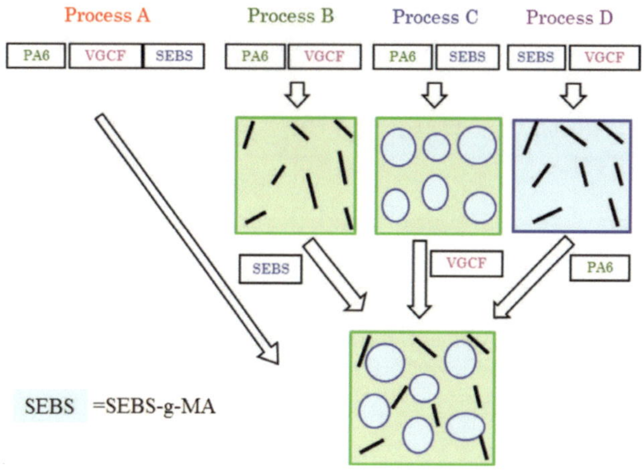

Figure 17. Schematic diagram of four different processing sequences for ternary composites (VGCF/PA6/SEBS-g-MA).

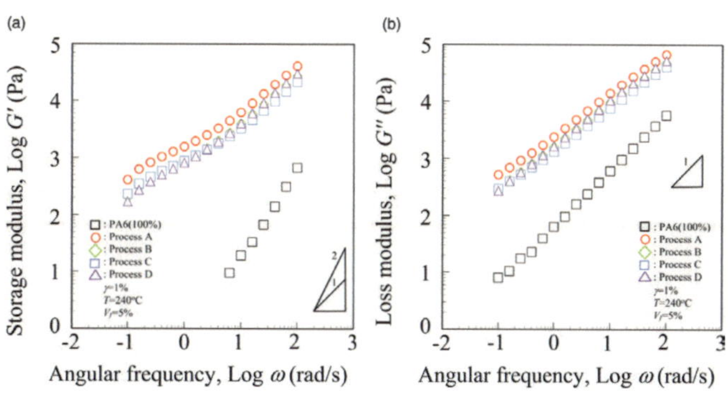

Figure 18. Dynamic viscoelastic properties as a function of angular frequency for various processing sequence of ternary composites (VGCF/PA6/SEBS-g-MA) composites at 1% strain and 240°C. (a) Storage modulus. (b) Loss modulus.

to the processing sequences and the volume fraction of fiber. The G'_r of these various composites with V_f= 1 vol% decreases in the following orders: Process A > Process C > Process D > Process B. On the other hand, that of V_f= 5 vol% decreases in the following orders: Process A > Process C > Process B > Process D. These tendencies may be attributed to the change in the internal structure by the number of mixing as mentioned earlier. However, it is not clearly explained by the order of decrease G'_r changing according to V_f. Therefore, it is necessary to observe the internal structure for better understanding the relationships between processing sequences and dynamic viscoelastic properties of these ternary composites. This will be discussed in the next sections.

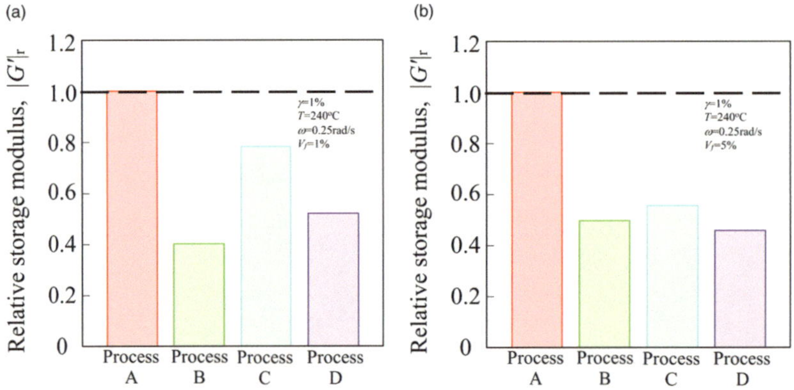

Figure 19. Relative storage modulus for various processing sequence of ternary composites (VGCF/PA6/SEBS-g-MA) composites at 1% strain and 240°C. (a) VGCF 1vol.%. (b) VGCF 5vol.%.

4.4. Morphology

Furthermore, to clarify the relationships between processing sequences and rheological properties of ternary composites (VGCF/PA6/SEBS-g-MA), we observed the internal structure, which is the dispersibility of SEBS particles and VGCF in these ternary composites. **Figure 20** shows the SEM photographs of cryogenically fractured surfaces, which were etched by toluene in order to remove the dispersed SEBS-g-MA particles, of various ternary composites prepared by different processing sequences at the same magnification rate of 20,000. Each ternary composite indicates the good dispersion of VGCF and dispersed SEBS-g-MA particles. In addition, it was not observed for VGCF in toluene solution for etching in order to remove the SEBS particles. Thus, there are VGCFs only in PA6 matrix domains, and are not presented in the dispersed SEBS particle domain. In particular, in the case of process D, VGCF should be present in the SEBS domain; nevertheless, VGCF was mixed with SEBS. However, VGCF was not found in the SEBS domain after mixing the second time in process D. Thus, VGCF is transferred from SEBS domain to the PA6 continuous matrix domain in process D, i.e., the localization of VGCF is changed. The reason for changing the localization of VGCF in process D can be explained by the following mechanisms: VGCF is selected by the PA6 domain, because the viscosity of PA6 is lower than that of SEBS-g-MA at the mixing temperature of

240°C. This is due to the stronger interactions of fiber-matrix polymer, which is wettability and chemical affinity, between VGCF and PA6 than those of VGCF and SEBS-g-MA. In addition, these detail mechanisms, which are filler migration inside a molten heterogeneous medium, were suggested by Fenouillot et al., Elias et al., and Baudouin et al. [61–63].

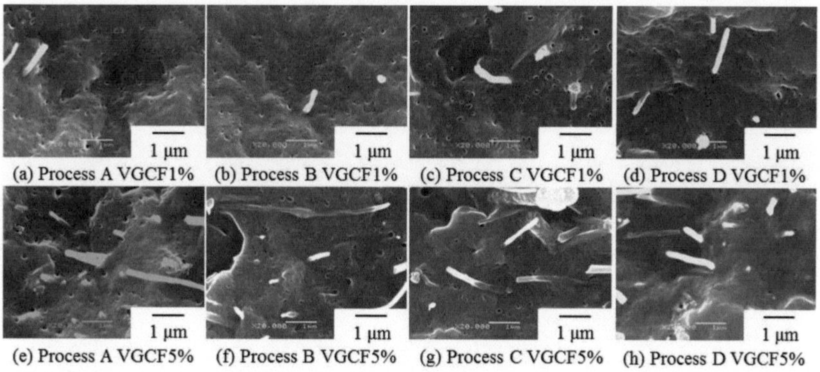

Figure 20. SEM micrographs of etched fracture surface for various processing sequence of the ternary composites (VGCF/PA6/SEBS-g-MA) (SEBS-g-MA content is 20 wt% and etched with toluene for 24 h to remove the dispersed phases): (a) Process A VGCF 1%, (b) Process B VGCF 1%, (c) Process C VGCF 1%, (d) Process D VGCF 1%, (e) Process A VGCF 5%, (f) Process B VGCF 5%, (g) Process C VGCF 5%, and (h) Process D VGCF 5%.

On the other hand, the dispersed SEBS-g-MA particles demonstrate good dispersion in the PA6 continuous matrix domains. **Table 7** summarizes the various data of the dispersed SEBS-g-MA particles in each ternary composites prepared by different processing sequences, which are calculated by image processing from SEM photographs such as number average diameter d_n, the weight average diameter d_w, the volume average diameter d_v, and polydispersibity d_w/d_n. The sizes of dispersed SEBS-g-MA particles change with the processing sequences and V_f, and these decrease fundamentally in the following orders: Process A > Process B > Process C > Process D. Thus, the size of dispersed SEBS particles values is closely related to the number of mixed SEBS-g-MA, and double mixing methods of SEBS-g-MA, which are the remixing of SEBS-g-MA such as processes C and D, are more effective than unimixing methods such as processes A and B. These behaviors may be attributed to the change in the mixing time, the physical interaction between PA6 and SEBS-g-MA, which is called as compatibilization effect, and also the dispersion and localization of VGCF; however, it is difficult to establish the exact reasons. Next, the number average fiber length l_n, weight average fiber length l_w, fiber distribution l_w/l_n, and number average aspect ratio $(a_r)_N$ are listed in **Table 8**. The number average aspect ratio $(a_r)_N$ is defined as l_n/d, where d is the average fiber diameter of VGCF ($d = \phi 150$ nm). The fiber length values of l_n, l_w, and $(a_r)_N$ generally decrease with increasing V_f except for process C, and the effect of processing sequences on the fiber length values decreases in the following orders: Process C > Process A > Process B > Process D. These behaviors may be attributed to the number of VGCFs mixed as mentioned earlier. In the case of process D, which is the first mixing step where VGCF was mixed with higher viscosity SEBS rather than PA6, enhances the damage of fiber in melt mixing.

Material		d_n (nm)	d_w (nm)	d_v (nm)	d_w/d_n
PA6/SEBS-g-MA (VGCF = 0%)		**88**	**97**	**121**	**1.11**
Process A	VGCF 1 vol%	81	88	104	1.08
	VGCF 5 vol%	51	59	95	1.17
Process B	VGCF 1 vol%	78	89	117	1.14
	VGCF 5 vol%	72	80	98	1.10
Process C	VGCF 1 vol%	68	77	102	1.12
	VGCF 5 vol%	52	55	66	1.07
Process D	VGCF 1 vol%	65	73	92	1.11
	VGCF 5 vol%	53	57	67	1.07

Table 7. Dispersed SEBS particle size of ternary composites (VGCF/PA6/SEBS-g-MA) prepared by different processing sequences.

Material		l_n (μm)	l_w (μm)	l_w/l_n	$(a_r)_N$
Process A	VGCF 1 vol%	2.28	3.12	1.37	15.2
	VGCF 5 vol%	2.22	2.96	1.33	14.8
Process B	VGCF 1 vol%	1.78	2.54	1.43	11.9
	VGCF 5 vol%	1.56	2.46	1.58	10.4
Process C	VGCF 1 vol%	2.37	3.32	1.40	15.8
	VGCF 5 vol%	2.39	3.32	1.39	15.9
Process D	VGCF 1 vol%	1.63	2.08	1.28	10.9
	VGCF 5 vol%	1.41	1.79	1.27	9.4

Table 8. Fiber length, its distribution, and aspect ratio of ternary composites (VGCF/PA6/SEBS-g-MA) prepared by different processing sequences.

To further clarify whether relationships between morphologies of ternary composites prepared by different processing sequences and dynamic viscoelastic properties of ones vary or not in this study, it is necessary to investigate the influences of the morphology of ternary composites on the dynamic viscoelastic properties in detail. G' is replotted as a function of average weight fiber length l_w in **Figure 21(a)** and the average weight diameter d_w in **Figure 21(b)**. In the figure, the number after each processing sequence code shows the volume fraction of VGCF in the composites. Generally, the fiber length dependence of G' slightly increases with l_w, and the influence of l_n on G' is smaller than that of V_f. These tendencies are related to the processing sequence of remixing of VGCF (processes B and D). On the contrary, the diameter size dependence of G' increases with decreasing d_w. However, this relation is weaker than the processing sequence of remixing of SEBS-g-MA (processes C and D). It was found from these results that the dynamic viscoelastic properties of ternary composites prepared by different processes are closely related to their morphologies. In particular, processes B and D, which are the remixing methods of VGCF, have good dispersibility of VGCF which improves dynamic viscoelastic properties. However, the dynamic viscoelastic properties are not simply correlated with the fiber length of VGCF and diameter size of dispersed SEBS-g-MA particles. Therefore, further studies of how the relationships among dynamic viscoelastic properties, processing sequences, and morphologies of these ternary composites are determined should be conducted. The findings would not only contribute to the research of these ternary composites but also to other high performance polymer matrix composites.

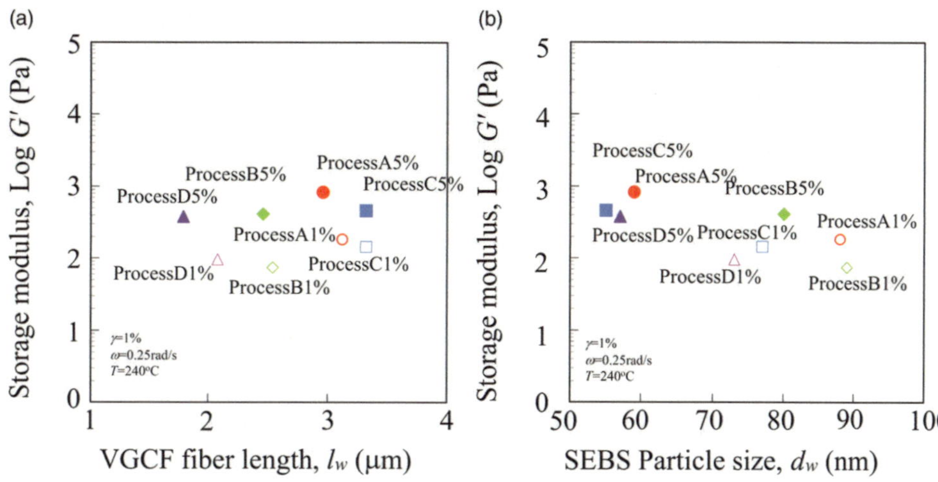

Figure 21. Influence of internal structure parameters such as SEBS particles size and fiber length on the storage modulus of various processing sequence of ternary composites (VGCF/PA6/SEBS-g-MA). (a) VGCF fiber length. (b) SEBS particle size.

5. Conclusion

We investigated the dynamic viscoelastic properties of carbon nanofiber (CNF)-filled polyamide composites and the blend of these composites and thermoplastic elastomer (TPE) in the molten state. In particular, this study discussed the effect of the addition of vapor grown carbon fiber (VGCF), which is a type of CNF, addition of TPE, and processing sequence on the dynamic viscoelastic properties in the molten state. It was found that VGCF has a stronger influence on the dynamic viscoelastic properties of the composites in the molten state. VGCF also has a high sensitive effect on the strain of transition from linear to nonlinear viscoelastic behavior, and the critical strain values at the transition decrease logarithmically with increasing VGCF contents. Composites with lower VGCF content (1 vol%) showed lower dynamic viscoelastic properties (both storage and loss moduli) than those of pure PA66. However, the viscoelastic properties of composites with higher contents above 5 vol% increased rapidly with increasing VGCF content. From the results of various rheological behaviors, rheological percolation thresholds seem to exist between 1 and 5 vol% of VGCF contents. On the other hand, the effect of the addition of TPE on the dynamic viscoelastic properties of VGCF/PA6 composites in the molten state differed according to each viscoelastic value. It was clarified that the dynamic viscoelastic properties of VGCF/PA6/SEBS-g-MA ternary composites are higher than those of VGCF/PA6/SEBS ones. Furthermore, the influence of processing sequences on the dynamic viscoelastic properties of VGCF/PA6/SEBS-g-MA composites in the molten state differed according to mixing steps by materials. In particular, the viscoelastic properties of these ternary composites prepared by different processing sequences in the low angular frequency region changed with the processing sequences, which decrease in the following order: Process A > Process C > Process B > Process D. In particular, processes B and D are remixing methods of VGCF where VGCF is mixed twice, and are therefore more effective than unimixing methods such as processes A and C, and have good VGCF dispersibility which improves dynamic viscoelastic properties. These may be attributed to the change in the internal structure due to the addition of TPE, type of SEBS, and processing sequences.

Acknowledgements

We would like to thank the Functional Microstructured Surfaces Research Center (FMS, MEXT, Japan) of Kogakuin University, the Project Research of Research Institute for Science and Technology of Kogakuin University, and the Ogasawara Foundation for the Promotion of Science & Engineering for funding this study, and partial support by the national budget of Czech Republic within the framework of the Centre of Polymer Systems project (reg. number: CZ. 1.05/2.1.00/03.0111). The authors thank Mr. T. Natio, Mr. S. Togashi, Mr. Y. Takenaka, Mr. S. Sano, Mr. Y. Osada, and many master course and undergraduate course students at the Polymeric Materials Laboratory of Kogakuin University for their assistance with the experiments.

Author details

Yosuke Nishitani[1*] and Takeshi Kitano[2]

*Address all correspondence to: at13152@ns.kogakuin.ac.jp

1 Kogakuin University, Hachioji, Tokyo, Japan

2 Tomas Bata University in Zlin, Zlín, Czech Republic

References

[1] Paul D. R., Robeson L. M. Polymer nanotechnology: Nanocomposites. Polymer. 2008;49:3187–3204. DOI: 10.1016/j.polymer.2008.04.017

[2] Kim H., Abdala A. A., Macosko C. W. Graphene/polymer nanocomposite. Macromolecules. 2010;43:6515–6530. DOI: 10.1021/ma1005

[3] Spitalsky Z., Tasis D., Papagelis K., Galiotis C. Carbon nanotube–polymer composites: Chemistry, processing, mechanical and electrical properties. Prog. Polym. Sci. 2010;35:357–401. DOI: 10.1016/j.progpolymsci.2009.09.003

[4] Han Z., Fina A. Thermal conductivity of carbon nanotubes and their polymer nanocomposites: A review. Prog. Polym. Sci. 2011;35:914–944. DOI: 10.1016/j.progpolymsci.2010.11.004

[5] Gary G. T., Max L. L., Karla L. S., Brian P. R. A review of the fabrication and properties of vapor-grown carbon nanofiber/polymer composites: A review. Compos. Sci. Technol. 2007;67:1709–1718. DOI: 10.1016/j.compscitech.2006.06.015

[6] Iijima S. Helical microtubules of graphitic carbon. Nature. 1991;354:56–58. DOI: 10.1038/354056a0

[7] Endo M., Kim Y. A., Hayashi T., Nishimura K., Matusita T. Vapor-grown carbon fibers (VGCFs) basic properties and their battery applications. Carbon. 2001;39:1287–1297. DOI: 10.1016/S0008-6223(00)00295-5

[8] Caldeira G., Maia J. M., Carneiro O. S., Covas J. A., Bernardo C. A. Production and characterization of innovative carbon fiber–polycarbonate composites. Polym. Compos. 1998;19:147–151. DOI: 10.1002/pc.10085

[9] Hattum F. W. J. V., Bernardo C. A., Finegan J. C., Tibbetts G. G., Alig R. L., Lake M. L. A study of the thermomechanical properties of carbon fiber-polypropylene composites. Polym. Compos. 2000;21:970–977. Doi: 10.1002/pc.10391. 2000;21:970-977. DOI: 10.1002/pc.10391

[10] Carneiro O. S., Maia J. M. Rheological behavior of (short) carbon fiber/thermoplastic composites. Part I: The influence of fiber type, processing conditions and level of incorporation. Polym. Compos. 2000;21:960–969. DOI: 10.1002/pc.10249

[11] Carneiro O. S., Maia J. M. Rheological behavior of (short) carbon fiber/thermoplastic composites. Part II: The influence of matrix type. Polym. Compos. 2000;21:970–977. DOI: 10.1002/pc.10250

[12] Enomoto K., Yasuhara T., Ohtake N., Kato K. Injection molding of polystyrene matrix composites filled with vapor grown carbon fiber. JSME Intl. J. Series A. Solid Mech. Mater. Eng. 2003;46:353–358. DOI: 10.1299/jsmea.46.353

[13] Choi Y. K., Sugimoto K., Song S. M., Endo M. Mechanical and thermal properties of vapor-grown carbon nanofiber and polycarbonate composite sheets. Matter. Lett. 2005;59:3514–3520. DOI: 10.1016/j.matlet.2005.05.082

[14] Al-Saleh M. H., Sundararaj U. A review of vapor grown carbon nanofiber/polymer conductive composites. Carbon. 2009;47:2–22. DOI: 10.1016/j.carbon.2008.09.039

[15] Nishitani Y., Hirano Y., Ishii C., Kitano T., Sekiguchi I. Tribological properties of vapor grown carbon fiber filled polybutylene terephthalate composites. Mat. Technol. 2008;26:114–122.

[16] Nishitani Y., Ito J., Ishii C., Kitano T., Sekiguchi I. Tribological properties of vapor grown carbon fiber and carbon fiber reinforced polyimide composites. J. Mat. Testing Res. Assoc. Japan. 2009;54:12–23.

[17] Nishitani Y., Naito T., Kitano T., Sekiguchi I. Effects of addition of thermoplastic elastomers on tribological and mechanical properties of vapor grown carbon fiber filled polybutylene terephthalate composites. In: Proceedings of World Tribology Congress (WTC2009); 6–11 September; Kyoto 421. 2009.

[18] Naito T., Nishitani Y., Sekiguchi I., Ishii C., Kitano T., Sekiguchi I. Influence of addition of thermoplastic elastomers on mechanical and tribological properties of vapor grown carbon fiber filled polybutylene terephthalate composites. Seikei-Kakou (J. Japan Soc. Polymer Process.). 2010;22:35–47.

[19] Nishitani Y., Sekiguchi I., Kitano T. Rheological properties of various carbon fibers filled PBT composites. In: Proceedings of the Polymer Processing Society 26th Annual Meeting (PPS-26); 4–9 July, Banff, Canada. 2010. R01–134:1–9.

[20] Nishitani Y., Sekiguchi I., Ishii C., Kitano T. Dynamic viscoelastic properties of carbon nanofiber filled polyamide 66 composites in molten state. Mat. Technol. 2010;28:114–122.

[21] Nishitani Y., Togashi S., Sekiguchi I., Ishii C., Kitano T. Tribological properties of carbon nanofiber filled polyamide 66 composites. Mat. Technol. 2010;28:292–302.

[22] Nishitani Y., Sano S., Takenaka Y., Kitano T., Sekiguchi I. Effect of processing sequence on the mechanical and tribological properties of ternary composites (VGCF/PA6/SEBS-g-MA). J. Mat. Testing Res. Assoc. Japan. 2015;60:164–171.

[23] Hausnerova B., Honkova N., Lengalova A., Kitano T., Saha P. Rheology and fiber degradation during shear flow of carbon-fiber-reinforced polypropylenes. Polym. Sci. Series A. 2006;48:951–960. DOI: 10.1134/S0965545X06090100

[24] Hausnerova B., Honkova N., Lengalova A., Kitano T., Saha P. Rheological behavior of fiber-filled polymer melts at low shear rate Part. I. Modeling of rheological properties. Polimery/Polymers. 2008;53:507–512.

[25] Nishitani Y., Ishii C., Kitano T. Rheological properties of surface treated glass fiber reinforced polypropylenes in molten state, polypropylene. Dr. Fatih Dogan (Ed.); 2012.

Available from: http://www.intechopen.com/books/polypropylene/rheological-properties-of-surface-treated-glass-fiber-reinforced-polypropylenes-in-molten-state. DOI: 10.5772/35785

[26] Wu G., Asai S., Sumita M., Yui H. Entropy penalty-induced self-assembly in carbon black or carbon fiber filled polymer blends. Macromolecules. 2002;35:945–951. DOI: 10.1021/ma0104940

[27] Sumita M., Sakata K., Asai S., Miyasaka K., Nakagawa H. Dispersion of fillers and the electrical conductivity of polymer blends filled with carbon black. Polym. Bull. 1991;25:265–271. DOI: 10.1007/BF00310802

[28] Chow W.S., Ishak Z. A. M., Karger-Kocsis J., Apostolov A. A., Ishiaku U. S. Compatibilizing effect of maleated polypropylene on the mechanical properties and morphology of injection molded polyamide 6/polypropylene/organoclay nanocomposites. Polymer. 2003;44:7427–7440. DOI: 10.1016/j.polymer.2003.09.006

[29] Gonzalez I., Eguiazabal J. I., Nazabal J. Rubber-toughened polyamide 6/clay nanocomposites. Compos. Sci. Technol. 2006;66:1833–1843. DOI: 10.1016/j.compscitech.2005.10.008

[30] Dasari A., Yu Z. Z., Yang M., Zhang Q. X., Xie X. L., Mai Y. W. Micro- and nano-scale deformation behavior of nylon 66-based binary and ternary nanocomposites. Compos. Sci. Technol. 2006;66:3097–3114. DOI: 10.1016/j.compscitech. 2005.10.008

[31] Nishitani Y., Ohashi K., Sekiguchi I., Ishi C., Kitano T. Influence of addition of styrene-ethylene/butylene-styrene copolymer on rheological, mechanical, and tribological properties of polyamide nanocomposites. Polym. Compos. 2010;31:68–76. DOI: 10.1002/pc.20767

[32] Nishitani Y., Yamada Y., Ishi C., Sekiguchi I., Kitano T. Effects of addition of functionalized SEBS on rheological, mechanical, and tribological properties of polyamide 6 nanocomposites. Polym. Eng. Sci. 2010;50:100–112. DOI: 10.1002/pen.21516

[33] Han C. D. Rheology and Processing of Polymeric Materials. In: Polymer Rheology, Oxford University Press. 2007; 1 Oxford, UK.

[34] Dasari A., Yu Z. Z., Mai Y. W. Effect of blending sequence on microstructure of ternary nanocomposites. Polymer. 2005;46:5986–5991. DOI: 10.1016/j.polymer.2005.05.145

[35] Gonzalez I., Eguiazabal J. I., Nazabal J. Effects of the processing sequence and critical interparticle distance in PA6-clay/mSEBS nanocomposites. Eur. Polym. J. 2008;44:287–299. DOI: 10.1016/j.eurpolymj.2007.11.027

[36] Tosaki M., Sakai H., Sasaki A., Takahashi Y., Matsubara M., Endo M. Rheological and mechanical properties of polypropylene filled with vapor-grown carbon fiber. Kobunsi Ronbunshu. 2005;62:585–590. DOI: 10.1295/koron.62.585

[37] Wang Y., Xu J., Bechtel S. E., Koelling K. W. Melt shear rheology of carbon nanofiber/polystyrene composites. Rheol. Acta. 2006;45:919–941. DOI: 10.1007/s00397-005-0077-8

[38] Durmus A., Kasgoz A., Macosko C. W. Linear low density polyethylene (LLDPE)/clay nanocomposites. Part I: Structural characterization and quantifying clay dispersion by melt rheology. Polymer. 2007;48:4492–4502. DOI: 10.1016/j.polymer.2007.05.074

[39] Wan T., Clifford M. J., Gao F., Bailey A. S., Gregory D. H., Somsunan R. Strain amplitude response and the microstructure of PA/clay nanocomposites. Polymer. 2005;46:6429–6436. DOI: 10.1016/j.polymer.2005.04.105

[40] Cassagnau P. Melt rheology of organoclay and fumed silica nanocomposites. Polymer. 2008;49:2183–2196. DOI: 10.1016/j.polymer.2007.12.035

[41] Ferry J. D. Viscoelastic Properties of Polymers. John Wiley & Sons, Incorporated; 1980 New York, USA.

[42] Nishitani Y., Sekiguchi I., Hausnerova B., Nagatsuka Y., Kitano T. Dynamic viscoelastic properties of long organic fibre reinforced polypropylene in molten state. Polym. Polym. Compos. 2001;9:199–211.

[43] Kitano T., Kataoka T. The effect of the mixing methods on viscous properties of polyethylene melts filled with fibers. Rheol. Acta. 1980;19:753–763. DOI: 10.1007/BF01521868

[44] Kitano T., Kataoka T., Nagatsuka Y. Shear flow rheological properties of vinylon- and glass-fiber reinforced polyethylene melts. Rheol. Acta. 1984;23:20–30. DOI: 10.1007/BF01333872

[45] Kitano T., Kataoka T., Nagatsuka Y. Dynamic flow properties of vinylon fibre and glass fiber reinforced polyethylene melts. Rheol. Acta. 1984;23:408–416. DOI: 10.1007/BF01329193

[46] Petra P., Fornes T. D., Paul D. R. Rheological behavior of multiwalled carbon nanotube/polycarbonate composites. Polymer. 2002;43:3247–3255. DOI: 10.1016/S0032-3861(02)00151-9

[47] Potschke P., Abdel-Goad M., Alig I., Dudkin S., Lellinger D. Rheological and dielectrical characterization of melt mixed polycarbonate-multiwalled carbon nanotube composites. Polymer. 2004;45:8863–8870. DOI: 10.1016/j.polymer.2004.10.040

[48] Kenneth S. C., Robert H. C. Dispersion and absorption in dielectrics I. alternating current characteristics. J. Chem. Phys. 1941;9:341. DOI: 10.1063/1.1750906

[49] Wu D., Wu L., Zhang M., Zhao Y. Viscoelasticity and thermal stability of polylactide composites with various functionalized carbon nanotubes. Polym. Degrad. Stab. 2008;93:1577–1584. DOI: 10.1016/j.polymdegradstab.2008.05.001

[50] Nishitani Y., Sekiguchi I., Hausnerova B., Zdrazilova N., Kitano T. Rheological properties of aminosilane surface treated short glass fibre reinforced polypropylenes.

Part 1: Steady shear and oscillatory flow properties in molten state. Polym. Compos. 2007;15:111–120.

[51] Kayatin M. J., Davis V. A. Viscoelasticity and shear stability of single-walled carbon nanotube/unsaturated polyester resin dispersions. macromolecules. 2009;42:6624–6632. DOI: 10.1021/ma901010d

[52] Van G. M., Palmen J. Time-temperature superposition for polymeric blends. Rheol. Bull. 1998;67:5–8.

[53] Meincke O., Kaempfer D., Weickmann H., Friedrich C., Vathauer M., Warth H. Mechanical properties and electrical conductivity of carbon-nanotube filled polyamide-6 and its blends with acrylonitrile/butadiene/styrene. Polymer. 2004;45:739–748. DOI: 10.1016/j.polymer.2003.12.013

[54] Trinkle S., Friedrich C. Van Gurp-Palmen-plot: A way to characterize polydispersity of linear polymers. Rheol. Acta. 2001;40:322–328. DOI: 10.1007/s003970000137

[55] Trinkle S., Walter P., Friedrich C. Van Gurp-Palmen Plot II? Classification of long chain branched polymers by their topology. Rheol. Acta. 2002;41:103–113. DOI: 10.1007/s003970200010

[56] Nishitani Y., Hasumi M., Kitano T. Influence of silane coupling agents on the rheological behavior of hemp fiber filled polyamide 1010 biomass composites in molten state [thesis]. In: AIP Conference Proceedings: 1664. Available from: http://dx.doi.org/10.1063/1.4918425

[57] Huang J. J., Keskkula H., Paul D. R. Rubber toughening of an amorphous polyamide by functionalized SEBS copolymers: Morphology and Izod impact behavior. Polymer. 2004;45:4203–4215. DOI: 10.1016/j.polymer.2004.04.002

[58] Balamurugan G. P., Maiti S. N. Influence of microstructure and deformation behavior on toughening of reactively compatibilized polyamide 6 and poly(ethylene-co-butyl acrylate) blends. Eur. Polym. J. 2007;43:1786–1805. DOI: 10.1016/j.eurpolymj.2007.02.035

[59] Nishitani Y., Kawahara T., Kitano T., Sekiguchi I. Effects of processing sequence on the physical properties of nanoomposites based on polyamide. In: Proceedings of Japan Society of Polymer Processing 2009 (Seikei-Kakou 09); 3–4 June 2009; Tokyo. 2009. pp. 181–182 (in Japanese).

[60] Tsurubuchi J., Naito T., Nishitani Y., Gotoh Y., Kitano T., Sekiguchi I. Effects of processing sequence on tribological properties of carbon nanofiber filled thermoplastic ternary composites. In: Proceedings of International Tribology Conference (ITC2011); 30 October–3 November 2011; Hiroshima. 2011. pp. P05–03:1.

[61] Fenouillot F., Cassagnau P., Majeste J.-C. Uneven distribution of nanoparticles in immiscible fluids: Morphology development in polymer blends. Polymer. 2009;50:1333–1350. DOI: 10.1016/j.polymer.2008.12.029

[62] Elias L., Fenouillot F., Majeste J.-C., Martin G., Cassagnau P. Migration of nanosilica particles in polymer blends. J. Polym. Sci. Part B: Polym. Phys. 2008;46:1976–1983. DOI: 10.1002/polb.21534

[63] Baudouin A.-C., Devaux J., Bailly C. Localization of carbon nanotubes at the interface in blends of polyamide and ethylene acrylate copolymer. Polymer. 2010;51:1341–1354. DOI: 10.1016/j.polymer.2010.01.050

Chapter 7

Viscoelastic Properties of Cardiovascular Tissues

Zhijie Wang, Mark J. Golob and Naomi C. Chesler

Additional information is available at the end of the chapter

http://dx.doi.org/10.5772/64169

Abstract

The aims of this chapter are to review the current state of knowledge regarding the viscoelastic behavior of cardiovascular tissues. We begin with a brief, general discussion of measurement and modeling of cardiovascular tissue viscoelasticity. We then review known viscoelastic behavior of arteries, veins, capillaries, blood components, the heart, and lymphatics. For each tissue type, we highlight tissue-specific measurement methods, the cellular and extracellular components responsible for tissue viscoelasticity, and the clinical implications of energy loss due to viscoelasticity. We conclude with a summary and suggestions for future research.

Keywords: viscoelasticity, energy loss, blood vessel, heart, lymph system, hemodynamics

1. Measurement and modeling

1.1. Experimental measurement approaches

Cardiovascular tissues are viscoelastic, exhibiting behaviors that combine features of elastic solids and viscous fluids. Elasticity, viscosity, and viscoelasticity can be quantified from mechanical testing techniques that relate the dynamics of a tissue's deformation to an applied load. For example, linear elastic materials subjected to an applied load exhibit a time-independent stress that is linearly proportional to strain (**Figure 1a**). A common metric of material elasticity is the elastic modulus, E, which is the slope of the stress-strain curve. Materials that are nonlinearly elastic respond differently to different levels of strain and remain time-independent (**Figure 1b**). In this case, elastic moduli can be defined at any point along the stress-strain curve. In arteries, for example, it is often convenient to quantify the behavior in low and high strain regions separately and to calculate low and high strain moduli (E_{low} and E_{high}), respectively [1, 2], which is discussed in Section 2.3.

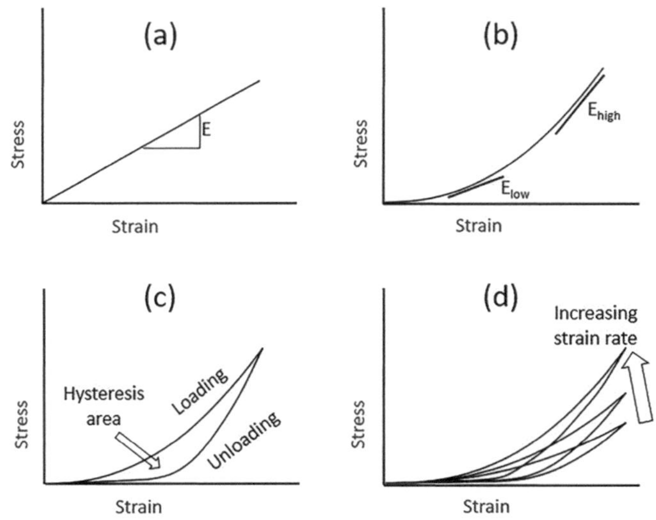

Figure 1 (a) Stress-strain curve for a linear elastic material. (b) Stress-strain curve for a typical elastic material displaying nonlinear behavior. Low-strain and high-strain behavior can be quantified by fitting elastic moduli to those regions of the stress-strain curve. (c) Typical stress-strain curve of nonlinear viscoelastic cardiovascular tissue exhibiting energy dissipation resulting in distinct nonlinear loading and unloading curves. (d) Typical hysteresis loops of nonlinear viscoelastic cardiovascular tissue exhibiting strain rate dependence.

Linear and nonlinear elastic materials do not dissipate energy after deformation or exhibit time-dependent behavior; therefore, stress-strain behavior of these materials is not different between loading and unloading. In contrast, viscoelastic materials dissipate energy upon deformation, which can be observed through hysteresis in the stress-strain curve (**Figure 1**c). The energy lost during a loading cycle is equal to the hysteresis area between the loading and unloading curves. Due to this energy dissipation, loading and unloading behaviors are not identical and loading moduli can be determined separately from unloading moduli. Since viscoelastic material behavior is also time-dependent, the loading-unloading stress-strain behavior also depends on strain rate (**Figure 1d**).

To measure viscoelasticity, the strain rate-, frequency-, or time-dependent mechanical behavior of a material must be measured. Often, a sinusoidal input (stress or strain) is applied to tissue, and an output signal (the corresponding strain or stress) is measured. The output signal is in phase with the input signal for a purely elastic material and out of phase for a viscoelastic material. Alternatively, viscoelastic behavior can be quantified with creep and stress relaxation experiments. Creep testing is performed by subjecting a material to a constant load and recording the time-dependent changes in strain. Stress relaxation is conducted by applying a constant strain and measuring the time-dependent stress reduction. Typically, preconditioning is performed before data collection, which is described briefly in the following section.

As a consequence of viscoelastic behavior, the mechanical response to an initial load (force, deformation, stretch, etc.) may differ from the response to a subsequent load. Fung implemented a preconditioning procedure to cyclically load and unload soft tissue to ensure a more consistent mechanical response [3]. The biological basis for the procedure is that with each subsequent cyclic load, the internal structure aligns with the direction of loading, and the tissue dissipates less energy [4]. The goal of preconditioning is to induce a pseudoelastic state, in which the tissue structure no longer changes with cyclic loading, resulting in a consistent mechanical response to a load or deformation. For instance, preconditioning has been implemented for up to 40 [5] or 80 [6] cycles in heart valve tissue before stable mechanical responses were observed. If the magnitude of the applied load changes, the internal material structure will change, so the tissue must undergo a new preconditioning protocol [4]. However, it is unclear whether tissues require additional preconditioning if the loading frequency is changed (while keeping the magnitude constant). In mechanical testing of cardiovascular tissues, preconditioning is nearly universally reported in arterial [7–12], heart valve [13, 14], and cardiac [15, 16] tissue. It is important to precondition tissues in order to more accurately estimate viscoelastic properties from single-valued elastic constants (i.e., modulus).

Though preconditioning is performed to reduce sample variability, repeated cyclic loading can cause a sample "memory" problem. That is, mechanical behavior during testing can be affected by loading from previous cycles (i.e., testing after preconditioning cycles). To reduce the sample memory problem, researchers have induced rest periods (24 hours vs. 15 seconds) between sets of preconditioned valve samples [17]. This can be accomplished by allowing tissues a recovery period following preconditioning cycles. In heart valve tissues, protocols with a 24-hour rest period between cyclic preconditioning sets had the lowest errors between the predicted model and experimental data [17]. However, the rest periods encompassed several orders of magnitude (in seconds), and it is unknown if an intermediate time is sufficient to reset the strain history. Another study on heart valves reported reduced hysteresis and a stable mechanical response after only a 60-second rest period between two sets of preconditioning cycles [6]. Therefore, while it is clear that preconditioning is necessary, protocols for cardiovascular tissue viscoelasticity measurements have not yet been standardized.

1.2. Empirical models

Using the experimental techniques described above, stress-strain data can be fit to empirical models, which are often useful for predicting viscoelastic material mechanical behavior. Classical models of viscoelastic materials use combinations of spring and dashpot elements to characterize stress-strain behavior. Spring elements represent elastic behavior, where the spring constant of proportionality, k, directly relates the applied force, F, to the resulting deformation, x. Dashpot elements represent viscous behavior, where the applied force is related to the rate of deformation, $\dot{x}$, by the viscosity, μ. Spring and dashpot elements are arranged in series in Maxwell models and in parallel in Kelvin-Voight models (**Figure 2**). Since the elements are arranged in parallel, force in a Kelvin-Voight model is a sum of the two individual element forces, and elements share the same deformation. Conversely, the elements in a Maxwell model share the same force, and the total deformation is the sum of the individual element deformations. The combination of spring and dashpot elements in these two models can be used to construct more complex viscoelastic solid models. Additional details on these models and their use are available in Ref. [18].

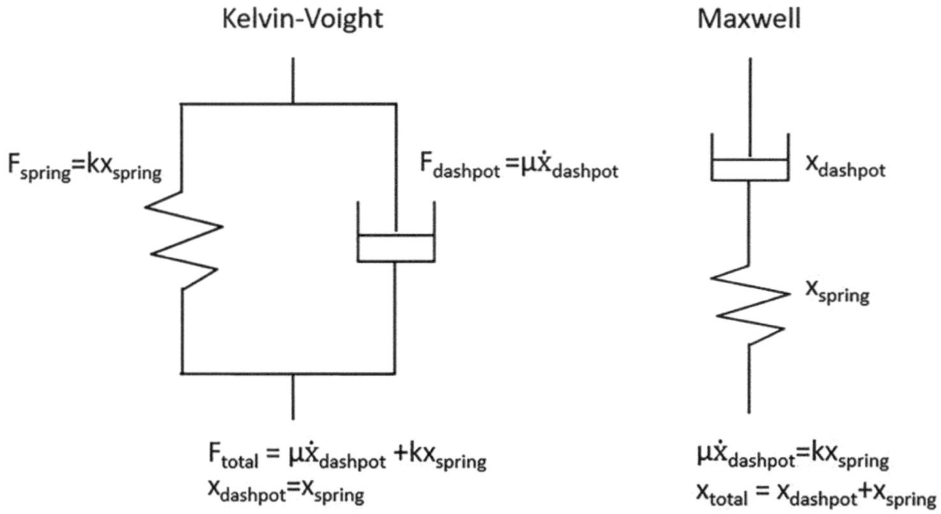

Figure 2 Maxwell and Kelvin Voight models composed of dashpots and springs. Elements in parallel (Kelvin-Voight model) have the same deformation and separate forces, whereas elements in series (Maxwell model) have the same force and separate deformations.

2. Arteries

2.1. Viscoelastic characteristics of arteries

The importance of arterial viscoelasticity is supported by the findings that arterial morphometry correlates more strongly with pulsatile rather than steady pressure [19] and that dynamically measured mechanical properties are different from statically measured mechanical properties [20–22] (**Table 1**). Measuring the viscoelastic behavior of isolated blood vessels can be accomplished by quantifying dynamic length-tension relationships in tissue strips or rings, or pressure-diameter relationships in intact segments either with sinusoidal pressurization [23–25] or step-wise increases in pressure (**Figure 3**). However, because step-wise pressurization does not accurately mimic the pulsatile blood pressure waveform that arteries experience in vivo (**Figure 3**), sinusoidal pressurization is a better method for characterizing arterial viscoelasticity. Arterial viscosity, or the damping capacity, can then be obtained from the hysteresis loop [1, 26]. Additional details on arterial mechanical testing methods are available in recent reviews [27, 28]. Arterial viscoelastic properties also depend on the function of the artery (e.g., conduit or muscular), species, and health status (**Table 1**).

		Pulmonary artery [22]	Carotid artery [25]	Aorta*	Carotid artery [29]	Carotid artery [21]
	Species, age	**Mouse, young**	**Mouse, aged**	**Mouse, aged**	**Dog, young**	**Human, aged**
		Elastic modulus	**Elastic modulus**	**Elastic modulus**	**Young's modulus**	**Distensibility**
Elasticity	**Static modulus**	14±2 kPa	116±118 kPa	52±8 kPa	690±48 kPa	$3.8 \pm 1.4 \cdot 10^{-3}$ $mmHg^{-1}$
	Dynamic modulus	61±3 kPa (at 1 Hz)	118±35 kPa (at 1 Hz)	39±9 kPa (at 1 Hz)	1100±100 kPa (at 2 Hz)	$2.1 \pm 0.9 \cdot 10^{-3}$ $mmHg^{-1}$
Viscosity	**Phase angle φ (radians)**	0.14±0.01 (at 1 Hz)	0.14±0.05 (at 1 Hz)	0.06±0.01 (at 1 Hz)	~0.11 (at 2 Hz)	~0.1 [30] (at 2 Hz)

The viscosity of an artery is presented as the phase difference between the force (stress) and deformation (strain).*Unpublished data. Mean ± SE shown for [22], [25], [29], and *. Mean ± SD shown for [21].

Table 1. Static and viscoelastic circumferential properties measured in conduit arteries.

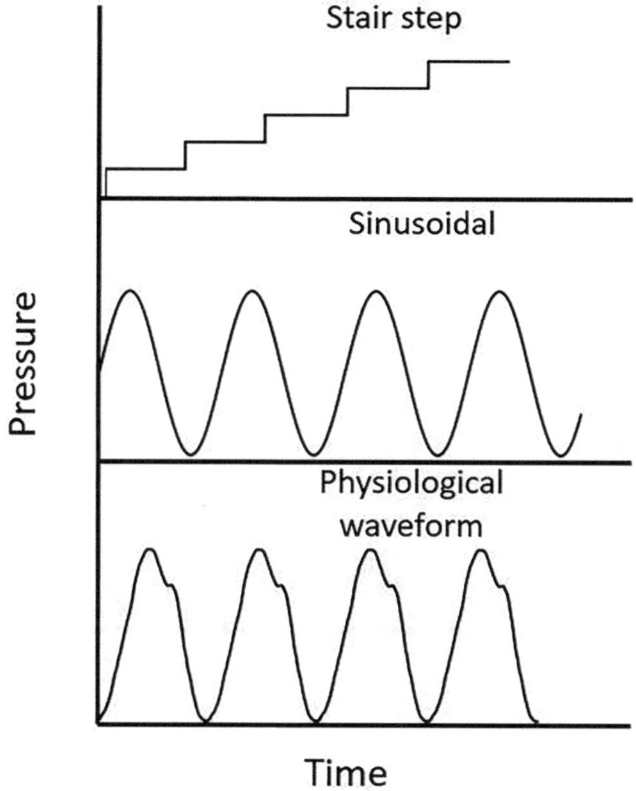

Figure 3 Stair-step and sinusoidal pressure inputs compared to a physiological waveform.

In healthy arteries, viscoelastic properties are frequency-dependent. In arteries from humans and large animals such as dogs, the dynamic elastic modulus increases over a low-frequency range (<3 Hz) and then remains nearly constant at higher frequencies (up to 10 Hz) [29–32]. The frequency-dependent behavior is different for small animals such as rats or mice; in mouse pulmonary arteries, the dynamic elastic modulus remains constant up to 5 Hz and then starts increasing with increasing frequency (up to 20 Hz) [22]. This discrepancy may be explained by the physiological frequency of the normal heart beat rate, which is largely species size-dependent. In large animals and humans, the normal heart rate is ~1 Hz, whereas in small animals, the heart rate is ~5–10 Hz. Therefore, if scaled to the physiological frequency, we can conclude that the arterial dynamic modulus increases rapidly at 2- to 3-times the natural heart rate and then plateaus at higher frequencies (10- to 20-times higher than the natural heart rate) [22]. In terms of viscous properties, the phase lag of energy dissipation increases as frequency increases in all species [22, 29, 30].

2.2. Contribution of smooth muscle cells

While it is generally agreed that elastin and collagen are responsible for the nonlinear elasticity of arteries, the cellular and molecular determinants of arterial viscoelasticity remain incompletely understood. Early experimental data comparing carotid and femoral arteries showed a higher viscosity in femoral arteries, and the investigators hypothesized that the higher smooth muscle cell (SMC) content in the femoral artery is responsible [29, 30]. However, there was no measurement of SMC content or activity in these studies. Recent evidence suggests that SMC activity is key to arterial viscoelasticity because removal of SMC tone reduces damping [33, 34] and activation of SMC contraction by phenylephrine or renovascular hypertension in dogs and sheep increases damping [35–37]. Therefore, both SMC content and SMC tone play important roles in arterial viscosity and viscoelasticity.

2.3. Contribution of extracellular matrix

The role of extracellular matrix (ECM) components in arterial mechanical behavior is originally evidenced in the nonlinear elasticity curve, which arises from differential load-bearing contributions of ECM proteins. Specifically, the protein elastin dominates load-bearing in the low-strain region whereas at higher strains, collagen fibers, which impart strength and stiffness, are engaged. The transition region in between occurs because increasing strain causes collagen fiber alignment and recruitment resulting in a transition from increasing collagen engagement to collagen-dominated load-bearing. Therefore, ECM proteins are critical to arterial nonlinear elasticity.

ECM proteins and cell-ECM interactions may also play important roles in arterial viscoelasticity. Collagen and proteoglycans are known to affect energy loss in tendon and cartilage [38–41]. In the first measurements of mouse pulmonary arterial viscoelasticity, Wang et al. [22] found supporting evidence that collagen and proteoglycans affect frequency-dependent changes in arterial damping capacity. Similarly, in carotid artery strips, Garcia et al. [42] found evidence that elastin is an important contributor to stress relaxation.

The contribution of ECM components to tissue viscoelasticity is best studied through single ECM protein degradation or tissue decellularization. The viscoelastic behavior of elastin was studied using a cyanogen bromide treatment, which removes cells and all ECM components except elastin [43]. In this experiment, stress-relaxation was highest in intact aortas, intermediate in decellularized ECM, and lowest in cyanogen bromide-treated aortas, i.e., arterial elastin. Interestingly, the creep behavior of the aorta in all the three aforementioned groups was negligible.

Our group has found that increased expression levels of collagen and proteoglycans are associated with increased pulmonary arterial stiffness and decreased damping capacity [22]. Since we did not investigate the independent effects of proteoglycan and collagen content on pulmonary arterial viscoelasticity, both proteins may be involved. Furthermore, the concomitant changes in these ECM proteins and arterial viscoelasticity and lack of changes in SMC content, coupled with the absence of SMC tone, suggest that these ECM proteins are critical to arterial viscoelasticity in a way that has not been previously described. Future investigations should clarify the roles of individual ECM protein and the ECM-fibril interaction or cell-ECM interaction in arterial viscoelasticity.

2.4. Effects of aging and hypertension

With aging or hypertension, arteries become less distensible [44, 45]. This is mainly associated with changes in the ECM proteins collagen and elastin. It is well known that fragmentation of elastin and degeneration of collagen occurs with age in arteries [45]. In addition, the degree of cross-linking of the ECM proteins increases. These changes lead to a higher elastic modulus. Hypertension and arterial stiffening are closely associated with age; however, arterial stiffness is an independent prognostic factor for cardiovascular outcomes [46]. Indeed, recent investigations have reported that arterial stiffening precedes blood pressure elevation [45]. The cause-and-effect relationship between hypertension development and arterial stiffening remains a key knowledge gap in current investigations [47, 48].

The changes in conduit arterial viscosity during aging and hypertension remain incompletely understood. From in vitro isobaric experiments on human aorta or mouse pulmonary artery, viscosity of the arterial wall has been found to decrease in these conditions [22, 30]. However, inconsistent findings are reported elsewhere. In vitro studies on rat aorta found increased viscosity with aging [49]. In another case, in vivo measurements have shown increased pulse damping (or viscosity) as hypertension develops [33, 34, 37]. In these in vivo studies, the dynamic responses of healthy and hypertensive arteries were measured under different pressure ranges due to the developing disease. It is known that arterial viscoelasticity is dependent on the pressure level: as the pressure increases gradually, arterial stiffness and viscosity both increase [50]. Therefore, the shift and deformation of the stress-strain hysteresis loop are a combined effect of changes in intrinsic mechanical properties and extrinsic pressure/stretch ranges, which must be taken into account when interpreting dynamic mechanical testing data.

2.5. Hemodynamic and cardiac consequences

The impact of large, conduit arterial viscoelastic properties on cardiovascular hemodynamics has not been well studied. As noted above, arterial stiffening (i.e., decreased elasticity) is a useful prognostic indicator of cardiovascular events [51, 52] and is well known to contribute to increased ventricular afterload [53–55]. Increased stiffness may also impair wave reflections, which further augment the ventricular afterload [56, 57].

The energy dampening function achieved through arterial viscosity is beneficial in normal physiological conditions because it absorbs the energy from the wave reflections. In sheep with acute pulmonary hypertension, an SMC activation-mediated increase in pulmonary arterial viscosity led to a decrease in wave reflections and increase in characteristic impedance (the vascular impedance that represents the opposition to pulsatile flow), thus reducing the fraction of oscillatory to total right ventricular hydraulic power and improving hemodynamic function [58]. In the mouse extralobar pulmonary arteries, our group found a decrease in arterial damping capacity as pulmonary hypertension developed [22]. We speculated that the reduced viscosity may be in part responsible for the increased pulse wave velocity and pulse pressure during pulmonary hypertension progression [59], which eventually increased pulsatile right ventricular afterload.

Additional clinical studies are needed to elucidate the implications of arterial viscosity for cardiovascular hemodynamics. Also, elasticity and viscosity are coupled behaviors in an artery, yet the relationship between them is rarely reported. The investigation of the independent and interdependent impacts of arterial elasticity and viscosity on cardiovascular function could be an important area for future research.

2.6. Use of an arterial viscoelasticity index for clinical diagnosis

A novel index of arterial viscoelasticity, which was measured in a large scale, clinical study, is worthy of special mention. Taniguchi et al. [60] used noninvasive methods to assess carotid artery viscoelasticity in 383 patients. The authors defined a nondimensional parameter I^* derived from the ratio of the gradients of vascular wall deformation rates during deflation and inflation over a cardiac cycle, in which $I^* < 0$ indicates healthy viscoelasticity whereas $I^* > 0$ indicates abnormal viscoelasticity. A positive I^* was found in female, elderly (>60 yr), and hypertensive (blood pressure > 140 mmHg) subjects and was a significant, independent risk factor for coronary artery disease based on univariate and multivariate analyses [60]. While this parameter is only an indirect measure of arterial viscoelastic behavior and its physical meaning requires further investigation, it is noteworthy that this index can be measured noninvasively through ultrasonic Doppler [60, 61]. In order to better understand the clinical significance of arterial viscoelasticity, it is crucial to assess arterial viscoelasticity in more patient populations, with additional measurement parameters. Research like this may establish new, useful viscoelasticity indices that aid in clinical diagnosis and prognosis.

3. Veins

Vein biomechanics are understudied compared to arterial biomechanics, and there is a general lack of understanding of venous viscoelasticity. That said, the venous wall is structurally similar to the arterial wall with less muscle and elastic tissue, resulting in thinner walls that are more compliant and collapse easily [62]. Thus, venous viscoelasticity likely shares several features in common with arterial viscoelasticity, including a dependence on SMC content and tone and the ECM proteins collagen, proteoglycans, and elastin. The measurement methods for venous viscoelasticity are identical to those for arterial viscoelasticity. Vein viscoelasticity is clinically relevant to coronary artery bypass grafting, which often uses saphenous veins to replace diseased coronary arteries, since coronary artery perfusion is critical to myocardial health. Venous viscoelasticity is an important and largely unexplored area for future research.

4. Capillaries

Capillaries are very thin microvessels that are composed of a single layer of endothelial cells. Capillary beds serve as the location for the exchange of gases and nutrients between blood and tissues. Because of the delicate structure of the capillary wall, it is impossible to isolate the capillaries without injury. To obtain the stress-strain relationship for capillaries, capillary dimensions can be measured by microscopy imaging while the local intravascular pressure can be altered and measured by micropuncture [63], occlusion of upstream (arterial) and downstream (venous) vessels [64] or controlled perfusion [65]. Like other types of blood vessels, the capillaries exhibit nonlinear stress-strain behavior [64] and viscoelastic features like creep and relaxation [65]. In normal conditions, the distensibility of the capillary vessels is in between the small arteries (arterioles) and small veins (venules), and the viscosity is the least in capillary vessels in a passive creep test in rats (i.e., without smooth muscle tone in the arterioles and venules) [65] (see **Table 2**). Like venous viscoelasticity, capillary viscoelasticity is a largely unexplored area for future research with unknown clinical relevance.

	Arcade arterioles	**Capillary**	**Arcade venules**
$\alpha 1$ ($\times 10^2$ mmHg)	6.34±3.59	7.42±3.00	9.52±3.24
β ($\times 10^4$ mmHg s)	4.25±4.17	3.00±2.25	5.52±3.16

The experimental data were fitted with a three-element linear solid viscoelasticity model. Coefficients $\alpha 1$ and β represent the elastic and viscous characteristics, respectively. Mean ± SD shown. Adapted with permission from Skalak et al. [65].

Table 2. Viscoelastic properties measured in rat arterioles, capillary vessels, and venules using a single-step creep test (pressure = 50 mmHg).

5. Blood components

Blood is a non-Newtonian, viscoelastic fluid made up of cellular components, including erythrocytes (red blood cells, RBCs), leucocytes (white blood cells), and thrombocytes (platelets), and a fluidic component called plasma. The volume percentage of all blood cells in whole blood is about 45% in adult men. The viscoelasticity of blood depends on the hematocrit (volume fraction of RBCs) in whole blood, plasma viscosity, and aggregation and mechanical properties of blood cells. The deformability of blood cells plays an important role in their main functions as well as the blood rheology and hemodynamics [66, 67]. While the viscoelasticity of blood cells, including RBCs [66–68] and white blood cells [69, 70], has been studied, here we will discuss only RBCs since they have the highest concentration in whole blood and are the most relevant to hemodynamics under most physiological and pathological conditions.

The RBC is a simply structured biological component that consists of a bilayer membrane and thin cytoskeleton of spectrin filaments [68]. Mammalian RBCs lack a cell nucleus. RBCs exhibit a unique deformability, which enables them to change shape reversibly in response to an external force (e.g., under the stress applied by the capillary wall). Despite the structural simplicity of RBCs, the understanding of their mechanical properties is still incomplete. The viscoelastic properties of RBCs, which are key determinants of RBC deformability, can be classified into elastic (or storage) and viscous (loss) moduli: the shear modulus that describes the uniaxial elongation property, the area expansion modulus that describes the changes in RBC membrane area, the bending modulus that describes the bilipid layer associated with resting shape changes, and the viscosity that describes the rate of deformation of RBC membrane [71, 72].

Typical methods for viscoelasticity measurement of single or multiple RBCs were recently reviewed [71] and include micropipette aspiration, atomic force microscopy, optical tweezers, fluid or microfluidic filtration, and laser diffractometry.

RBC viscoelasticity is closely related to blood rheology and hemodynamics [66, 67]. For example, reduced RBC viscoelasticity leads to a significant increase in microvascular flow resistance and blood viscosity [67, 71]. Exchanging native RBCs with RBCs hardened with glutaraldehyde causes a doubling of the filtration resistance and reduced flow in rats [67]. Altered RBC viscoelasticity is frequently reported in pathological conditions such as diabetes [73, 74], hyperglycemia [74, 75], and sickle cell disease [76]. While cardiovascular complications including arterial stiffening, heart failure, or stroke are often associated with these diseases [77, 78], it is possible that the altered RBC mechanics may exert an impact on overall hemodynamics, which is a critical contributor to cardiovascular health. Further understanding of the influence of RBC viscoelasticity on cardiovascular hemodynamics could inspire novel therapies that target the biomechanical mechanism of the disease.

6. The heart

6.1. Heart valves

Viscoelastic behavior has been demonstrated to different degrees in native or pericardial-based heart valve tissues, which are usually made up of highly organized ECM and valve interstitial cells. Though direct comparisons cannot be made between studies with different testing methods, viscoelastic characteristics including strain-rate dependence, stress relaxation, and creep have been observed after loading of valvular tissue from *ex vivo* mechanical testing. Similar to arterial tissue discussed above, interestingly, mechanisms of stress relaxation appear to be distinct from those of creep. For example, stress relaxation observed in valvular tissue was accompanied by negligible creep over a 3-hour timeframe [79]. Therefore, use of multiple modalities for valve tissue viscoelastic property measurement is recommended.

As with arterial tissue, valvular viscoelastic behavior is dependent on ECM components. In heart valves, collagen in particular has a complex fiber alignment pattern that appears to generate more viscoelasticity in the circumferential direction compared to the radial direction [80]. Imaging techniques including scanning electron microscopy and bright field or polarized light microscopy have demonstrated that collagen fibers are highly aligned in the circumferential direction but more randomly distributed in the radial direction. Therefore, a more highly aligned fiber structure likely contributes to larger degree of viscoelastic behavior. Also, the effect of preconditioning has been shown to be direction-dependent with the circumferential direction displaying a more consistent mechanical response after a fewer number of cycles [80]. This indicates that an initially aligned fiber structure will achieve a consistent mechanical response faster than randomly oriented fibers.

Like arteries, heart valves exhibit less creep than stress relaxation [81]. This behavior seems to be unique to collagen-rich soft biological tissues like ligament [82] and cornea [83]. The underlying mechanism is not fully understood. Thornton et al. [84] attribute it to different mechanisms involved in the two tests. In particular, these authors suggest that the stress relaxation response is determined by a discrete group of ECM fibers recruited at constant elongation, whereas the creep response is determined by different fibers being progressively recruited at constant stress. Another hypothetical mechanism is the 'fibril-locking' mechanism that collagen fibrils maintain under a constant stress, whereas under a constant strain the fibril stress decreases [81]. The interactions between collagen and proteoglycans also have been proposed to affect tissue viscoelasticity in heart valves [79, 85].

Not surprisingly, the chemical treatments of valvular tissue that are used prior to bioartificial heart valve replacements have been shown to affect viscoelastic behavior. Glutaraldehyde fixation reduces the immune response and prevents leaflet degradation but can induce ECM cross-links and other structural changes [86, 87] that reduce radial stiffness and increase extensibility [86, 88] as well as reduce creep and stress relaxation [13, 86, 88, 89]. Fixation also increases the number of preconditioning cycles required to achieve a consistent mechanical response in dynamic mechanical testing [88]. These effects can be minimized with pressure fixation, which results in more similar viscoelastic behavior in fixed tissues compared to untreated tissues [88].

The decreased stress-relaxation that occurs in heart valve tissues treated with glutaraldehyde has important implications for bioprosthetic heart valves because the industry standard of accelerated wear testing (AWT) is used to determine the absolute fatigue life of bioprosthetic heart valves. In AWT, valves are cycled at frequencies higher than the normal heart rate (up to ~30 Hz). With reduced stress-relaxation, the valve may not have time to relax to its natural state, resulting in a higher preload for each subsequent cycle [13]. A better understanding of heart valve viscoelasticity and the molecular basis for tissue viscoelastic properties could improve bioprosthetic valve design and testing.

6.2. Myocardium

Ventricular myocardial tissue exhibits viscoelastic characteristics, which are changed in diseased states. Viscoelastic properties have been examined in the healthy swine myocardium using noninvasive shear wave velocity techniques [90]. Briefly, external actuators are used to create and propagate waves in the myocardial wall; the velocity is measured at several frequencies, and these data are fit to empirical viscoelastic models described in Section 1.2. Human myocardium tissue tested ex vivo exhibited multiple viscoelastic characteristics including directional-dependent hysteresis and stress relaxation, and rate-dependent stress-strain curves [16]. The subjects from whom tissues were harvested had various causes of death, so the data cannot be interpreted in the context of healthy versus diseased states. However, several studies have shown changes in viscoelastic properties with disease. The use of a Kelvin-Voight model demonstrated that elastic and viscous damping constants increased in the pressure-overloaded feline right ventricle [91]. Similarly, elastic and viscous constants increased in the pressure-overloaded rat left ventricle [92]. An important future direction is to assess the viscoelastic behavior in healthy versus diseased myocardium and determine the impact on cardiac performance.

The structural basis of myocardial viscoelasticity can be attributed to a combination of cardiac cells and ECM proteins, but the nature and extent of each component's contributions to viscoelasticity are still debated. Myocardial tissue is composed of cardiac cells or myocytes. Myocytes are composed of repeating units called sarcomeres, which contain titin, actin, and myosin proteins. By isolating cardiomyocytes from surrounding ECM structures, viscoelastic properties can be measured. Uniaxial testing of myocytes revealed force-length hysteresis [93], indicating they are one source of viscoelasticity. Titin functions as a spring, and titin-actin interactions may contribute to viscoelastic behavior [94]. Cardiac cells are surrounded and consequently interconnected by ECM proteins to provide structural integrity to the heart. Evidence that collagen degradation is accompanied by decreased elastic stiffness and viscous damping suggests that collagen plays an important role in myocardial viscoelasticity [91]. Collagen accumulation (or fibrosis) is a characteristic of failing ventricles and with hypertrophic remodeling the collagen content can increase by ~50% [95]. Therefore, it may be important to investigate the effect of collagen accumulation on myocardium viscoelasticity during heart failure development.

Importantly, the majority of myocardial viscoelastic properties are reported from the left ventricle, and differences in viscoelastic behavior between ventricles are unknown. The left ventricle differs embryologically, geometrically, and structurally from the right ventricle [96–98], so results from the one cannot be extrapolated to the other. Indentation testing ex vivo showed similar baseline levels of elastic and viscous constants between both ventricles in rats [92]. Differences in viscoelastic properties between ventricles will be important directions for future research to understand the role of viscoelasticity in healthy and diseased myocardial function.

7. The lymphatic system

The lymphatic system is the third circulation in the human body and is complementary to the first two: the systemic and pulmonary circulations. It consists of branched lymphatic vessels that collect and transport lymph fluid as well as organs (e.g., lymph nodes, spleen, and thymus) that assist with lymph transport. When pressure is greater in the interstitial fluid than the tiny, closed-end lymphatic capillaries, lymph flows in [99] and is transported to gradually larger lymphatic vessels, collected and filtered through lymph nodes, and finally enters the lymphatic duct where it is Reintroduced into the bloodstream. The ontogenesis of lymphatic vessels is not fully understood but some evidence suggests that they are derived from the veins [62]. Lymphatics are often compared to veins because both are thin walled, valved structures and their main function is to transport fluid (blood or lymph) for nutrition and immune purposes, respectively.

Despite of their structural similarity to blood vessels, lymphatics are different in several aspects. First, at a similar distance from the heart, lymphatics are larger and thinner than veins [62]. Second, lymphatics are subjected to lower pressures than veins and have more compliant walls [62, 100, 101]. Third, unlike the veins in which the blood is conducted only passively to the heart as a result of valve action and in combination with the intermittent compression by adjacent tissue, lymphatics have both extrinsic and intrinsic pumping mechanisms [101, 102]. Both experimental studies and mathematical modeling have been done to characterize lymphatic pump function [103–105]. For instance, incorporating experimental measurements of a mesenteric lymph vessel, Bertram and Moore developed a model to capture intrinsic pumping function [102]. Similar to arteries and veins, chronic changes in pressure and flow can cause remodeling of lymphatics [101]. For example, lymphatic pumping weakens in response to venous hypertension [106]. The active pumping function can be impaired due to metabolic disorders, local immune-compromise, and lymphedema [104].

Few studies have been performed on the viscoelasticity of lymph vessels. Ohhashi [100] compared the viscoelastic properties of bovine mesenteric lymph vessel, which is considered as muscular lymphatics because of the rich and well-developed SMC content, to the canine thoracic duct, which is considered a fibrous lymphatic because of the lower SMC content and higher elastin and collagen content in the wall. They found that the muscular lymphatic vessel is more compliant, and its hysteresis loop is wider than that of the fibrous thoracic duct, suggesting a larger viscosity due to more layers of SMCs. Because lymphatic pumping is critical to lymph transport and lymphatic viscoelasticity likely affects lymphatic pumping, understanding the determinants of lymphatic viscoelasticity and its impact on pumping function may unveil new mechanisms for lymphatic dysfunction.

The viscoelastic properties of another main component in the lymph system – the lymph node – have been recently reported. McClain et al. found that lymph node tissue becomes stiffer and has increased energy loss with cancerous tumor development in mice [107]. While the finding is novel and exciting, it raises more questions such as how changes in lymph node viscoelastic properties relate to tumor regression, antitumor immune response, or metastatic colonization of the lymph node.

8. Summary

An improved understanding of cardiovascular tissue viscoelastic properties and their dependence on cardiovascular tissue structure will no doubt provide valuable insights into the functional behavior of these tissues. More importantly, despite the long-standing recognition that cardiovascular tissues are nonlinearly elastic and viscoelastic materials, the clinical implications of cardiovascular tissue viscoelasticity remain poorly understood. In the example of arterial viscoelasticity, inconsistent findings of arterial viscoelastic changes with disease and a lack of tools for simple and easily accessible clinical measurement have likely contributed to a lack of information on the impact of arterial viscoelasticity on cardiovascular hemodynamics and cardiac function. Therefore, there is a pressing need to elucidate the clinical implications of cardiovascular tissue viscoelasticity in disease progression in future research.

We recommend the following as promising and impactful future areas of research:

1. Protocols for preconditioning: It is well accepted that preconditioning is necessary for mechanical property measurements. The most appropriate protocols for preconditioning tissues subjected to different testing frequencies and different strain ranges is unclear, however. Also, the sample 'memory' behaviour should be investigated to standardize mechanical testing protocols for cardiovascular tissue viscoelasticity measurements and reduce sample variation.

2. Structural models that include viscoelasticity: The rapid development of arterial structural models has advanced the understanding of the structural-functional relationship of arteries in ways that empirical, phenomenological models cannot. However, current models predict static mechanical behavior only. Future structural models should incorporate elements that are responsible for viscous behavior and predict cardiovascular tissue dynamic mechanical behavior.

3. Creep vs. stress relaxation behaviours: The mechanisms that lead to differences in creep and stress relaxation behaviours are not fully understood. Investigating these character-

istics in different cardiovascular tissues would improve our understanding of tissue viscoelastic behavior.

4. Structural determinants of viscoelasticity: The contributions of cells (e.g., SMCs, endothelial cells, or myocytes), ECM (e.g., elastin, collagen, or proteoglycans), and cell-ECM interactions to tissue viscoelasticity are not fully understood. Modern high spatial and temporal resolution imaging techniques (e.g., multiphoton microscopic imaging) will provide insight into the ways in which cardiovascular tissue structures generate viscoelastic function.

5. Hemodynamic and cardiac impact of viscoelasticity: Arterial stiffening is known to occur in aging and hypertension and predict heart failure. Future research should investigate the impact of arterial viscoelasticity on cardiovascular disease development and progression.

Author details

Zhijie Wang, Mark J. Golob and Naomi C. Chesler*

*Address all correspondence to: naomi.chesler@wisc.edu

University of Wisconsin – Madison, Madison, Wisconsin, USA

References

[1] Wang Z , Chesler NC. Role of collagen content and cross-linking in large pulmonary arterial stiffening after chronic hypoxia. Biomech Model Mechanobiol. 2012;11(1–2): 279–289.

[2] Tabima DM , Chesler NC. The effects of vasoactivity and hypoxic pulmonary hypertension on extralobar pulmonary artery biomechanics. J Biomech. 2010;43(10):1864–1869.

[3] Fung YC , Fronek K, Patitucci P. Pseudoelasticity of arteries and the choice of its mathematical expression. Am J Physiol. 1979;237(5):H620–H631.

[4] Humphrey JD , DeLange S. An Introduction to Biomechanics: Solids and Fluids, Analysis and Design: Springer-Verlag New York; 2004.

[5] Pham T , Sun W. Material properties of aged human mitral valve leaflets. J Biomed Mater Res A. 2014;102(8):2692–2703.

[6] Martin C , Sun W. Biomechanical characterization of aortic valve tissue in humans and common animal models. J Biomed Mater Res A. 2012;100(6):1591–1599.

[7] Huang W , Delgado-West D, Wu JT, Fung YC. Tissue remodeling of rat pulmonary artery in hypoxic breathing. II. Course of change of mechanical properties. Ann Biomed Eng. 2001;29(7):552–562.

[8] Lawrence AR , Gooch KJ. Transmural pressure and axial loading interactively regulate arterial remodeling ex vivo. Am J Physiol Heart Circ Physiol. 2009:475–484.

[9] Deguchi JO , Huang H, Libby P, Aikawa E, Whittaker P, Sylvan J, . Genetically engineered resistance for MMP collagenases promotes abdominal aortic aneurysm formation in mice infused with angiotensin II. Lab Invest. 2009;89(3):315–326.

[10] Amin M , Le VP, Wagenseil JE. Mechanical testing of mouse carotid arteries: from newborn to adult. J Vis Exp. 2012(60).

[11] Ferruzzi J , Bersi MR, Humphrey JD. Biomechanical phenotyping of central arteries in health and disease: advantages of and methods for murine models. Ann Biomed Eng. 2013;41(7):1311–1330.

[12] Tian L , Henningsen J, Salick MR, Crone WC, Gunderson M, Dailey SH, . Stretch calculated from grip distance accurately approximates mid-specimen stretch in large elastic arteries in uniaxial tensile tests. J Mech Behav Biomed Mater 2015;47:107–113.

[13] Vesely I , Boughner DR, Leeson-Dietrich J. Bioprosthetic valve tissue viscoelasticity: implications on accelerated pulse duplicator testing. Ann Thorac Surg. 1995;60(2 Suppl):S379–S382; discussion S83.

[14] Boronyak SM , Merryman WD. Development of a simultaneous cryo-anchoring and radiofrequency ablation catheter for percutaneous treatment of mitral valve prolapse. Ann Biomed Eng. 2012;40(9):1971–1981.

[15] Ghaemi H , Behdinan K, Spence AD. In vitro technique in estimation of passive mechanical properties of bovine heart part I. Experimental techniques and data. Med Eng Phys. 2009;31(1):76–82.

[16] Sommer G , Schriefl AJ, Andra M, Sacherer M, Viertler C, Wolinski H, . Biomechanical properties and microstructure of human ventricular myocardium. Acta Biomater. 2015;24:172–192.

[17] Carew EO , Barber JE, Vesely I. Role of preconditioning and recovery time in repeated testing of aortic valve tissues: validation through quasilinear viscoelastic theory. Ann Biomed Eng. 2000;28(9):1093–1100.

[18] Hiemenz PC , Lodge TP. Polymer Chemistry: CRC Press; 2007.

[19] Eberth JF , Gresham VC, Reddy AK, Popovic N, Wilson E, Humphrey JD. Importance of pulsatility in hypertensive carotid artery growth and remodeling. J Hypertens. 2009;27(10):2010–2021.

[20] Glaser E , Lacolley P, Boutouyrie P, Sacunha R, Lucet B, Safar ME, . Dynamic versus static compliance of the carotid artery in living Wistar-Kyoto rats. J Vasc Res. 1995;32(4): 254–265.

[21] Lenard Z , Fulop D, Visontai Z, Jokkel G, Reneman R, Kollai M. Static versus dynamic distensibility of the carotid artery in humans. J Vasc Res. 2000;37(2):103–111.

[22] Wang Z , Lakes RS, Golob M, Eickhoff JC, Chesler NC. Changes in large pulmonary arterial viscoelasticity in chronic pulmonary hypertension. PLoS One. 2013;8(11):e78569.

[23] Chesler NC , Thompson-Figueroa J, Millburne K. Measurements of mouse pulmonary artery biomechanics. J Biomech Eng. 2004;126(2):309–314.

[24] Liu A , Tian L, Golob M, Eickhoff JC, Boston M, Chesler NC. 17Beta-estradiol attenuates conduit pulmonary artery mechanical property changes with pulmonary arterial hypertension. Hypertension. 2015;66(5):1082–1088.

[25] Golob MJ , Tian L, Wang Z, Zimmerman TA, Caneba CA, Hacker TA, . Mitochondria DNA mutations cause sex-dependent development of hypertension and alterations in cardiovascular function. J Biomech. 2015;48(3):405–412.

[26] Tian L , Wang Z, Lakes RS, Chesler NC. Comparison of approaches to quantify arterial damping capacity from pressurization tests on mouse conduit arteries. J Biomech Eng. 2013;135(5):54504.

[27] Tian L , Chesler NC. In vivo and in vitro measurements of pulmonary arterial stiffness: a brief review. Pulm Circ. 2012;2(4):505–517.

[28] Wang Z , Chesler NC. Pulmonary vascular wall stiffness: an important contributor to the increased right ventricular afterload with pulmonary hypertension. Pulm Circ. 2011;1(2):212–223.

[29] Bergel DH . The dynamic elastic properties of the arterial wall. J Physiol. 1961;156(3): 458–469.

[30] Learoyd BM , Taylor MG. Alterations with age in the viscoelastic properties of human arterial walls. Circ Res. 1966;18(3):278–292.

[31] Cox RH . Viscoelastic properties of canine pulmonary arteries. Am J Physiol. 1984;246(1 Pt 2):H90–H96.

[32] Gamero LG , Armentano RL, Barra JG, Simon A, Levenson J. Identification of arterial wall dynamics in conscious dogs. Exp Physiol. 2001;86(4):519–528.

[33] Armentano RL , Barra JG, Santana DB, Pessana FM, Graf S, Craiem D, . Smart damping modulation of carotid wall energetics in human hypertension: effects of angiotensin-converting enzyme inhibition. Hypertension. 2006;47(3):384–390.

[34] Stefanadis C , Dernellis J, Vlachopoulos C, Tsioufis C, Tsiamis E, Toutouzas K, . Aortic function in arterial hypertension determined by pressure-diameter relation: effects of diltiazem. Circulation. 1997;96(6):1853–1858.

[35] Armentano RL , Barra JG, Levenson J, Simon A, Pichel RH. Arterial wall mechanics in conscious dogs. Assessment of viscous, inertial, and elastic moduli to characterize aortic wall behavior. Circ Res. 1995;76(3):468–478.

[36] Santana DB , Barra JG, Grignola JC, Gines FF, Armentano RL. Pulmonary artery smooth muscle activation attenuates arterial dysfunction during acute pulmonary hypertension. J Appl Physiol. 2005;98(2):605–613.

[37] Barra JG , Levenson J, Armentano RL, Cabrera Fischer EI, Pichel RH, Simon A. In vivo angiotensin II receptor blockade and converting enzyme inhibition on canine aortic viscoelasticity. Am J Physiol. 1997;272(2 Pt 2):H859–H868.

[38] Elliott DM , Robinson PS, Gimbel JA, Sarver JJ, Abboud JA, Iozzo RV, . Effect of altered matrix proteins on quasilinear viscoelastic properties in transgenic mouse tail tendons. Ann Biomed Eng. 2003;31(5):599–605.

[39] Li LP , Herzog W. The role of viscoelasticity of collagen fibers in articular cartilage: theory and numerical formulation. Biorheology. 2004;41(3–4):181–194.

[40] Lujan TJ , Underwood CJ, Jacobs NT, Weiss JA. Contribution of glycosaminoglycans to viscoelastic tensile behavior of human ligament. J Appl Physiol. 2009;106(2):423–431.

[41] Dourte LM , Pathmanathan L, Jawad AF, Iozzo RV, Mienaltowski MJ, Birk DE, . Influence of decorin on the mechanical, compositional, and structural properties of the mouse patellar tendon. J Biomech Eng. 2012;134(3):031005.

[42] Garcia A , Martinez MA, Pena E. Viscoelastic properties of the passive mechanical behavior of the porcine carotid artery: influence of proximal and distal positions. Biorheology. 2012;49(4):271–288.

[43] Zou Y , Zhang Y. The orthotropic viscoelastic behavior of aortic elastin. Biomech Model Mechanobiol. 2011;10(5):613–625.

[44] Tomiyama H , Hashimoto H, Matsumoto C, Odaira M, Yoshida M, Shiina K, . Effects of aging and persistent prehypertension on arterial stiffening. Atherosclerosis. 2011;217(1):130–134.

[45] Sun Z. Aging, arterial stiffness, and hypertension. Hypertension. 2015;65(2):252–256.

[46] Payne RA , Wilkinson IB, Webb DJ. Arterial stiffness and hypertension: emerging concepts. Hypertension. 2010;55(1):9–14.

[47] Jacobs DR Jr., Duprez DA, Shimbo D. Invited commentary: hypertension and arterial stiffness-origins remain a dilemma. Am J Epidemiol. 2016.

[48] Franklin SS . Arterial stiffness and hypertension: a two-way street? Hypertension. 2005;45(3):349–351.

[49] Antonov P , Antonova M, Nikolova N, Antonova N, Vlaskovska M, Kasakov L. Age dependent changes of arterial wall viscoelasticity. Clin Hemorheol Microcirc. 2008;39(1–4):63–68.

[50] Langewouters GJ , Wesseling KH, Goedhard WJ. The pressure dependent dynamic elasticity of 35 thoracic and 16 abdominal human aortas in vitro described by a five component model. J Biomech. 1985;18(8):613–620.

[51] Steppan J , Barodka V, Berkowitz DE, Nyhan D. Vascular stiffness and increased pulse pressure in the aging cardiovascular system. Cardiol Res Pract. 2011;2011:263585.

[52] Laurent S , Boutouyrie P, Asmar R, Gautier I, Laloux B, Guize L, . Aortic stiffness is an independent predictor of all-cause and cardiovascular mortality in hypertensive patients. Hypertension. 2001;37(5):1236–1241.

[53] Fourie PR , Coetzee AR, Bolliger CT. Pulmonary artery compliance: its role in right ventricular-arterial coupling. Cardiovasc Res. 1992;26(9):839–844.

[54] Pagnamenta A , Dewachter C, McEntee K, Fesler P, Brimioulle S, Naeije R. Early right ventriculo-arterial uncoupling in borderline pulmonary hypertension on experimental heart failure. J Appl Physiol. 2010;109(4):1080–1085.

[55] Kass DA . Ventricular arterial stiffening: integrating the pathophysiology. Hypertension. 2005;46(1):185–193.

[56] O'Rourke MF , Kelly RP. Wave reflection in the systemic circulation and its implications in ventricular function. J Hypertens. 1993;11(4):327–337.

[57] Manisty C , Mayet J, Tapp RJ, Parker KH, Sever P, Poulter NR, . Wave reflection predicts cardiovascular events in hypertensive individuals independent of blood pressure and other cardiovascular risk factors: an ASCOT (Anglo-Scandinavian Cardiac Outcome Trial) substudy. J Am Coll Cardiol. 2010;56(1):24–30.

[58] Grignola JC , Gines F, Bia D, Armentano R. Improved right ventricular-vascular coupling during active pulmonary hypertension. Int J Cardiol. 2007;115(2):171–182.

[59] Schreier DA , Hacker TA, Hunter K, Eickoff J, Liu A, Song G, . Impact of increased hematocrit on right ventricular afterload in response to chronic hypoxia. J Appl Physiol. 2014;117(8):833–839.

[60] Taniguchi R , Hosaka A, Miyahara T, Hoshina K, Okamoto H, Shigematsu K, . Viscoelastic deterioration of the carotid artery vascular wall is a possible predictor of coronary artery disease. J Atheroscler Thromb. 2015;22(4):415–423.

[61] Yokobori AT , Ohmi T, Monma R, Tomono Y, Inoue K, Owa M, . Correlation between the characteristics of acceleration and visco elasticity of artery wall under pulsatile flow

conditions (physical meaning of I* as a parameter of progressive behaviors of athero-sclerosis and arteriosclerosis). Biomed Mater Eng. 2013;23(1–2):75–91.

[62] Deng X , Guidoin R. Arteries, Veins and Lymphatic Vessels. In: Black J, Hastings G, editors. Handbook of Biomaterial Properties. London: Chapman & Hall; 1998. p. 81–105.

[63] Gore RW . Pressures in cat mesenteric arterioles and capillaries during changes in systemic arterial blood pressure. Circ Res. 1974;34(4):581–591.

[64] Lee J , Schmid-Schonbein GW. Biomechanics of skeletal muscle capillaries: hemodynamic resistance, endothelial distensibility, and pseudopod formation. Ann Biomed Eng. 1995;23(3):226–246.

[65] Skalak TC , Schmidschonbein GW. Viscoelastic properties of microvessels in rat spinotrapezius muscle. J Biomech Eng. 1986;108(3):193–200.

[66] Baskurt OK , Meiselman HJ. Blood rheology and hemodynamics. Semin Thromb Hemost. 2003;29(5):435–450.

[67] Chien S. Red cell deformability and its relevance to blood flow. Annu Rev Physiol. 1987;49:177–192.

[68] Yoon YZ , Kotar J, Yoon G, Cicuta P. The nonlinear mechanical response of the red blood cell. Phys Biol. 2008;5(3):036007.

[69] Dong C , Skalak R. Leukocyte deformability: finite element modeling of large viscoelastic deformation. J Theor Biol. 1992;158(2):173–193.

[70] Chien S , Schmid-Schonbein GW, Sung KL, Schmalzer EA, Skalak R. Viscoelastic properties of leukocytes. Kroc Found Ser. 1984;16:19–51.

[71] Kim J , Lee HY, Shin S. Advances in the measurement of red blood cell deformability: a brief review. J Cell Biotechnol. 2015;1:63–79.

[72] Puig-de-Morales-Marinkovic M , Turner KT, Butler JP, Fredberg JJ, Suresh S. Viscoelasticity of the human red blood cell. Am J Physiol Cell Physiol. 2007;293(2):C597–C605.

[73] Brown CD , Ghali HS, Zhao Z, Thomas LL, Friedman EA. Association of reduced red blood cell deformability and diabetic nephropathy. Kidney Int. 2005;67(1):295–300.

[74] Singh M , Shin S. Changes in erythrocyte aggregation and deformability in diabetes mellitus: a brief review. Indian J Exp Biol. 2009;47(1):7–15.

[75] Babu N , Singh M. Influence of hyperglycemia on aggregation, deformability and shape parameters of erythrocytes. Clin Hemorheol Microcirc. 2004;31(4):273–280.

[76] Lei H , Karniadakis GE. Quantifying the rheological and hemodynamic characteristics of sickle cell anemia. Biophys J. 2012;102(2):185–194.

[77] Grundy SM , Benjamin IJ, Burke GL, Chait A, Eckel RH, Howard BV, . Diabetes and cardiovascular disease: a statement for healthcare professionals from the American Heart Association. Circulation. 1999;100(10):1134–1146.

[78] Belizna C , Loufrani L, Ghali A, Lahary A, Primard E, Louvel JP, . Arterial stiffness and stroke in sickle cell disease. Stroke. 2012;43(4):1129–1130.

[79] Stella JA , Liao J, Sacks MS. Time-dependent biaxial mechanical behavior of the aortic heart valve leaflet. J Biomech. 2007;40(14):3169–3177.

[80] Lee JM , Courtman DW, Boughner DR. The glutaraldehyde-stabilized porcine aortic valve xenograft. I. Tensile viscoelastic properties of the fresh leaflet material. J Biomed Mater Res. 1984;18(1):61–77.

[81] Liao J , Yang L, Grashow J, Sacks MS. The relation between collagen fibril kinematics and mechanical properties in the mitral valve anterior leaflet. J Biomech Eng. 2007;129(1):78–87.

[82] Hingorani RV , Provenzano PP, Lakes RS, Escarcega A, Vanderby R Jr. Nonlinear viscoelasticity in rabbit medial collateral ligament. Ann Biomed Eng. 2004;32(2):306–312.

[83] Boyce BL , Jones RE, Nguyen TD, Grazier JM. Stress-controlled viscoelastic tensile response of bovine cornea. J Biomech. 2007;40(11):2367–2376.

[84] Thornton GM , Oliynyk A, Frank CB, Shrive NG. Ligament creep cannot be predicted horn stress relaxation at low stress: a biomechanical study of the rabbit medial collateral ligament. J Orthop Res. 1997;15(5):652–656.

[85] Grashow JS , Sacks MS, Liao J, Yoganathan AP. Planar biaxial creep and stress relaxation of the mitral valve anterior leaflet. Ann Biomed Eng. 2006;34(10):1509–1518.

[86] Vesely I , Noseworthy R. Micromechanics of the fibrosa and the ventricularis in aortic valve leaflets. J Biomech. 1992;25(1):101–113.

[87] Golomb G , Schoen FJ, Smith MS, Linden J, Dixon M, Levy RJ. The role of glutaraldehyde-induced cross-links in calcification of bovine pericardium used in cardiac valve bioprostheses. Am J Pathol. 1987;127(1):122–130.

[88] Lee JM , Boughner DR, Courtman DW. The glutaraldehyde-stabilized porcine aortic valve xenograft. II. Effect of fixation with or without pressure on the tensile viscoelastic properties of the leaflet material. J Biomed Mater Res. 1984;18(1):79–98.

[89] Rousseau E , Sauren A, Van Hout M, Van Steenhoven A. Elastic and viscoelastic material behaviour of fresh and glutaraldehyde-treated porcine aortic valve tissue. J Biomech. 1983;16(5):339–348.

[90] Urban M , Pislaru C, Nenadic IZ, Kinnick RR, Greenleaf J. Measurement of viscoelastic properties of in vivo swine myocardium using lamb wave dispersion ultrasound vibrometry (LDUV). IEEE Trans Med Imaging. 2013;32(2):247–261.

[91] Stroud JD , Baicu CF, Barnes MA, Spinale FG, Zile MR. Viscoelastic properties of pressure overload hypertrophied myocardium: effect of serine protease treatment. Am J Physiol Heart Circ Physiol. 2002;282(6):H2324–H2335.

[92] Rubiano A , Qi Y, Guzzo D, Rowe K, Pepine C, Simmons C. Stem cell therapy restores viscoelastic properties of myocardium in rat model of hypertension. J Mech Behav Biomed Mater. 2015;59:71–77.

[93] Helmes M , Trombitas K, Centner T, Kellermayer M, Labeit S, Linke WA, . Mechanically driven contour-length adjustment in rat cardiac titin's unique N2B sequence: titin is an adjustable spring. Circ Res. 1999;84(11):1339–1352.

[94] Granzier HL , Labeit S. The giant protein titin: a major player in myocardial mechanics, signaling, and disease. Circ Res. 2004;94(3):284–295.

[95] Schreier D , Hacker T, Song G, Chesler N. The role of collagen synthesis in ventricular and vascular adaptation to hypoxic pulmonary hypertension. J Biomech Eng. 2013;135(2):021018.

[96] Golob M , Moss RL, Chesler NC. Cardiac tissue structure, properties, and performance: a materials science perspective. Ann Biomed Eng. 2014;42(10):2003–2013.

[97] Bellofiore A , Chesler NC. Methods for measuring right ventricular function and hemodynamic coupling with the pulmonary vasculature. Ann Biomed Eng. 2013;41(7): 1384–1398.

[98] Sacks MS , Chuong CJ. Biaxial mechanical properties of passive right ventricular free wall myocardium. J Biomech Eng. 1993;115(2):202–205.

[99] Scallan J , Huxley VH, Korthuis RJ. The Lymphatic Vasculature. Capillary Fluid Exchange: Regulation, Functions, and Pathology. Integrated Systems Physiology: From Molecule to Function to Disease. San Rafael, CA: Morgan & Claypool Life Sciences, 2010.

[100] Ohhashi T. Comparison of viscoelastic properties of walls and functional characteristics of valves in lymphatic and venous vessels. Lymphology. 1987;20(4):219–223.

[101] Niklason LE , Koh J, Solan A. Tissue engineering of the lymphatic system. Ann N Y Acad Sci. 2002;979:27–34; discussion 5–8.

[102] Bertram CD , Macaskill C, Davis MJ, Moore JE Jr. Development of a model of a multi-lymphangion lymphatic vessel incorporating realistic and measured parameter values. Biomech Model Mechanobiol. 2014;13(2):401–416.

[103] Wilson JT , van Loon R, Wang W, Zawieja DC, Moore JE Jr. Determining the combined effect of the lymphatic valve leaflets and sinus on resistance to forward flow. J Biomech. 2015;48(13):3593–3599.

[104] Munn LL . Mechanobiology of lymphatic contractions. Semin Cell Dev Biol. 2015;38:67–74.

[105] Nipper ME , Dixon JB. Engineering the lymphatic system. Cardiovasc Eng Technol. 2011;2(4):296–308.

[106] Dongaonkar RM , Nguyen TL, Quick CM, Heaps CL, Hardy J, Laine GA, . Mesenteric lymphatic vessels adapt to mesenteric venous hypertension by becoming weaker pumps. Am J Physiol Regul Integr Comp Physiol. 2015;308(5):R391–R399.

[107] McClain J , Tuell SL, Thomas SN. Tumors Change the Elastic and Viscoelastic Properties of Draining Lymph Node Tissues. Proceedings of the ASME Summer Bioengineering Conference – 2013, Pt B. 2014.

Chapter 8

Nonviscous Modes of Viscoelastically Damped Vibrating Systems

Mario Lázaro, César F. Casanova and Carlos Lázaro

Additional information is available at the end of the chapter

http://dx.doi.org/10.5772/64205

Abstract

Nonviscously damped vibrating systems are characterized by dissipative mechanisms depending on the time history of the response velocity, introduced in the physical models using convolution integrals involving hereditary kernel functions. One of the most used damping viscoelastic models is Biot's model, whose hereditary functions are assumed to be exponential kernels. The free-motion equations of these types of nonviscous systems lead to a nonlinear eigenvalue problem enclosing certain number of the so-called nonviscous modes with nonoscillatory nature. Traditionally, the nonviscous modes (eigenvalues and eigenvectors) for nonproportional systems have been computed using the state-space approach, computationally expensive. This number of real eigenvalues is directly related to the rank of the damping matrices associated with the exponential kernels. The state-space approach has traditionally been used up to now as the only method to compute the nonviscous modes for nonproportionally damped systems. Motivated by this open problem, we propose in this chapter to describe the available numerical methods for classically damped systems and present the recent methods for nonclassically damped systems. It is shown that the problem of finding the nonviscous modes can be reduced to solve as a set of linear eigenvalue problems. The presented methods are compared through a numerical example.

Keywords: vibrating systems, nonviscous damping, eigenvalues and eigenvectors, nonproportional systems, numerical methods

1. Introduction and background

It has been always very difficult to model the physical fundamentals of damping in structural dynamics. In general, the proposed models depend on several parameters, which must be fitted according to experimental results. The viscous model, proposed by Rayleigh [1], is the most used representation of dissipative forces for vibrating systems as it predicts an exponential decay rate of displacements, something that can be observed experimentally in a great variety of structural materials such as metals, concrete, wood, glass, or masonry. However, damping models need to be updated for the mathematical modeling of the real behavior of the so-called viscoelastic damping materials, widely used for vibration control and energy dissipation devices. Although the term *viscoelastic damping* has traditionally been used, in the last years the concept *nonviscous damping* is also found in the bibliography, since this behavior can be considered as a generalization of the classic viscous damping. These materials, used in different areas of engineering as mechanical, civil, industrial, or aeronautics, are formed by polymer derivatives, rubbers, and rubber-like materials, and are characterized by a time-dependent constitutive model and by frequency-dependent Young and shear moduli.

Viscoelastic models of energy dissipation are introduced in the structure assuming that the damping forces are proportional to the history of the degrees-of-freedom (dof) velocities via kernel hereditary functions. These functions, also named damping functions, are the terms of the viscoelastic damping matrix in time domain, denoted by $\mathcal{G}(t) \in \mathbb{R}^{n \times n}$. The dynamic balance of internal forces yields to the system of motion differential equations for a viscoelastically damped structure, with the form

$$\begin{aligned} &\mathbf{M}\ddot{\mathbf{u}} + \int_0^t \mathcal{G}(t-\tau)\dot{\mathbf{u}}(\tau)\, d\tau + \mathbf{K}\mathbf{u} = \mathbf{f}_e(t) \\ &\mathbf{u}(0) = \mathbf{v}_0\,, \quad \dot{\mathbf{u}}(0) = \mathbf{u}_0 \end{aligned} \tag{1}$$

where the dofs' time-domain response is represented by $\mathbf{u}(t) \in \mathbb{R}^n$. $\mathbf{M}, \mathbf{K} \in \mathbb{R}^{n\times n}$ are the mass and stiffness matrices. In general, we do not assume symmetry in these matrices although the mass matrix will be assumed to be non-singular. Under these conditions, the modes of the system can be obtained as the nontrivial solutions of the free-motion problem obtained considering $\mathbf{f}_e(t)=\mathbf{v}_0=\mathbf{u}_0=\mathbf{0}$ in Eq. (1). Thus, checking functions of the form $\mathbf{u}(t)=\mathbf{u}e^{st}$ we obtain

$$\left[s^2\mathbf{M} + s\mathbf{G}(s) + \mathbf{K}\right]\mathbf{u} \equiv \mathbf{D}(s)\mathbf{u} = \mathbf{0} \tag{2}$$

where $\mathbf{D}(s)$: $\mathbb{C}\rightarrow\mathbb{C}^{n\times n}$ is the so-called *dynamic stiffness matrix*.

In this chapter, we will analyze Biot's damping model with N exponential kernels, a restriction commonly assumed in engineering applications. The expressions of the normalized damping functions in time and in frequency domain are, respectively

$$\mathcal{G}(t) = \sum_{k=1}^{N} \mathbf{C}_k \mu_k e^{-\mu_k t} \tag{3}$$

http://dx.doi.org/10.5772/64205

$$\mathbf{G}(s) = \mathcal{L}\{\mathcal{G}(t)\} = \sum_{k=1}^{N} \frac{\mu_k}{s + \mu_k} \mathbf{C}_k \tag{4}$$

where $\mu_k > 0$ with $1 \leq k \leq N$ are the relaxation or nonviscous parameters and $C_k \in \mathrm{R}^{n \times n}$ are the damping matrices (in general asymmetric) of the limit viscous model, obtained if the relaxation parameters tend to infinite, that is

$$\sum_{k=1}^{N} \mathbf{C}_k = \lim_{\mu_1 \cdots \mu_N \to \infty} \mathbf{G}(s) \tag{5}$$

The coefficients μ_k control the time and frequency dependence of the damping model while the spatial location and the level of damping are modeled via the matrices $\mathbf{C}_k$. From this property, it is not strange to find them also as the damping coefficients of Biot's model. The level of damping is closely related to the magnitude of these limit-damping matrices, while the relaxation parameters give information on how far is our nonviscous model from a viscous behavior [2]. It is also easy to demonstrate that the limit viscous damping and the time-domain kernel function are related by

$$\sum_{k=1}^{N} \mathbf{C}_k = \int_0^{\infty} \mathcal{G}(t)\, dt \tag{6}$$

Eqs. (2) and (4) clearly show the frequency dependence of the damping matrix, characteristic in this type of systems. This fact leads to a nonlinear eigenvalue problem whose eigenvalues are the roots of the equation

$$\det\left[\mathbf{D}(s)\right] = 0 \tag{7}$$

In general, the damping matrix $\mathbf{G}(s)$ admits a rational representation, so that the polynomial of the denominator is at least of one order less than that of the numerator [2]. Thus, the determinant can be written as a polynomial, whose order is greater than $2n$ and, therefore, the total number of roots of Eq. (7) can be expressed as $2n + r$ and arranged as

$$\{s_1, s_1^*, \ldots, s_n, s_n^*\}_{j=1}^{n} \cup \{\sigma_1, \ldots, \sigma_r\} \tag{8}$$

where s_i, s_i^* are n complex conjugate pairs and $\sigma_1, \ldots, \sigma_r$ are r negative real numbers named *nonviscous* eigenvalues. The name is chosen precisely because they are characteristic of nonviscous or viscoelastic models. The number of these nonviscous eigenvalues will depend on the nature of the damping function, particularly on the number of hereditary exponential kernels. The complex conjugate pair forces the solution to be oscillatory, whereas the other eigenvalues are associated with overdamped, nonoscillatory modes. The latter modes decay rapidly and in general are not important for the system response.

The representation of the hereditary behavior was originally introduced by Boltzman [3] at the end of the nineteenth century. Its application to viscoelastic materials and to damping of vibrating systems was studied by different authors in the middle of the twentieth century. Among them, it is worth mentioning specially Biot [4, 5] whose multi-exponential hereditary model has widely been used for modeling viscoelastic damping materials. The fundamentals of viscoelasticity, a thorough study on the time-dependence constitutive models, and its application for modeling damping materials can be found in books such as Fluegge [6], Nashif [7], and Jones [8]. Although this chapter is closely related to Biot's damping model, we must not forget the other viscoelastic models based on the fractional derivatives and widely used for representing the frequency-dependent behavior of damping materials. This model allows to use less parameters than exponential-based models [9], although the mathematical treatment is more difficult to implement, especially in the time domain, which is computationally more expensive [10].

This chapter is focused on the study of the r nonviscous modes of a nonviscously damped vibrating system. It is known that the effect of not considering these modes in the time-domain response is not important [2, 11]. Additionally, the exact calculation requires the use of the state-space approach, significantly increasing the computational effort and losing the physical insight of the involved internal variables [12, 13]. Maybe, these two reasons put together explain why they have not been analyzed in detail in the literature. Recently, Lázaro [14] derived a numerical approximated method to extract the nonviscous modes avoiding the state-space approach. In this chapter, we present a review of the nonviscous modes with nonoscillatory nature, giving their characteristics, mathematical properties, and the current available numerical methods for their computation.

2. Single degree-of-freedom systems

A single dof nonviscously damped vibrating system is dynamically characterized by a mass m, a linear stiffness k, and a nonviscous hereditary damping function $\mathcal{G}(t)$. The motion equation is

$$m\ddot{\mathrm{u}} + \int_{-\infty}^{t} \mathcal{G}(t-\tau)\,\dot{\mathrm{u}}(\tau)\,\mathrm{d}\tau + k\mathrm{u} = \mathfrak{f}_e(t) \tag{9}$$

where $\mathrm{u}(t)$ represents the degree of freedom and $\mathfrak{f}_e(t)$ represents the applied force in time domain. In this chapter, we analyze the nonviscous modes associated to Biot's damping model with N exponential kernels. The expressions of the normalized damping functions in time and in frequency domain for single dof are, respectively

$$\mathcal{G}(t)=\sum_{k=1}^{N} c_k\mu_k\, e^{-\mu_k t}, \qquad G(s)=\mathcal{L}\{\mathcal{G}(t)\}=\sum_{k=1}^{N}\frac{c_k\mu_k}{s+\mu_k} \tag{10}$$

where $\mu_k > 0$ with $1 \le k \le N$ are the relaxation or nonviscous parameters. Eqs. (5) and (6) can be particularized for single dof resulting

$$\lim_{\mu_1\cdots\mu_N\to\infty} G(s)=\sum_{k=1}^{N} c_k, \quad \int_0^{\infty}\mathcal{G}(t)dt=\sum_{k=1}^{N} c_k \tag{11}$$

Checking solutions of the form $u(t)=u\,e^{st}$, we can derive the characteristic equation

$$\text{ms}^2+sG(s)+k=\text{ms}^2+\sum_{k=1}^{N}\frac{\text{sc}_k\mu_k}{s+\mu_k}+k=0 \tag{12}$$

Multiplying this expression by $\prod_{k=1}^{N}(s+\mu_k)$, it results in a $2+N$-order polynomial. If the system is lightly or moderately damped, the set of eigenvalues presents the form $\{s_0, s_0^*, \sigma_1, \ldots, \sigma_N\}$, where s_0, s_0^* are a pair of conjugate-complex eigenvalues representing the modes of oscillatory nature. The rest N roots are negative real numbers representing the nonviscous eigenvalues of nonoscillatory nature (overcritically damped). In this point, we focus on giving a mathematical characterization of these eigenvalues and to provide efficient methods to approximate the nonviscous eigenvalues avoiding solving the polynomial equation.

2.1. Mathematical characterization of eigenvalues

Let us see that the damping function evaluated at a nonviscous eigenvalue must always verify certain inequality related to the dynamic properties of the system, say mass m and stiffness k. Eq. (12) is rewritten in terms of the undamped natural frequency $\omega_n=\sqrt{k/m}$ and of a new dimensionless damping function denoted by

$$J(s)=\frac{G(s)}{2m\omega_n} \tag{13}$$

The characteristic Eq. (11) now becomes

$$s^2 + 2sJ(s)\omega_n + \omega_n^2 = 0 \tag{14}$$

Reordering this equation, we can express it as

$$s^2 + 2sJ(s)\omega_n + \omega_n^2 = \left[s + J(s)\omega_n\right]^2 + \left[1 - J^2(s)\right]\omega_n^2 = 0 \tag{15}$$

Let $\sigma \in \mathbb{R}^-$ be any real nonviscous eigenvalue. Since $J(s) \in \mathbb{R}$ for all $s \in \mathbb{R}$, it can be ensured that

$$1 - J^2(\sigma) = -\left[\frac{\sigma}{\omega_n} + J(\sigma)\right]^2 \leq 0 \tag{16}$$

equivalent to $|J(\sigma)| \geq 1$. This inequality always holds for any real eigenvalue of Eq. (12). This result is a generalization of the well-known relationship between the dynamic parameters m, k, and c of a single dof viscously damped oscillator for having real eigenvalues: $c \geq c_{cr} = 2\sqrt{m\,k}$ (condition for critical damping).

As a direct consequence, we can define the following set:

$$\mathcal{B} = \left\{s \in \mathbb{R}^- : |G(s)| \geq 2\sqrt{m\,k}\right\} \tag{17}$$

assuring that every real eigenvalue of Eq. (12) lies inside $\mathcal{B}$. Lázaro and Pérez-Aparicio [15] derived the necessary condition expressed as $|J(\sigma)| \geq 1$ and calculated approximate limits for the set $\mathcal{B}$, denoted as *nonviscous set*.

2.2. Numerical computation

It is known that the influence in the response of the nonviscous modes is much less important than that of the oscillatory complex modes [2, 16, 17]. For this reason, it is reasonable to look for closed-form approaches, avoiding the computational effort needed for solving the characteristic polynomial. Two methods based on the hypothesis of light damping can be found in the literature. They allow to approximate the nonviscous eigenvalues using closed-form formulas as function of the dynamic and damping parameters. The first one due to Adhikari and Pascual [18] approximates the nonviscous eigenvalues with the first iteration of Newton's method applied to the characteristic polynomial. The second one, developed by Lázaro in his PhD Thesis [19] and published in the paper [20], is a perturbation-based approach. Both methods will be described in detail below and can be applied for both single dof systems and multiple dof systems with proportional (or classical) damping.

2.2.1. Adhikari and Pascual's method

Let us denote by $\zeta_k = c_k / 2m\omega_n$ to the damping ratio associated to the jth exponential kernel. Introducing these parameters, the characteristic equation can be written as

$$s^2 + 2s\omega_n \sum_{k=1}^{N} \frac{\zeta_k \mu_k}{s+\mu_k} + \omega_n^2 = 0 \tag{18}$$

As mentioned before, the characteristic polynomial can be obtained multiplying the above equation by $\prod_{j=1}^{N}(s+\mu_j)$, resulting the $2+N$-order polynomial

$$P(s) = \left(s^2 + \omega_n^2\right)\prod_{j=1}^{N}(s+\mu_j) + 2s\omega_n \sum_{k=1}^{N} \zeta_k \mu_k \prod_{\substack{j=1 \\ j\neq k}}^{N}(s+\mu_j) \tag{19}$$

The method of Adhikari and Pascual [18] is based on the application of the first iteration of Newton's method with $s = -\mu_j$ as the initial point. Indeed, assuming that $-\mu_j + \Delta_j$ is close to the solution, Δ_j can be explicitly calculated from the first-order expansion of $P(s)$ around the initial point

$$0 \approx P(-\mu_j + \Delta_j) \approx P(-\mu_j) + \frac{\partial P(-\mu_j)}{\partial s}\Delta_j \tag{20}$$

After some simplifications, the expressions of Adhikari and Pascual published in Ref. [18] can be rewritten in terms of the current notation as

$$\sigma_j \approx -\mu_j - \frac{P(-\mu_j)}{\dfrac{\partial P(-\mu_j)}{\partial s}} = -\mu_j + \frac{\mu_j \zeta_j p_j}{\zeta_j (p_j - \mu_j q_j) - r_j + \dfrac{p_j}{2\omega_n \mu_j}\left(\mu_j^2 + \omega_n^2\right)} \tag{21}$$

where

$$p_j = \prod_{\substack{k=1 \\ k\neq j}}^{N}(\mu_k - \mu_j), \qquad q_j = \sum_{\substack{k=1 \\ k\neq j}}^{N} \prod_{\substack{r=1 \\ r\neq j,k}}^{N}(\mu_r - \mu_j), \qquad r_j = \sum_{\substack{k=1 \\ k\neq j}}^{N} \zeta_k \mu_k \prod_{\substack{r=1 \\ r\neq j,k}}^{N}(\mu_r - \mu_j) \tag{22}$$

Under the hypothesis of light damping $\zeta_j \ll 1$, the nonviscous eigenvalue lies close to $-\mu_j$; therefore, it is expected that the solution from Eq. (21) accurately estimates the exact solution.

2.2.2. *Lázaro's method*

Lázaro's method [19, 20] is based on considering the jth nonviscous eigenvalue σ_j, $1 \le j \le N$ as a function of the jth associated damping ratio ζ_j. The damping ratio ζ_j can be interpreted as a perturbation parameter of Eq. (18). Thus, we can write $\sigma_j = \sigma_j(\zeta_j)$ and Eq. (18) can be written for this eigenvalue as

$$\sigma_j^2(\zeta_j) + 2\sigma_j(\zeta_j)\omega_n \frac{\zeta_j \mu_j}{\sigma_j(\zeta_j) + \mu_j} + 2\sigma_j(\zeta_j)\omega_n \sum_{\substack{k=1 \\ k \neq j}}^{N} \frac{\zeta_k \mu_k}{\sigma_j(\zeta_j) + \mu_k} + \omega_n^2 = 0 \tag{23}$$

Now, multiplying this equation by $\sigma_j(\zeta_j) + \mu_j$, we obtain

$$\left(\sigma_j(\zeta_j) + \mu_j\right)\left[\sigma_j^2(\zeta_j) + 2\sigma_j(\zeta_j)\omega_n \mathcal{Z}_j\left(\sigma_j(\zeta_j)\right) + \omega_n^2\right] + 2\sigma_j(\zeta_j)\zeta_j \omega_n \mu_j = 0 \tag{24}$$

With this operation, the singularity associated to the jth nonviscous eigenvalue can be avoided. The function $\mathcal{Z}_j(s)$ introduced above is defined as

$$\mathcal{Z}_j(s) = \sum_{\substack{j=1 \\ j \neq k}}^{N} \frac{\zeta_j \mu_j}{s + \mu_j} \tag{25}$$

Eq. (24) explicitly defines σ_j as a function of ζ_j. Assuming light damping, we can expand $\sigma_j(\zeta_j)$ in terms of the damping parameter ζ_j, considering the latter as a perturbation parameter within the equation. Thus,

$$\sigma_j(\zeta_j) = \sigma_j(0) + \sigma_j'(0)\zeta_j + \sigma_j''(0)\frac{\zeta_j^2}{2} + \cdots \tag{26}$$

The value $\sigma_j(0)$ can be obtained evaluating Eq. (24) at $\zeta_j = 0$

$$\left(\sigma_j(0) + \mu_j\right)\left[\sigma_j^2(0) + 2\sigma_j(0)\omega_n \mathcal{Z}_j\left(\sigma_j(0)\right) + \omega_n^2\right] = 0 \tag{27}$$

The eigenvalue associated to the kth hereditary kernel lies closely to the kth relaxation parameter [15, 21]. Therefore, we are interested in the real solution $\sigma_j(0) = -\mu_j$. The first-order derivative can be calculated solving for $\sigma'_j(0)$ after taking derivatives with respect to ζ_j in Eq. (24). The rest of higher-order derivatives $\sigma''_j(0)$, $\sigma'''_j(0)$, ... are derived following the same procedure and using the previously calculated results. In general, it is sufficient to take up the second-order term since this approximation accurately estimates the nonviscous eigenvalues within a wide range of the damping ratios, including lightly and moderately damped structures [15, 20]. After obtaining the coefficients $\sigma'_j(0)$ and $\sigma''_j(0)$, the closed-form expression for σ_j remains as follows:

$$\begin{aligned}\sigma_j &\approx \sigma_j(0) + \frac{\partial \sigma_j(0)}{\partial \zeta_j}\zeta_j + \frac{\partial^2 \sigma_j(0)}{\partial \zeta_j^2}\zeta_j^2 \\ &= -\mu_j + \frac{2\mu_j^2\omega_n\zeta_j}{\mu_j^2+\omega_n^2-2\omega_n\mu_j\eta_{0j}} + 4\mu_j^3\omega_n^2\zeta_j^2\frac{\mu_j^2-\omega_n^2+2\omega_n\mu_j^2\eta_{1j}}{\left[\mu_j^2+\omega_n^2-2\omega_n\mu_j\eta_{0j}\right]^3}\end{aligned} \tag{28}$$

where

$$\eta_{0j} = \mathcal{Z}_j(-\mu_j) = \sum_{\substack{k=1\\k\neq j}}^{N} \frac{\zeta_k\mu_k}{\mu_k-\mu_j} \quad , \quad \eta_{1j} = \left.\frac{\partial \mathcal{Z}_j}{\partial s}\right|_{s=-\mu_j} = -\sum_{\substack{k=1\\k\neq j}}^{N} \frac{\zeta_k\mu_k}{(\mu_k-\mu_j)^2} \tag{29}$$

Both Lázaro's and Adhikari and Pascual's methods are presented as closed-form expressions. On one hand, numerical computation of polynomial roots is avoided, and on the other hand the analytical expressions allow to explicitly observe the dependence of the nonviscous eigenvalues as functions of the rest of the parameters of the problem.

3. Multiple degrees-of-freedom systems

This section deals with the properties of the nonviscous modes in asymmetric nonproportional viscoelastically damped vibrating systems. A generalization of the mathematical characterization proved for single dof systems in the previous point will be derived. Regarding numerical analysis, the available methods for computing nonviscous modes will also be presented. As mentioned in the introduction, we consider an n-dof vibrating structure with mass and stiffness matrices denoted $\mathbf{M}, \mathbf{K} \in \mathbb{R}^{n\times n}$. No restrictions with respect to the symmetry of these matrices are imposed and, additionally, it will be assumed that the mass matrix is not singular. The damping matrix $\mathcal{G}(t) \in \mathbb{R}^{n\times n}$ contains the hereditary functions of the viscoelastic dissipative model. There are also no restrictions on the symmetry of the damping matrices $\mathbf{C}_j$ $1 \leq j \leq N$ and therefore it will be considered that $\mathbf{C}_j \neq \mathbf{C}_j^T$. Thus, the eigenvalues are the roots of the nonlinear equation

$$\det\left[s^2\mathbf{M} + s\,\mathbf{G}(s) + \mathbf{K}\right] = 0 \tag{30}$$

The eigenvalues can be separated in n conjugate-complex pairs $\{s_l, s_l^*\}_{l=1}^{n}$ with oscillatory nature and the r real nonviscous eigenvalues $\{\sigma_j\}_{j=1}^{r}$. The number r of nonviscous eigenvalues and the range of the damping matrices $\mathbf{C}_j$ are directly related [12, 13]. In fact, Adhikari and Wagner [12] proved that $r = \sum_{j=1}^{N} \text{rank}(\mathbf{C}_j)$ in the context of the state-space approach, proof of which will be presented in this section.

3.1. Mathematical characterization of eigenmodes

It is assumed that the damping matrix is not proportional, that is, $\mathbf{G}(s)$ does not verify the necessary conditions to be diagonal in the modal space of the undamped problem [22]. As known, proportional damping matrices allow to reduce an n-dof system to n single dof systems due to the simultaneous decoupling capability. Thus, for these kinds of structures, the results of the previous point would apply. Assuming the nonproportionality, each nonviscous mode is characterized by a real eigenvalue $\sigma_j \in \mathbb{R}^-$, $1 \leq j \leq r$, and both right and left real eigenvectors are denoted by $\mathbf{u}_j$ and $\mathbf{v}_j$, respectively, so that

$$\begin{aligned} \left[\sigma_j^2 \mathbf{M} + \sigma_j \mathbf{G}(\sigma_j) + \mathbf{K}\right]\mathbf{u}_j &= \mathbf{D}(\sigma_j)\mathbf{u}_j = \mathbf{0} \quad 1 \leq j \leq r \\ \mathbf{v}_j^T\left[\sigma_j^2 \mathbf{M} + \sigma_j \mathbf{G}(\sigma_j) + \mathbf{K}\right] &= \mathbf{v}_j^T \mathbf{D}(\sigma_j) = \mathbf{0} \quad 1 \leq j \leq r \end{aligned} \tag{31}$$

We define the following expressions for each nonviscous eigenmode:

$$\mathfrak{M}_j = \mathbf{v}_j^T \mathbf{M}\,\mathbf{u}_j, \quad \mathfrak{K}_j = \mathbf{v}_j^T \mathbf{K}\mathbf{u}_j, \quad \Omega_j = \sqrt{\frac{\mathfrak{K}_j}{\mathfrak{M}_j}}, \tag{32}$$

These values can be interpreted as modal mass and stiffness, respectively, associated to the jth nonviscous mode. Using these new modal parameters, we can write that

$$\mathbf{v}_j^T \mathbf{D}(\sigma_j)\mathbf{u}_j = \mathfrak{M}_j \sigma_j^2 + \sigma_j \mathbf{v}_j^T \mathbf{G}(\sigma_j)\mathbf{u}_j + \mathfrak{K}_j \tag{33}$$

We introduce functions $\mathfrak{I}_j(s) : \mathbb{C} \to \mathbb{C}$ defined as

$$\mathfrak{I}_j(s) = \frac{\mathbf{v}_j^T \mathbf{G}(s)\mathbf{u}_j}{2\sqrt{\mathfrak{M}_j \mathfrak{K}_j}} \tag{34}$$

which can be interpreted as the dimensionless modal representation of the damping matrix at the jth real mode. Introducing this relationship in Eq. (33), we obtain

$$\sigma_j^2 + 2\sigma_j \mathfrak{J}_j(\sigma_j)\Omega_j + \Omega_j^2 = 0 \tag{35}$$

We can identify in this equality the same form as that of Eq. (14), derived for single dof oscillators. Therefore, and using identical mathematical manipulations, we can deduce that $\left|\mathfrak{J}_j(\sigma_j)\right| \geq 1$, or equivalently in terms of the damping matrix

$$\left|\mathbf{v}_j^T \mathbf{G}(\sigma_j)\mathbf{u}_j\right| \geq 2\sqrt{\mathfrak{M}_j \mathfrak{K}_j}\ , \quad 1 \leq j \leq r \tag{36}$$

expression of which represents the generalization for multiple dof systems of the necessary condition derived for single dof systems in the previous point, Eq. (16). Additionally, Eq. (36) can also be considered as a generalization of the result published by Lázaro and Pérez-Aparicio [15] for symmetric systems.

3.2. The state-space approach

In this section, the general state-space representation of the dynamic problem will be described. This methodology allows to transform the general n-dof system of integro-differential equations into a system of $m>2n$ first-order differential equations through the introduction of internal variables. It was developed by Wagner and Adhikari [12] for symmetric systems and by Adhikari and Wagner [13] for asymmetric system.

It turns out that the final size m of the extended state-space formulation is directly related to the rank of the damping matrices $\mathbf{C}_j$, $1 \leq j \leq N$. Because of that, it is appropriate to introduce the algebra associated to the matrix $\mathbf{C}_j$. Let us assume that $\text{rank}(\mathbf{C}_j) = r_j \leq n$, then there exist two matrices $\mathbf{X}_j, \mathbf{Y}_j \in \mathbb{R}^{n\times n}$, such that

$$\mathbf{Y}_j^T \mathbf{C}_j \mathbf{X}_j = \begin{bmatrix} \mathbf{d}_j & \mathbf{O}_{1j} \\ \mathbf{O}_{1j}^T & \mathbf{O}_{2j} \end{bmatrix} \tag{37}$$

where $\mathbf{d}_j \in \mathbb{R}^{r_j \times r_j}$ is a diagonal block matrix with the nonzero eigenvalues of $\mathbf{C}_j$ and the blocks $\mathbf{O}_{1j}$ and $\mathbf{O}_{2j}$ are null matrices of size $r_j \times (n - r_j)$ and $(n - r_j)\times(n - r_j)$, respectively. The columns of matrices $\mathbf{X}_j$ and $\mathbf{Y}_j$ form two different bases of space R^n, hence both matrices can be written in the form

$$\mathbf{X}_j = \left[\mathbf{x}_{j1} \cdots \mathbf{x}_{j,r_j}\, \mathbf{x}_{j,r_j+1} \cdots \mathbf{x}_{j,n}\right]$$
$$\mathbf{Y}_j = \left[\mathbf{y}_{j1} \cdots \mathbf{y}_{j,r_j}\, \mathbf{y}_{j,r_j+1} \cdots \mathbf{y}_{j,n}\right] \quad (38)$$

where $\mathbf{x}_{jk}$ and $\mathbf{y}_{jk}$ for $k=1, \ldots, r_j$ are the right and left eigenvectors of the nonzero eigenvalues of $\mathbf{C}_k$. These two bases have special relevance in the developments of the state-space method and it is convenient to group them in the two rectangular matrices

$$\mathbf{R}_j = \left[\mathbf{x}_{j1} \cdots \mathbf{x}_{j,r_j}\right] \in \mathrm{R}^{n\times r_j}$$
$$\mathbf{L}_j = \left[\mathbf{y}_{j1} \cdots \mathbf{y}_{j,r_j}\right] \in \mathrm{R}^{n\times r_j} \quad (39)$$

so that the following relations are straightforward:

$$\mathbf{L}_j^T \mathbf{C}_j \mathbf{R}_j = \mathbf{d}_j \quad (40)$$

Let us return now to the system of integro-differential equations presented in Eq. (1) written in terms of the dof $\mathbf{u}(t)$,, and let us introduce a set of $N+1$ internal variables denoted by $\mathbf{v}(t)$ and $\mathbf{w}_j(t)$, $1 \le j \le N$ and defined as

$$\mathbf{v}(t) = \dot{\mathbf{u}}(t) \quad \text{and} \quad \mathbf{w}_j(t) = \int_0^t \mu_j e^{-\mu_j(t-\tau)} \dot{\mathbf{u}}(\tau)\, d\tau\,, \quad 1 \le j \le N \quad (41)$$

For our purposes, we need the time derivative $\dot{\mathbf{w}}_j$, which can be calculated using Leibniz's rule for differentiation of an integral, yielding

$$\dot{\mathbf{w}}_j = -\mu_j^2 \int_0^t e^{-\mu_j(t-\tau)} \dot{\mathbf{u}}(\tau)\, d\tau + \mu_j \dot{\mathbf{u}}(t) = -\mu_j \mathbf{w}_j(t) + \mu_j \mathbf{v}(t) \quad (42)$$

With these new variables, Eq. (1) can be expressed as

$$\mathbf{M}\dot{\mathbf{v}} + \sum_{j=1}^{N} \mathbf{C}_j \mathbf{w}_j + \mathbf{K}\mathbf{u} = \mathbf{f}_e(t) \quad (43)$$

In the above expression, the vector $\mathbf{C}_j\mathbf{w}_j \in \mathbb{R}^n$ represents the image via the linear mapping defined by the matrix $\mathbf{C}_j$. The kernel of this mapping is a subspace of $\mathbb{R}^n$ with dimension $n-r_j$ and characterized by Ker $(\mathbf{C}_j) = \{\mathfrak{q} \in \mathbb{R}^n : \mathbf{C}_j\mathfrak{q} = \mathbf{0}\}$. The vectors $\mathbf{x}_{jk}, k=r_j+1,\ldots,n$ are a basis of this subspace. Therefore, only the r_j projections of $\mathbf{w}_j$ onto the rest of eigenvectors, say $\mathbf{x}_{jk}, k=1, \ldots, r_j$, are representative. Consequently, we can defined the r_j internal variables $\mathfrak{w}_j(t) \in \mathbb{R}^{r_j}$ from the rectangular transformation matrix $\mathbf{R}_j$

$$\mathbf{w}_j(t) = \mathbf{R}_j \mathfrak{w}_j(t) \tag{44}$$

Introducing this transformation into Eq. (43) and premultiplying by $\mathbf{M}^{-1}$

$$\dot{\mathfrak{v}} = -\sum_{j=1}^{N} \mathbf{M}^{-1}\mathbf{C}_j \mathbf{R}_j \mathfrak{w}_j(t) - \mathbf{M}^{-1}\mathbf{K}\,\mathfrak{u} + \mathbf{M}^{-1}\mathfrak{f}_e(t) \tag{45}$$

Now, in order to complete the extended linear system, we need to relate the variables $\mathrm{w}_j(t)$ and their time derivatives. For that, let us combine Eq. (44) with (42) resulting in

$$\mathbf{R}_j \dot{\mathfrak{w}}_j = -\mu_j \mathbf{R}_j \mathfrak{w}_j(t) + \mu_j \mathfrak{v}(t) \tag{46}$$

Premultiplying by matrix $\mathbf{L}_j^T$ and denoting by $\mathbf{T}_j=[\mathbf{L}_j^T\mathbf{R}_j]^{-1}\mathbf{L}_j^T$, we can write after some operations

$$\dot{\mathfrak{w}}_j = -\mu_j \mathfrak{w}_j(t) + \mu_j \mathbf{T}_j \mathfrak{v}(t)\,, \quad 1 \le j \le N \tag{47}$$

Eqs. (45) and (47) and the direct relations $\mathfrak{v}(t) = \dot{\mathfrak{u}}(t)$ can be put in order in the following extended linear system of ordinary differential equations:

$$\dot{\mathfrak{z}} = \mathbf{A}\,\mathfrak{z} + \mathfrak{r}(t) \tag{48}$$

where

$$\mathbf{A}=\begin{bmatrix} \mathbf{O}_{n,n} & \mathbf{I}_n & \mathbf{O}_{n,r_1} & \mathbf{O}_{n,r_2} & \cdots & \mathbf{O}_{n,r_N} \\ -\mathbf{M}^{-1}\mathbf{K} & \mathbf{O}_{n,n} & -\mathbf{M}^{-1}\mathbf{C}_1\mathbf{R}_1 & -\mathbf{M}^{-1}\mathbf{C}_2\mathbf{R}_2 & \cdots & -\mathbf{M}^{-1}\mathbf{C}_N\mathbf{R}_N \\ \mathbf{O}_{r_1,n} & \mu_1\mathbf{T}_1 & -\mu_1 I_{r_1} & \mathbf{O}_{r_1,r_2} & \cdots & \mathbf{O}_{r_1,r_N} \\ \mathbf{O}_{r_2,n} & \mu_2\mathbf{T}_2 & \mathbf{O}_{r_2,r_1} & -\mu_2\mathbf{I}_{r_2} & \cdots & \mathbf{O}_{r_1,r_N} \\ \vdots & \vdots & \vdots & \vdots & \ddots & \vdots \\ \mathbf{O}_{r_2,n} & \mu_N\mathbf{T}_N & \mathbf{O}_{r_N,r_1} & \mathbf{O}_{r_N,r_2} & \cdots & -\mu_N\mathbf{I}_{r_N} \end{bmatrix} \in \mathbb{R}^{m\times m} \tag{49}$$

$$\mathfrak{z}(t)=\left\{\mathfrak{u}^T(t),\mathfrak{v}^T(t),\mathfrak{w}_1^T(t),\ldots,\mathfrak{w}_N^T(t)\right\}^T \in \mathbb{R}^m, \quad \mathfrak{r}(t)=\left\{\mathbf{0}_n^T,(\mathbf{M}^{-1}\mathfrak{f}_e)^T,\mathbf{0}_{r_1}^T,\ldots,\mathbf{0}_{r_N}^T\right\}^T \in \mathbb{R}^m \tag{50}$$

In these expressions, $\mathbf{O}_{p,q} \in \mathbb{R}^{p\times q}$ and $\mathbf{0}_p \in \mathbb{R}^p$ represent the null matrix and vector (in column) of their respective spaces and $\mathbf{I}_{r_j}$ the identity matrix of $\mathbb{R}^{r_j\times r_j}$.. Since $\mathbf{T}_j$ represents a full-rank matrix of order r_j, then the total order of the system is

$$m = 2n + \sum_{j=1}^{n} r_j \tag{51}$$

showing that the extra order of the state-space formulation of a nonviscously damped vibrating system is governed by the rank of the damping matrices. Hence, the total number of nonviscous eigenvalues is given by $r=\sum_{j=1}^{n}$ rank $(\mathbf{C}_j)$. As known, checking solutions of the form $\mathbf{z}(t)=\mathbf{z}\, e^{st}$ in the free-motion equations ($\mathfrak{f}_e \equiv \mathbf{0}$) leads to the linear eigenvalue problem

$$\left(\mathbf{A} - s\,\mathbf{I}_m\right)\mathbf{z} = \mathbf{0}_m \tag{52}$$

The complete solution of this problem allows to construct the spectral set of nonviscously damped systems. On one hand, we have $2n$ complex modes with oscillatory nature and, on the other hand, the r nonviscous eigenmodes with their respective eigenvectors. A detailed study of the eigenvalue problem of Eq. (52) has been described in the work of Adhikari and Wagner [13]. From a mathematical point of view, the problem of calculating the eigenmodes is totally solved. However, we can expose two reasons why it is worth to deepen in the numerical problem of the nonviscous modes: (a) to solve a linear eigenvalue problem as that shown in Eq. (52) requires in general $\mathcal{O}(m^3)$ operations, something that affects the computational efficiency of the problem as we increase the number of hereditary kernels and the number of degrees of freedom. (b) The physical insight of the problem is somewhat lost with the introduction of new internal variables in the state-space method. Due to these two arguments, several numerical methods have been proposed in the bibliography to obtain the n complex modes with oscillatory nature (see for instance references [2, 18, 23–27]). On the contrary, the nonviscous modes have not been studied with so much detail since, obviously, their influence in the response is much less important. Lázaro published a research focusing on nonviscous modes of symmetric systems [14] trying, on one hand, to reduce the computational complexity of computing the nonviscous modes and, on the other hand, to supply a physical interpretation of the significance of these kinds of modes, closely related to the properties of the damping model and to the matrices $\mathbf{C}_j$, $j=1, \ldots, N$. In the next point, Lázaro's method will be described including a generalization for asymmetric systems.

3.3. Approximate numerical method

As described above, we derive here the numerical method proposed by Lázaro [14] for the computation of nonviscous modes. We work under the generally accepted assumption of light damping, something that allows to predict that the nonviscous eigenvalues are close to the relaxation parameters $\{-\mu_j\}_{j=1}^N$. Let us consider the following decoupling of the damping matrix in the Laplace domain associated to the jth relaxation parameter, μ_j

$$\mathbf{G}(s) = \frac{\mu_j}{s+\mu_j}\mathbf{C}_j + \sum_{\substack{k=1 \\ k\neq j}}^{N} \frac{\mu_k}{s+\mu_k}\mathbf{C}_k \equiv \frac{\mu_j}{s+\mu_j}\mathbf{C}_j + \mathbf{G}_j(s) \tag{53}$$

Something similar can be made for the dynamic stiffness matrix, yielding

$$\mathbf{D}(s) = s^2\mathbf{M} + s\mathbf{G}(s) + \mathbf{K} = s^2\mathbf{M} + s\mathbf{G}_j(s) + \mathbf{K} + \frac{s\mu_j}{s+\mu_j}\mathbf{C}_j \equiv \mathbf{D}_j(s) + \frac{s\mu_j}{s+\mu_j}\mathbf{C}_j \tag{54}$$

where $\mathbf{D}_j(s) = s^2\mathbf{M} + s\mathbf{G}_j(s) + \mathbf{K} \in \mathbb{C}^{n\times n}$ denotes the dynamic stiffness matrix without the jth hereditary damping function. Note that under this manipulation, the function $\mathbf{D}_j(s)$ is now continuous and with continous derivatives at $s=-\mu_j$. Let us denote by $\sigma_j \in \mathbb{R}^-$ to any nonviscous eigenvalue associated to μ_j and by $\mathbf{x}_j, \mathbf{y}_j \in \mathbb{R}^n$ the right and left eigenvectors associated to σ_j, respectively. The following relations hold:

$$\mathbf{D}(\sigma_j)\mathbf{x}_j = \left[\mathbf{D}_j(\sigma_j) + \frac{\sigma_j\mu_j}{\sigma_j+\mu_j}\mathbf{C}_j\right]\mathbf{x}_j = \mathbf{0} \tag{55}$$

$$\mathbf{D}^T(\sigma_j)\mathbf{y}_j = \left[\mathbf{D}_j^T(\sigma_j) + \frac{\sigma_j\mu_j}{\sigma_j+\mu_j}\mathbf{C}_j^T\right]\mathbf{y}_j = \mathbf{0} \tag{56}$$

In order not to have to repeat every step for the right and left eigenvalues, the developments will be carried out only for Eq. (55). Thus, multiplying Eq. (55) by $\sigma_j + \mu_j$

$$\left[(\sigma_j + \mu_j)\mathbf{D}_j(\sigma_j) + \sigma_j \mu_j \mathbf{C}_j\right]\mathbf{x}_j = \mathbf{0} \tag{57}$$

Let us define the matrix

$$\mathbf{A}_j(s) = (s + \mu_j)\mathbf{D}_j(s) \tag{58}$$

and Eq. (57) can be written as

$$\left[\mathbf{A}_j(\sigma_j) + \sigma_j \mu_j \mathbf{C}_j\right]\mathbf{x}_j = \mathbf{0} \tag{59}$$

Since the damping is assumed to be light, σ_j is close to $-\mu_j$ and consequently there exists certain $\lambda_j \in \mathrm{R}$, such that $\sigma_j = -\mu_j + \lambda_j$, with $|\lambda_j / \mu_j| \ll 1$. Expanding the matrix $\mathbf{A}_j(\sigma)$ around $\lambda_j = 0$ and neglecting second-order terms, we obtain

$$\mathbf{A}_j(\sigma_j) = \mathbf{A}_j(-\mu_j + \lambda_j) \approx \mathbf{A}_j(-\mu_j) + \mathbf{A}'_j(-\mu_j)\lambda_j \tag{60}$$

where $(\bullet)' = \partial(\bullet)/\partial s$. From the definition of $\mathbf{A}_j(s)$, we have that

$$\mathbf{A}_j(-\mu_j) = \mathbf{0}, \quad \mathbf{A}'_j(-\mu_j) = \mathbf{D}_j(-\mu_j) \tag{61}$$

Substituting this result together with $\sigma_j = -\mu_j + \lambda_j$ in Eq. (59) and rearranging

$$\left[\mathbf{C}_j - \lambda_j \mathbf{B}_j\right]\mathbf{x}_j = \mathbf{0} \tag{62}$$

where

$$\mathbf{B}_j = \frac{\mathbf{C}_j}{\mu_j} + \frac{\mathbf{D}_j(-\mu_j)}{\mu_j^2} \in \mathbb{R}^{n \times n} \tag{63}$$

Following the same steps for the left eigenvectors from Eq. (56), we obtain the following relation between λ_j and $\mathbf{y}_j \in \mathbb{R}^n$ is fulfilled:

$$[\mathbf{C}_j^T - \lambda_j \mathbf{B}_j^T]\mathbf{y}_j = \mathbf{0} \tag{64}$$

From Eqs. (62) and (64), λ_j, $\mathbf{x}_j$, and $\mathbf{y}_j$ represent an eigensolution of the generalized linear asymmetric eigenvalue problem of matrix $\mathbf{C}_j$ with respect to $\mathbf{B}_j$. Denoting by r_j=rank ($\mathbf{C}_j$)≤n, then $\lambda_j=0$ is eigenvalue of Eq. (26) with multiplicity $n-r_j$. Consequently, there exist other r_j no-null eigenvalues, which will be named $\lambda_{j,1}, \ldots, \lambda_{j,r_j}$. Hence, the complete spectral set of problem (64) can be listed as

$$\left\{\lambda_{j,1},\ldots,\lambda_{j,r_j},0,\overset{n-r_j}{\ldots},0\right\} \tag{65}$$

and the r_j nonviscous modes associated to the jth relaxation parameter can be denoted by

$$\left\{-\mu+\lambda_{j,k};\mathbf{x}_{j,k};\mathbf{y}_{j,k}\right\}_{k=1}^{r_j} \tag{66}$$

We highlight two interesting results from this method: (i) the computation of the nonviscous modes has been reduced to solve N linear eigenvalue problems of order n and (ii) there is no need to previously calculate neither the modal space of the undamped model nor the eigenproblem of the matrices $\mathbf{C}_j$. We find a limitation because a hypothesis of light damping has been used in the linearization (60). For vibrating problems under a higher level of damping, the method can be adapted just taking the second-order term in the expansion of $\mathbf{A}_j(-\mu_j+\lambda_j)$, that is

$$\mathrm{A}_j(-\mu_j+\lambda_j) \approx \mathrm{A}_j(-\mu_j) + \mathrm{A}_j'(-\mu_j)\lambda_j + \mathrm{A}_j''(-\mu_j)\frac{\lambda_i^2}{2!} = \mathrm{D}_j(-\mu_j)\lambda_j + \mathrm{D}_j'(-\mu_j)\lambda_j^2 \tag{67}$$

Introducing this expression in Eq. (59) and after some manipulations the resulting right and left eigenvalue problems are

$$\left[\mathbf{H}_j - \lambda_j \mathbf{F}_j\right]\mathbf{z}_j = \mathbf{0}, \qquad \left[\mathbf{H}_j^T - \lambda_j \mathbf{F}_j^T\right]\hat{\mathbf{z}}_j = \mathbf{0} \tag{68}$$

where

$$\mathbf{H}_j = \begin{bmatrix} \mathbf{C}_j & \mathbf{O}_n \\ \mathbf{O}_n & \mathbf{I}_n \end{bmatrix},\ \mathbf{F}_j = \begin{bmatrix} \mathbf{B}_j & \mathbf{D}_j'(-\mu_j)/\mu_j^2 \\ \mathbf{I}_n & \mathbf{O}_n \end{bmatrix},\ \mathbf{z}_j = \begin{Bmatrix} \mathbf{x}_j \\ \lambda_j \mathbf{x}_j \end{Bmatrix},\ \hat{\mathbf{z}}_j = \begin{Bmatrix} \mathbf{y}_j \\ \lambda_j \mathbf{y}_j \end{Bmatrix} \tag{69}$$

In general, the second-order approximation will lead to better approximations, although in this case a larger problem must be solved; this will be confirmed in the numerical example. The reader who wants to deepen in detail in higher-order approximations and their associated computational cost can refer to the work of Lázaro [14]. In this paper, it is proved that, from a computational point of view, it is profitable to increase the order of approximation up to certain limit order after which it is better to use the state-space approach. That limit value of the approximation order is $(2+N)/\sqrt[3]{N}$, is the number of hereditary damping kernels.

3.4. Numerical example

In this numerical example, the presented computational methods to calculate the nonviscous modes will be compared. For that, we use a five-degree-of-freedom discrete system with viscoelastic dampers, shown in **Figure 1**. Each dof represents the displacement of a mass $m=10^3$ kg. The linear stiffness between the masses is $k=10^5$ N/m. Two nonviscous dashpots are located between ground and dof 1 and between dofs 3, 4, and 5, whose constitutive relationships are expressed as the sum of exponential kernels.

$$\mathcal{G}_a(t)=c_a\left(\mu_1 e^{-\mu_1 t}+\mu_2 e^{-\mu_2 t}\right),\quad \mathcal{G}_b(t)=c_b\mu_3 e^{-\mu_3 t} \tag{70}$$

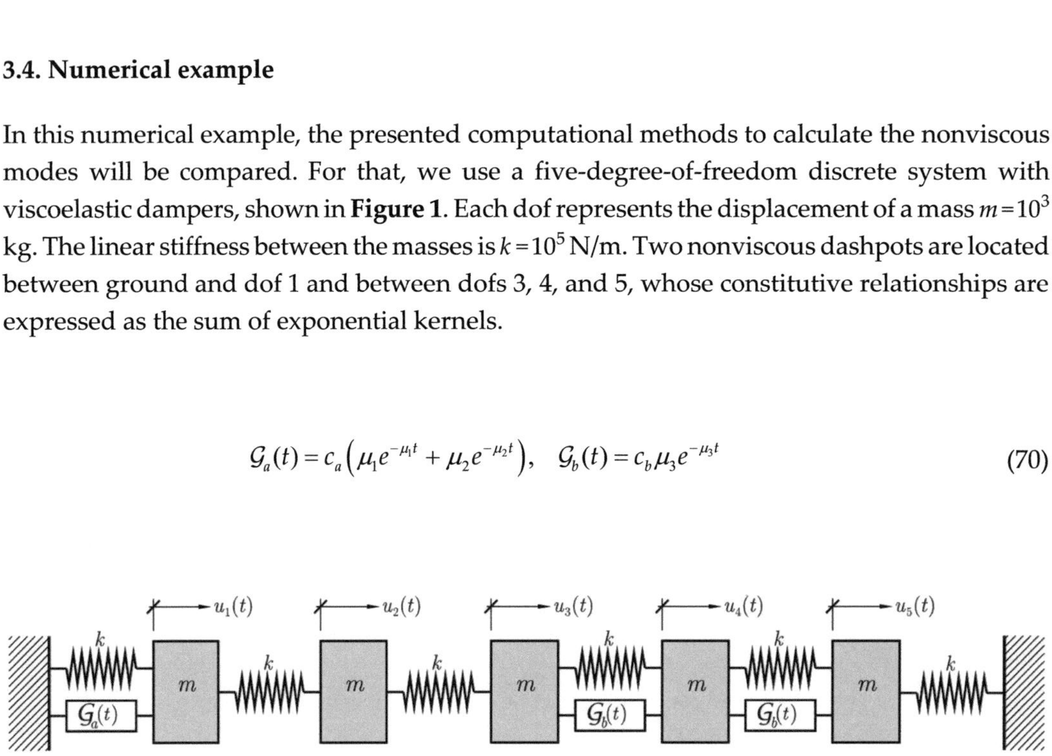

Figure 1. Numerical example: a five-degrees-of-freedom lumped mass system with viscoelastic dampers based on exponential kernels.

The damping coefficients are $c_a=600$ Nm^{-1}s and $c_b=200$ Nm^{-1}s and the relaxation parameters $\mu_j=\{10, 25, 45\}$ rad/s. Since we have three relaxation parameters, the damping matrix in time domain yields

$$\mathcal{G}(t)=\mathbf{C}_1\mu_1 e^{-\mu_1 t}+\mathbf{C}_2\mu_2 e^{-\mu_2 t}+\mathbf{C}_3\mu_3 e^{-\mu_3 t} \tag{71}$$

and according to the dashpots and rigidities distribution, the damping matrix coefficients and the stiffness matrix are

http://dx.doi.org/10.5772/64205

$$\mathbf{K}=\begin{bmatrix}2k & -k & 0 & 0 & 0\\ -k & 2k & -k & 0 & 0\\ 0 & -k & 2k & -k & 0\\ 0 & 0 & -k & 2k & -k\\ 0 & 0 & 0 & -k & 2k\end{bmatrix},\ \mathbf{C}_1=\mathbf{C}_2=c_a\begin{bmatrix}1 & 0 & 0 & 0 & 0\\ 0 & 0 & 0 & 0 & 0\\ 0 & 0 & 0 & 0 & 0\\ 0 & 0 & 0 & 0 & 0\\ 0 & 0 & 0 & 0 & 0\end{bmatrix},\ \mathbf{C}_3=c_b\begin{bmatrix}0 & 0 & 0 & 0 & 0\\ 0 & 0 & 0 & 0 & 0\\ 0 & 0 & 1 & -1 & 0\\ 0 & 0 & -1 & 2 & -1\\ 0 & 0 & 0 & -1 & 1\end{bmatrix} \tag{72}$$

	EIGENVALUES			
	$\mu_1=10,\ r_1=1$	$\mu_2=25,\ r_2=1$	$\mu_3=45,\ r_3=2$	
	$\sigma_{1,1}$	$\sigma_{2,1}$	$\sigma_{3,1}$	$\sigma_{3,2}$
Exact	−9,762536252	−24,539682264	−44,480104306	−44,817181343
1st order approx.	−9,767255360	−24,552342078	−44,490259517	−44,818527825
(error, %)	*(0,04834)*	*(0,05159)*	*(0,02283)*	*(0,00300)*
2nd order approx.	−9,762478350	−24,539554707	−44,480042888	−44,817178561
(error, %)	*(0,00059)*	*(0,00052)*	*(0,00014)*	*(0,00001)*
	EIGENVECTORS			
	$\mu_1=10,\ r_1=1$	$\mu_2=25,\ r_2=1$	$\mu_3=45,\ r_3=2$	
	$\mathbf{x}_{1,1}$	$\mathbf{x}_{2,1}$	$\mathbf{x}_{3,1}$	$\mathbf{x}_{3,2}$
Exact	0,920866633	0,991941304	0,000819770	0,001425705
	0,359382674	0,125658720	0,018202462	0,032082433
	0,140415976	0,016087932	0,395717303	0,707139978
	0,053098476	0,001862474	−0,829159074	0,000761150
	0,017681720	0,000209935	0,394424954	−0,706343512
1st order approx.	0,924263325	0,992396067	0,000786787	0,001405038
	0,353130865	0,122132136	0,017832254	0,031844687
	0,135129270	0,015194058	0,395980865	0,707139245
	0,050069699	0,001700855	−0,828890371	0,000749666
	0,016401313	0,000185500	0,394742065	−0,706355058
(error, %)	*(0,94544)*	*(0,36701)*	*(0,06167)*	*(0,02392)*
2nd order approx.	0,920823393	0,991936757	0,000819965	0,001425786
	0,359460764	0,125693455	0,018204709	0,032082906
	0,140482958	0,016096765	0,395715707	0,707139980
	0,053137115	0,001864208	−0,829160702	0,000761173
	0,017698131	0,000210211	0,394423029	−0,706343488
(error, %)	*(0,01192)*	*(0,00362)*	*(0,00037)*	*(0,00005)*

Table 1. Numerical example: results of nonviscous eigenvalues (rad/s) and eigenvectors.

The rank of these matrices can easily be calculated obtaining

$$r_1 = \text{rank}(\mathbf{C}_1) = 1, \quad r_2 = \text{rank}(\mathbf{C}_2) = 1, \quad r_3 = \text{rank}(\mathbf{C}_3) = 2 \tag{73}$$

The number of nonviscous eigenvalues of this system is $r = r_1 + r_2 + r_3 = 4$. The results of the four nonviscous eigenvalues and eigenvectors are shown in **Table 1**. Exact solutions based on the state-space approach are shown in the first rows. Below, we find the approximated solutions calculated with Lázaro's method using both the first- and the second-order approximation (see Eqs. (62) and (68), respectively). The relative error is also shown below each result (in brackets) for both eigenvalues and eigenvectors. For the latter, the relative error is calculated in terms of the vector norms. Note that in general, the eigenvalues are calculated more accurately than eigenvectors. Indeed, the relative error of the former is one order of magnitude lower than that of the latter. As expected, the second-order approximation improves notably the solution, decreasing the relative errors two or three orders of magnitude respect to those computed from the first-order approximation. In general, since the effect of the nonviscous modes in the response is not relevant, it is justified to use the first-or second-order approximations presented in this text, even for moderately or highly damped vibrating structures [14].

4. Conclusions

In this chapter, the mathematical modeling of damping materials has been presented. These materials are characterized by presenting dissipative forces depending on the history of degrees-of-freedom velocities via exponential kernel functions (or Biot's model). The free-motion vibration of these structural systems leads to a nonlinear eigenvalue problem. There exist two types of eigensolutions: on one hand, the complex eigenmodes, with oscillatory nature and considered as perturbations of the undamped natural modes, on the other hand, the so-called nonviscous modes, overcritically damped modes (without oscillatory nature), characteristic of the type of damping model. These latter modes are the main objective of the research of the present chapter.

The nonviscous modes behind a viscoelastic exponential-damping-based system are closely related to the relaxation parameter of the exponential functions. In general, their influence in the response of the system is several orders of magnitude less important than that of the complex modes. In this paper, we try to summarize some of the most relevant properties of these modes, both from a theoretical and from a numerical point of view. Nonviscous modes for both single and multiple dof systems are studied. For both cases, a necessary condition of nonviscous modes relating to eigenvector, eigenvalue, and dynamic matrices is provided. Additionally, numerical methods to extract nonviscous eigenvalues and eigenvectors, assuming asymmetric and nonproportional dynamic matrices, are reviewed. The results have been compared with a numerical example.

Author details

Mario Lázaro[1*], César F. Casanova[2] and Carlos Lázaro[1]

*Address all correspondence to: malana@mes.upv.es

1 Department of Continuum Mechanics and Theory of Structures, Polytechnic University of Valencia, Valencia, Spain

2 Faculty of Engineering, Piura University, Piura, Peru

References

[1] L. Rayleigh. Theory of Sound. Dover Publications, 2nd edition, London, 1945.

[2] S. Adhikari. Dynamics of non-viscously damped linear systems. *Journal of Engineering Mechanics*, 128(3):328–339, 2002.

[3] L. Boltzmann. Zur theorie der elastischen nachwirkung. *Annalen der Physik*, 241(11): 430–432, 1878.

[4] M.A. Biot. Theory of stress-strain relations in anisotropic viscoelasticity and relaxation phenomena. *Journal of Applied Physics*, 25(11):1385–1391, 1954.

[5] M.A. Biot. Variational principles in irreversible thermodynamics with application to viscoelasticity. *Physical Review*, 97(6):1463–1469, 1955.

[6] W. Flugge. Viscoelasticity. Springer-Verlag, 2nd edition, Berlin Heidelberg,1975.

[7] A.D. Nashif, D.I.G. Jones, and J.P. Henderson. Vibration Damping. John Wiley & Sons, New York, 1985.

[8] D.I.G. Jones. Handbook of Viscoelastic Vibration Damping. John Wiley & Sons, New York, 2001.

[9] T. Pritz. Analysis of four-parameter fractional derivative model of real solid materials. *Journal of Sound and Vibration*, 195(1):103–115, 1996.

[10] Y.A. Rossikhin and M.V. Shitikova. Application of fractional calculus for dynamic problems of solid mechanics: novel trends and recent results. *Applied Mechanics Reviews*, 63(1):010801(1)–010801(52), 2010.

[11] J. Woodhouse. Linear damping models for structural vibration. *Journal of Sound and Vibration*, 215(3):547–569, 1998.

[12] N. Wagner and S. Adhikari. Symmetric state-space method for a class of nonviscously damped systems. *AIAA Journal*, 41(5):951–956, 2003.

[13] S. Adhikari and N. Wagner. Analysis of asymmetric nonviscously damped linear dynamic systems. *Journal of Applied Mechanics (Transactions of ASME)*, 70(6):885–893, 2003.

[14] M. Lázaro. Nonviscous modes of nonproportionally damped viscoelastic systems. *Journal of Applied Mechanics (Transactions of ASME)*, 82(12):Art. 121011 (9pp), 2015.

[15] M. Lázaro and J.L. Pérez-Aparicio. Characterization of real eigenvalues in linear viscoelastic oscillators and the non-viscous set. *Journal of Applied Mechanics (Transactions of ASME)*, 81(2):Art. 021016–(14pp), 2014.

[16] S. Adhikari. Eigenrelations for non-viscously damped systems. *AIAA Journal*, 39(8): 1624–1630, 2001.

[17] J. Woodhouse. Viscous damping identification in linear vibration. *Journal of Sound and Vibration*, 203:475–500, 1998.

[18] S. Adhikari and B. Pascual. Eigenvalues of linear viscoelastic systems. *Journal of Sound and Vibration*, 325(4–5):1000–1011, 2009.

[19] M. Lázaro. *The Eigenvalue Problem in Linear Viscoelastic Structures: New Numerical Approach and the Equivalent Viscous Model*. PhD thesis, Polytechnic University of Valencia. Deparment of Continuum Mechanics & Theory of Structures. Valencia, Spain, 2013.

[20] M. Lázaro, C.F. Casanova, I. Ferrer, and P. Martín. Analysis of nonviscous oscillators based on the damping model perturbation. *Shock and Vibration*, 2016: Art. 368129 (20p), 2016.

[21] S. Adhikari and B. Pascual. Iterative methods for eigenvalues of viscoelastic systems. *Journal of Vibration and Acoustics*, 133(2):021002.1–001002.7, 2011.

[22] S. Adhikari. Classical normal modes in non-viscously damped linear systems. *AIAA Journal*, 39(5):978–980, 2001.

[23] F. Cortés and M.J. Elejabarrieta. Computational methods for complex eigenproblems in finite element analysis of structural systems with viscoelastic damping treatments. *Computer Methods in Applied Mechanics and Engineering*, 195(44–47):6448–6462, 2006.

[24] L. Duigou, E.M. Daya, and M. Potier-Ferry. Iterative algorithms for non-linear eigenvalue problems. Application to vibrations of viscoelastic shells. *Computer Methods in Applied Mechanics and Engineering*, 192(11–12):1323–1335, 2003.

[25] M. Lázaro. Eigensolutions of non-proportionally damped systems based on continuous damping sensitivity. *Journal of Sound and Vibration*, 363(C):532–544, 2016.

[26] M. Lázaro, J.L. Pérez-Aparicio, and M. Epstein. Computation of eigenvalues in proportionally damped viscoelastic structures based on the fixed-point iteration. *Applied Mathematics and Computation*, 219(8):3511–3529, 2012.

[27] M. Lázaro and J.L. Pérez-Aparicio. Multiparametric computation of eigenvalues for linear viscoelastic structures. *Computers & Structures*, 117:67–81, 2013.

Chapter 9

Viscoplastic Behaviour of Polyamides

Şerban Dan-Andrei

Additional information is available at the end of the chapter

http://dx.doi.org/10.5772/64563

Abstract

In this study, the viscoplastic behaviour of a polyamide-based thermoplastic polymer was investigated by performing a number of tests that highlighted the influence of time and temperature on the mechanical behaviour: strain-rate- and temperature-dependency tests, creep tests, dynamic mechanical analysis (DMA) tests, Mullins' effect tests, and low-cycle fatigue tests. The results are discussed and explanations are proposed regarding the particularities the investigated material exhibits during deformation.

Keywords: viscoplasticity, polyamide, thermoplastic polymers

1. Introduction

When designing assemblies for engineering applications, the material characteristics of various components are often considered to be characterized only by simple laws, such as linear elasticity and plasticity for solids and viscosity for fluids. In reality, all materials exhibit both elasticity and viscosity during deformation [1]. The elastic behaviour of fluids is often characterized by the bulk modulus, a measure of its stiffness during hydrostatic compression [2]. Viscous effects in solids can be observed in various loading scenarios, the most common being strain-rate dependency, temperature dependency and creep [3]. Although rarely taken into account for metals or concretes, viscous effects play a major part in defining the mechanical response of polymers [1, 4–6]. Their characteristic structure of long, covalently bonded chains of atoms, with very high molecular weights, is responsible for the mechanisms of inducing temporary damage, such as viscous flow, bond interchange, Thirion relaxation, etc. [2].

Throughout human history, polymers were used for a great variety of purposes. Natural polymers, such as cotton, silk, wool and resins, were employed for various household purposes as far back as 6000 years ago [5]. The extensive use of synthetic polymers began in the 1940s. Their advantages over conventional materials (metals, wood, textiles, etc.), namely low specific mass for fairly good mechanical properties, low production costs and good insulation properties, determined an exponential increase in the number of applications. For the past 20 years, the total production of polymers exceeded the combined metal production volume wise [5].

The unique time- and temperature-dependent properties of rubbers were observed and described by various scientists of the nineteenth century. Ludwig Boltzmann (the first to coin the term viscoelasticity), James Clerk Maxwell, Sir William Thompson and Lord Kelvin were the first to provide mathematical models in describing viscoelasticity [5]. Today, the basic models developed by the aforementioned scientist (the Maxwell fluid, the Kelvin solid coupled with the Boltzmann superposition principle) represent the backbone of various advanced models used in describing the mechanical behaviour of viscoelastic and viscoplastic materials [5, 6].

Breaking down the behaviour of viscoelastic materials into the solid-like and fluid-like response, several constitutive formulations can be considered. The elastic part is usually described by Hookean elasticity for brittle polymers (linear elastic response) and by various non-linear models for more compliant materials [5, 6]. Some thermosets exhibit yielding before fracture, an accurate description of their behaviour requiring the input of plasticity [2, 4]. On the other hand, thermoplastics exhibit a linear stress-strain response over a very short interval (under 1% deformation) and no clear yield point (the stress-strain curve gradually decreases its slope over a given strain interval) [1, 7, 8]. The stress-strain response of rubbers has a different shape (the characteristic 'S' curve) and can only be accurately modelled with the use of hyperelastic models, such as the Neo-Hookean formulation [9], Yeoh formulation [10], the Ogden formulation [11], the Arruda-Boyce formulation [12] or the implicit Marlow formulation [13]. These characteristics are presented in **Figure 1a**.

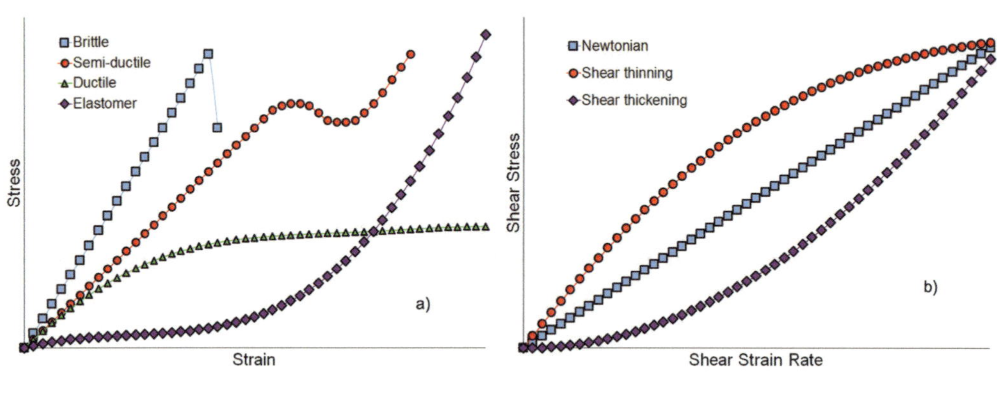

Figure 1. Mechanical behaviour of polymers: types of stress-strain response (a) and types of viscous flow (b).

Regarding the flow component, three models are usually employed to describe viscosity: the Newtonian fluid is the basic model in describing flow, where the shear stress increases linearly with the shear strain rate (i.e. water). For other types of fluids, the viscosity increases exponentially with the increase in strain rate (i.e. honey, quicksand, oobleck). They are called shear-thickening fluids or dilatants. Fluids that exhibit a logarithmic increase in shear stress with shear strain rate are called shear-thinning fluids or pseudoplastic fluids (i.e. blood, paint, ketchup).

In modelling viscoelastic materials, a common approach is the use of mathematical models that incorporate both elastic and viscous elements. The aforementioned Maxwell and Kelvin models used two components, namely an elastic element (spring with an elastic modulus E) and a damping element (dashpot with a damping coefficient η), connected in series (Maxwell fluid, **Figure 2a**) or in parallel (Kelvin solid, **Figure 2b**). Note that the elastic element can undergo both normal and tangential loadings along all directions. More advanced models increase the number of elements in order to capture the behaviour of the viscoelastic material more accurately. Examples of such models are the three-parameter fluid, the four-parameter solid (**Figure 2c**), the generalized Maxwell fluid or the generalized Kelvin solid (models which imply the use of a number of Maxwell and Kelvin elements connected in parallel or in series, respectively) [5, 6].

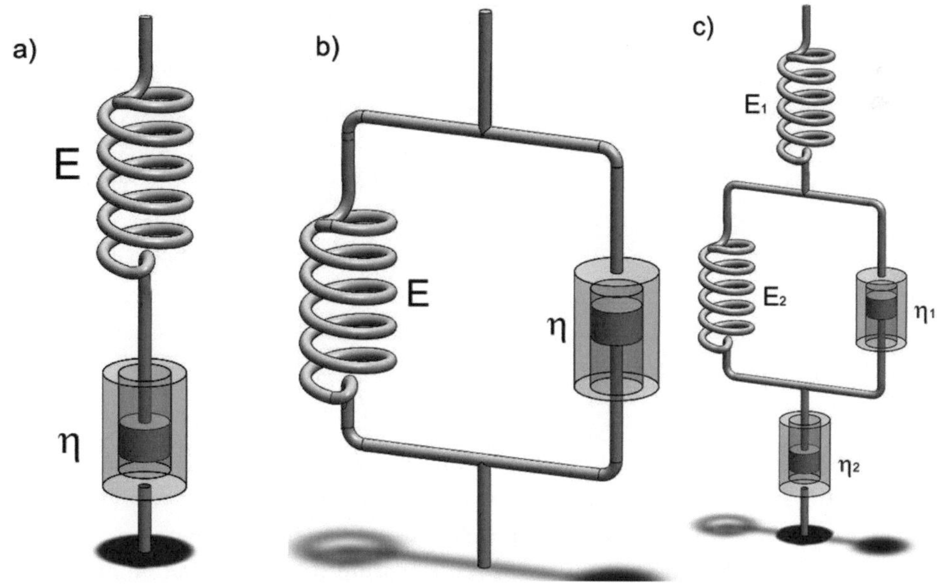

Figure 2. Mathematical models for the Maxwell fluid (a), Kelvin solid (b) and four-parameter solid (c).

For a higher degree of accuracy, the linear elastic response of the elastic element can be replaced with hyperelasticity, and the Newtonian fluid characteristic of the dashpot can be replaced with either a shear-thickening or a shear-thinning model. Viscoplasticity can be modelled by introducing a rate-independent flow component ('slip' element) in the presented mathematical models.

Other methods of modelling viscoplasticity consider a phenomenological approach. A notable example is the work of A.D. Drozdov and collaborators. Their constitutive models for various thermoplastics and elastomers consider the total deformation being composed of a sliding part and an elastic part and the stress-strain relation being governed by a strain-energy density function. The resulting constants and parameters are eventually fitted with the experimental data [14–16].

Considering the loading conditions, the same polymer can exhibit different behaviours: it can be characterized by either a glassy state (stiffer response, sometimes brittle) or a ductile state (a compliant response, yielding before fracture). The transition region between the two states is called the glass transition temperature (GTT) [3–6]. It is usually determined in temperature-sweep dynamic mechanical analysis (DMA) tests and can be extrapolated for strain rates using the time-temperature superposition.

One of the most notable examples of misestimating the mechanical properties of polymers is the case of the Challenger Space Shuttle disaster (28 January, 1986). The Presidential commission tasked with investigating the accident concluded that, due to the low temperatures recorded during the space shuttle launch, the rubber O-rings sealing the aft field joint, on the right solid rocket booster, became less resilient and ultimately failed. This allowed the pressurized hot gasses to make contact with the external tank, causing structural failure [17].

The investigated material of this study is a polyamide-based (PA) semi-crystalline thermoplastic polymer that was developed for low-temperature applications. Being established as the material used in manufacturing various components, it was observed that the given component's response to external loadings differed from the initial behaviour, after a certain period of service. Limited data were provided by the material supplier, so an extensive experimental plan was set up in order to determine the time and temperature dependency of the polyamide.

2. Experimental investigations

Considering the in-service conditions that the components manufactured from this compound are subjected to, an experimental plan was devised in order to capture the variation in mechanical properties with test parameters such as time (rate dependency and creep), temperature and humidity, as well as for several loading conditions (monotone and cyclic tests).

2.1. Monotone tests

2.1.1. Time dependency

The studied polyamide's variation of the mechanical properties with time was investigated for short-term (strain-rate-dependent tensile tests) and long-term scenarios (creep tests).

2.1.1.1. Strain-rate-dependency tests

Tensile tests represent the most facile experimental investigations used for the determination of mechanical properties of materials. With the evolution of universal testing machines over the past decades, materials of any class can be easily characterized in terms of their stress-strain behaviour, stiffness, strength and yield point.

For this study, strain-rate-dependent tests were performed on injected dumb-bell specimens, in accordance with ISO 527 [18]. The strain rates were chosen on a logarithmical scale, corresponding to the crosshead travel values suggested by the test standard: 0.00028, 0.0028, 0.028, 0.28 and 2.8 s^{-1} (corresponding for crosshead travel speeds ranging from 2 up to 20,000 mm/min).

Quasi-static tests were performed on two test systems: a ball-lead screw-driven electromechanical Zwick Z250 machine (for strain rates from 0.00028 up to 0.028 s^{-1}), equipped with a 10-kN load cell and on a servo-hydraulic Schenk PC63M (for a strain rate of 0.28 s^{-1}), equipped with a 40-kN load cell. In both cases, the strains were being recorded with an incremental extensometer on a gauge length of 50 mm.

The dynamic tests were performed on a servo-hydraulic Instron VHS 160/20 machine, equipped with a Kistler 9017A 400-kN piezoelectric load washer, mechanically preset at 200 kN. Strains were recorded through digital image correlation, using a high-speed camera at 190,000 frames per second, the images being processed using the ARAMIS system by GOM. Higher strain rates were also investigated but no clear results could be obtained due to system ringing, caused by the propagation of stress waves through the testing equipment [8, 19, 20].

The stress-strain curves for the strain-rate-dependent tests are presented in **Figure 3** [8]. The variation in elastic modulus and tensile strength with strain rate is presented in **Figure 4**. The elastic modulus was calculated using the recommendations from ISO 527 [18].

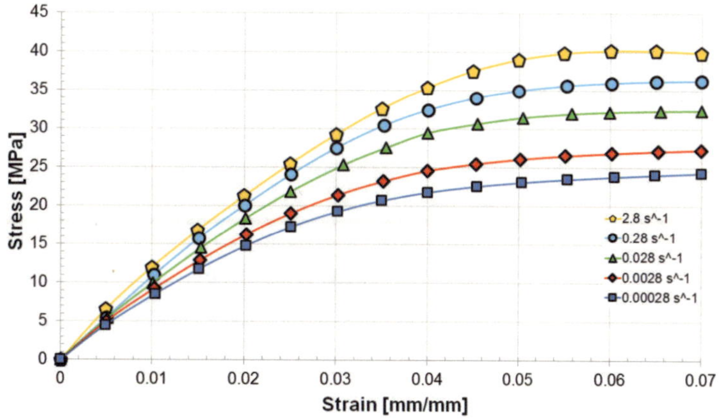

Figure 3. Stress-strain curves for strain-rate-dependency tests.

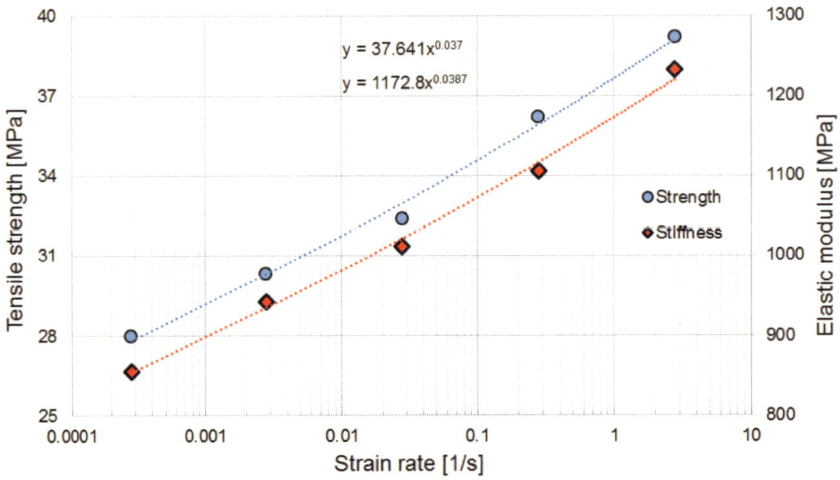

Figure 4. Variation of elastic modulus and tensile strength with strain rate.

Figure 4 shows that a correlation between the increase in tensile strength and stiffness with strain rate can be observed, both trend lines having similar exponents.

Regarding the stress-strain behaviour, **Figure 3** shows that the yield strain decreases with an increase in strain rate. Tests performed at low speeds, when the viscous effects become more significant, determine no clear yield point. By analysing the shape of the stress-strain curves, it can be observed that the linear region is very small, almost unnoticeable. An analysis of the variation of stiffness with strain, for the investigated strain rates, is presented in **Figure 5**.

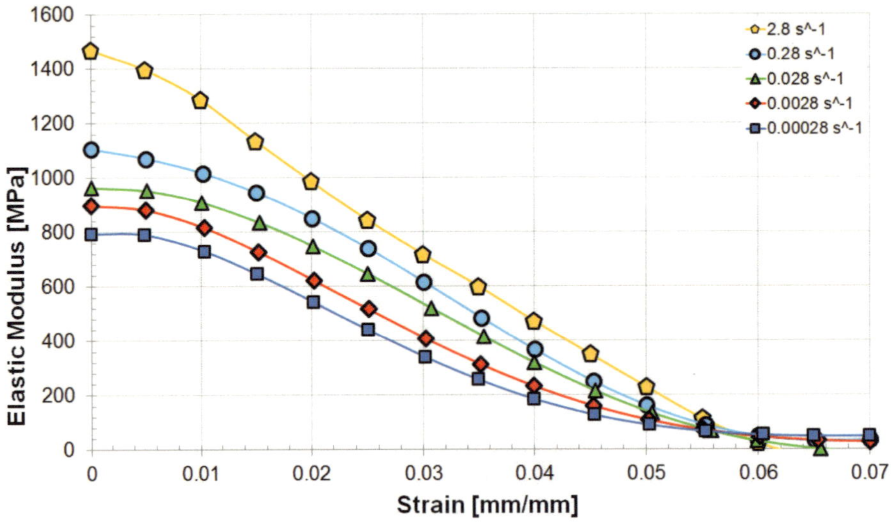

Figure 5. Variation of stiffness with strain, for the investigated strain rates.

2.1.1.2. Creep tests

Creep tests investigated the long-term influence of time on the mechanical properties of the polyamide, and were performed on a 5-kN Zwick/Roel machine. Three loading levels were used, corresponding to 30, 60 and 90% of the estimated yield stress [21], maintained for a period of 24 h. The normalized compliance is calculated with Eq. (1), test result being presented in **Figure 6**.

$$d(t)=\frac{D(t)}{D(0)}=\frac{\varepsilon(t)}{\sigma(0)}\cdot\frac{\sigma(0)}{\varepsilon(0)}=\frac{\varepsilon(t)}{\varepsilon(0)} \tag{1}$$

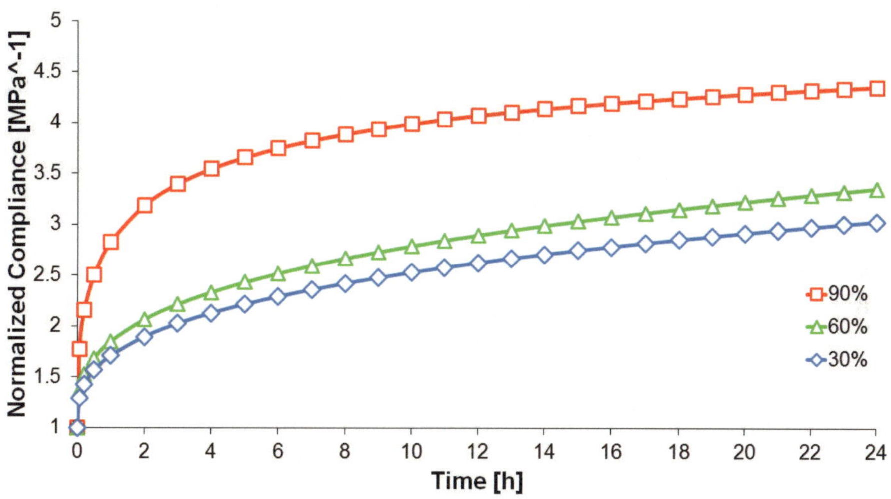

Figure 6. Creep tests at 30, 60 and 90% tensile strength.

A drastic increase in tensile compliance can be observed in the first hour of each test, after which the increase becomes gradually smaller. Given the non-linear behaviour of the material, the values for compliances differ for each level of loading, higher loads determining higher compliances.

Comparing the variation of the normalized tensile compliance with time, between the tests performed at high deformations (90% of the yield stress) and low deformations (60 and 30% of the yield strain), a significant difference is noticed. Although increased stress levels determine higher compliances, there is no linear correlation between the variations of the two parameters, but rather an exponential one. The observed phenomenon is caused by strain softening of the material. As the total strain grows towards the yield point, the role of the viscous and plastic components becomes significant. Given the fact that no clear plateau was recorded, one can conclude that the viscous strain did not reach saturation after 24 h of constant loading.

2.1.2. Temperature-dependency tests

The tests were performed with a crosshead travel speed of 20 mm/min on a ball-lead screw-driven electromechanical Zwick Z250 machine equipped with an environmental chamber. Strains were recorded with an incremental extensometer. The stress-strain curves are presented in **Figure 7** and the variation of tensile strength and elastic modulus with temperature is presented in **Figure 8**.

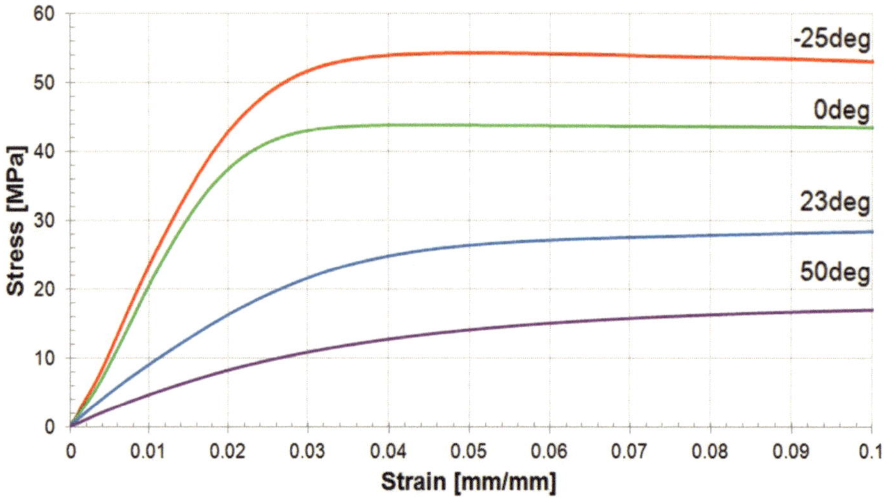

Figure 7. Stress-strain curves for temperature-dependency tests.

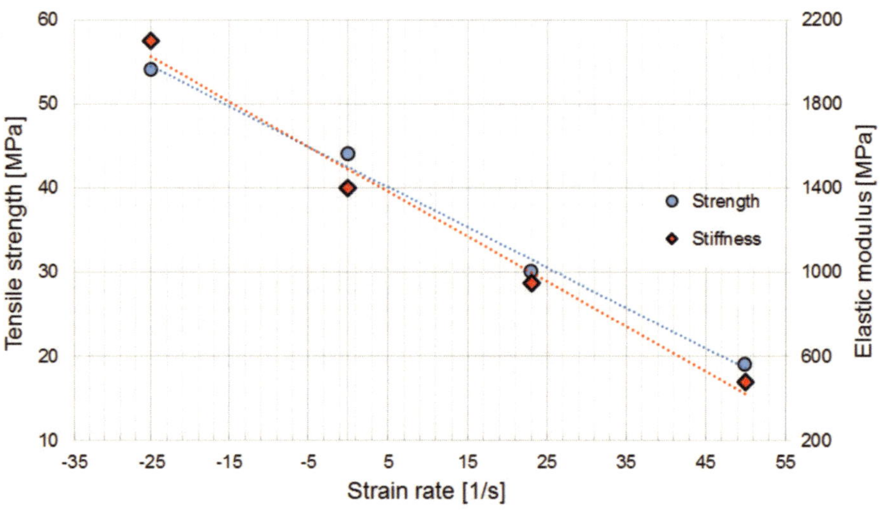

Figure 8. Variation of elastic modulus and tensile strength with temperature.

The same type of non-linearity as with strain-rate-dependent tests is observed; the instantaneous elastic modulus does not remain constant and varies with strain (**Figure 9**). For the low-temperatures tests, a toe region can be observed: the maximum elastic modulus does not coincide with the initial value, which occurs around 0.007-mm/mm deformation. After an abrupt decrease, the instantaneous modulus reaches values close to 0, for the plateau region (around 0.035 mm/mm strain). For higher temperatures, the variation is characterized by a uniform descent until the yield point, similar to ambient temperature tests.

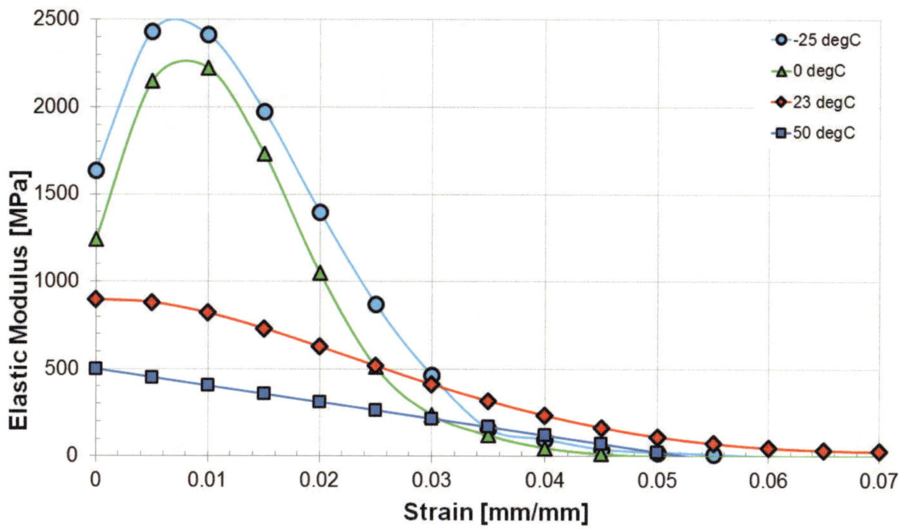

Figure 9. Variation of stiffness with strain for the investigated temperatures.

2.2. Cyclic tests

A more comprehensive look at the viscoplastic properties of the studied polyamide was obtained by performing cyclic tests, namely DMA tests, Mullins' effect tests and low-cycle fatigue (LCF) tests.

2.2.1. DMA tests

In this study, two types of DMA tests were performed. First tests consisted of temperature-sweep low-strain investigations (~0.1% strain), performed in single cantilever on a 'conventional' TA Instruments Q800 DMA machine. The second tests consisted of large-strain DMA tests (1.5–4.5% deformation), performed in tension on an electromagnetic Instron Electropuls machine.

The temperature sweep tests determined a glass transition temperature of 38.5°C at 1 Hz (**Figure 10**) and of 41.2°C at 10 Hz, which explains the large variation in stiffness and strength observed during the temperature-dependency tensile tests.

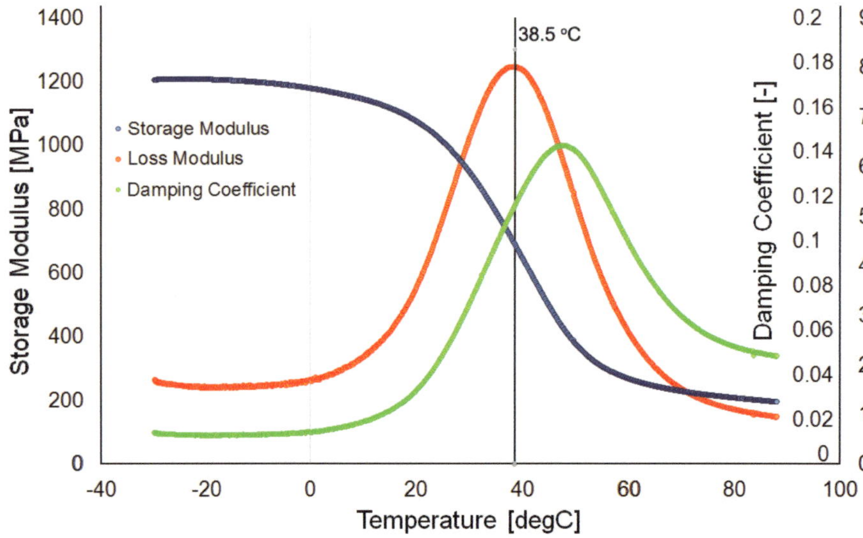

Figure 10. DMA results for a temperature sweep test performed at 1 Hz in single cantilever.

Considering that usual component undergoes higher deformations during service and that the instantaneous elastic modulus varies with strain (**Figure 5**), large-strain DMA tests were carried out in order to record the variation of the dynamic moduli with frequency and with amplitude.

For all the investigated frequencies and amplitudes, a similar trend was observed: the storage modulus increased while the damping coefficient and the loss modulus decreased with frequency (**Figure 11a**). When investigating the moduli variation with amplitude, it can be noted that the elastic potential decreases with amplitude. The loss modulus and damping coefficient increase with the amplitude until it reaches values of around 1.6 mm (equivalent of 4.3% total strain), when, due to the contribution of the remnant strain (plastic and viscous strains), the two parameters plateau (**Figure 11b**).

2.2.2. Mullins' effect tests

The cyclic-induced-softening behaviour of polyamides is determined by various deformation mechanisms that can occur. In literature, the non-recovered deformation observed when the loading stops is usually referred to as 'plastic strain' or 'permanent strain' [4, 5]. However, due to the viscous effects caused by the polyamide's structure, a part of that remnant deformation is recovered through relaxation (the stored potential energy causes some macromolecules to regain their initial position). The irreversible part of the deformation is caused by the slippage of the atomic planes in the crystalline phase (similar to metal plasticity [22, 23]) or by macro-molecular chain scission [1].

In consequence, the semi-crystalline structure of the polyamide causes a specific behaviour during deformation, the strain being divided into three categories: instantaneous elasticity (recovered immediately after the loadings stop), delayed elasticity (recovered after a certain time) and permanent deformation (caused by the irreversible change in structure). If the material is subjected to a constant load below the yield stress, the main deformation component that will be observed is the instantaneous elastic strain. If the loading is maintained for a longer time (creep tests), an increase in the total deformation is observed, due to the accumulation of viscous strain. If the specimen is strained towards yielding and beyond, the permanent deformation starts to become significant while the reversible strain begins to plateau [2, 7].

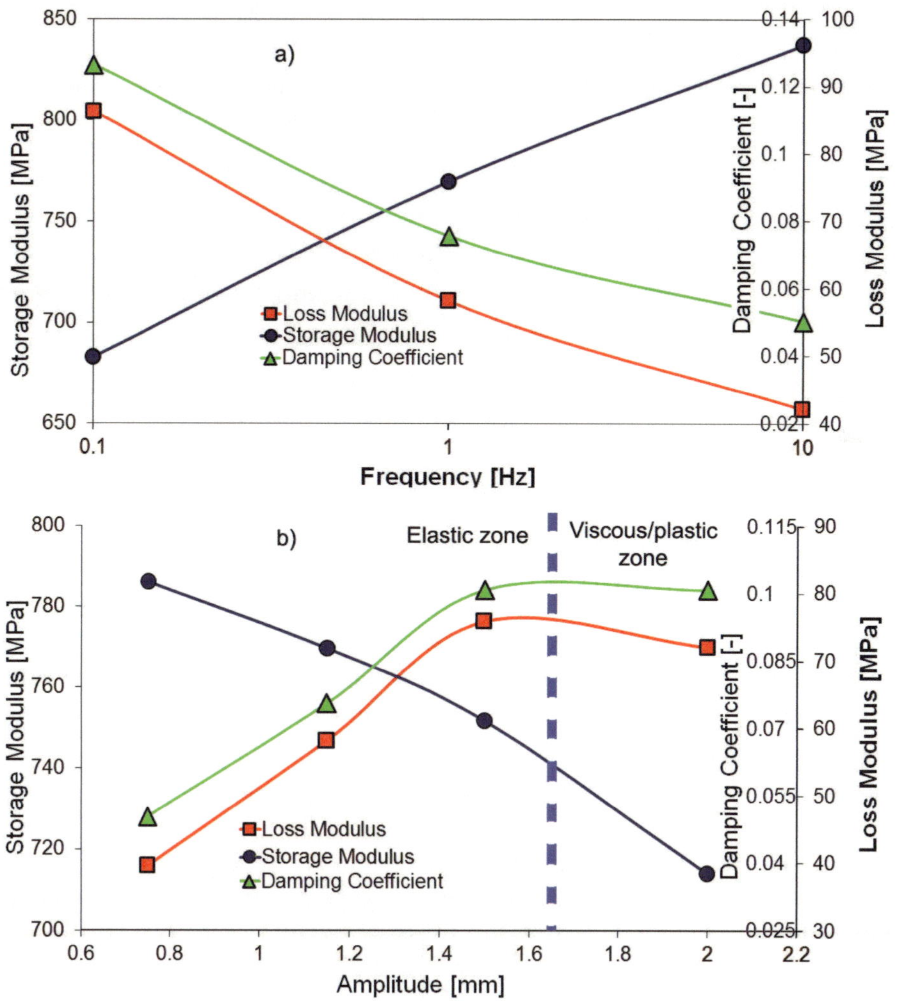

Figure 11. Results for DMA tests performed at high strains for frequency variation (a) and amplitude variation (b).

The same scenario is observed for cyclic loadings. **Figure 12** presents the strain decomposition for a sine loading. The total strain ε^{tot} is the sum of the instantaneous elastic strain ε^{el}, the viscous strain ε^{vsc} and the plastic strain ε^{pl}. In this example, the cyclic loadings are performed in strain control, thus ε^{tot} presents a constant sine variation in each cycle. The plastic strain ε^{pl} reaches a maximal value with the peak of the cycle while the viscous strain ε^{vsc} presents a steady increase until it reaches saturation (this point being considered the steady-state behaviour in cyclic loadings). The instantaneous elastic strain ε^{el} is obtained by subtracting ε^{pl} and ε^{vsc} from ε^{tot}.

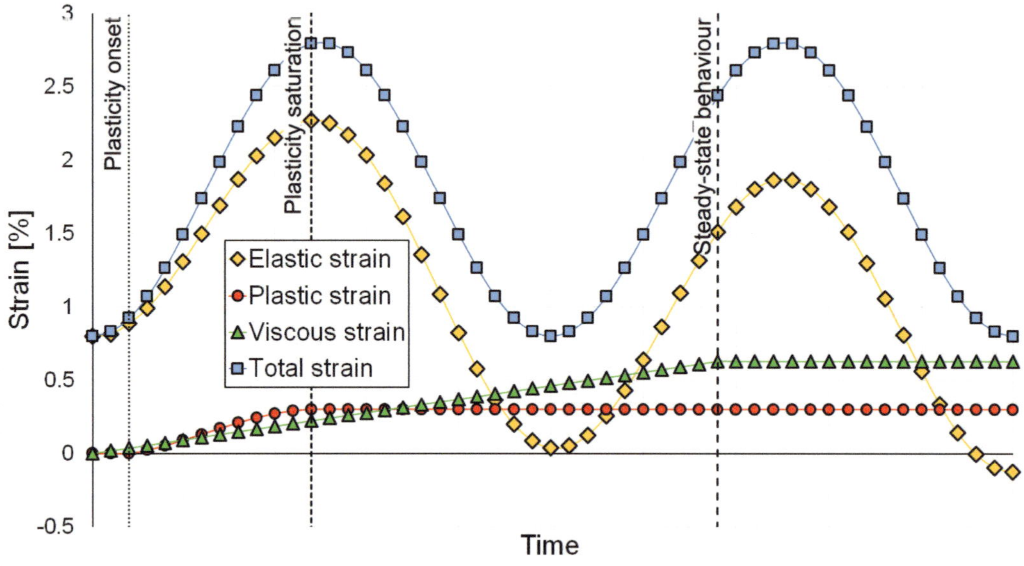

Figure 12. Strain decomposition for a tensile cyclic test on a viscoelastic material.

Mullins' effect tests were initially developed for rubber testing but were implemented for other types of polymers such as thermoplastic elastomers and semi-crystalline thermoplastic polymers [24–26]. Three main material characteristics are distinguishable for such tests: hysteresis of energy, strain softening and strain hardening [27–29].

The graphic representation depicting the first loading/unloading cycle of each amplitude is called the primary hyperelastic behaviour; the material's loading path exhibits strain hardening when loaded to stress values higher than the current cycle. If the material is subjected to several cycles with the same amplitude, it exhibits a strain-softening phenomenon, caused by the accumulation of viscous strain, as described above. After a certain number of cycles of the same amplitude, the behaviour stabilizes to a stress-strain curve of the same shape, due to the saturation of viscous strain, called the permanent set [30, 31].

The first set of Mullins' effect tests was performed on a 10-kN Walter + Bai servo-hydraulic fatigue machine. The experimental programme consisted of three blocks of different amplitudes with 10 cycles being performed in each block (**Figure 13a**). The machine was set in displacement control, with an equivalent crosshead travel of 25, 50 and 75% of the approximated yield stress. The test speed was held constant at 2 mm/min, results being presented in **Figure 13b** [31].

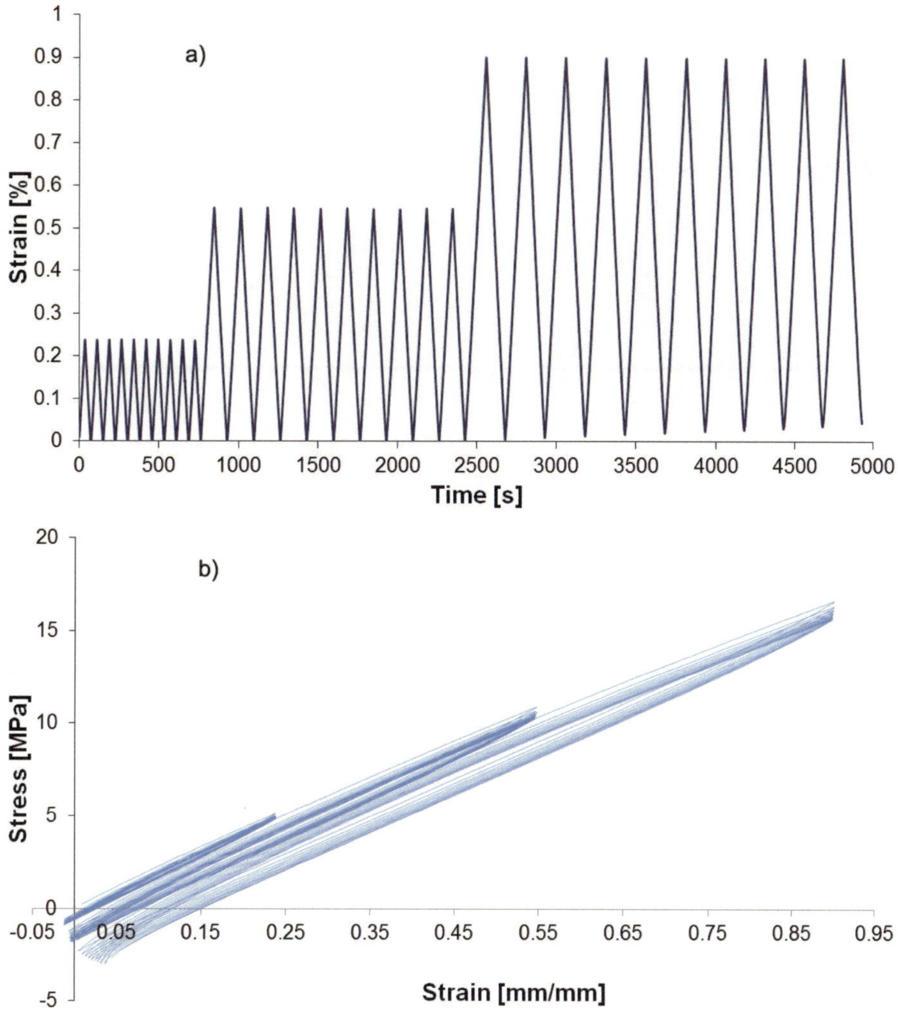

Figure 13. Program set-up (a) and experimental results (b) for the Mullins' effect test.

Considering the results presented in **Figure 13**, for Mullins' effect tests performed below the yield point, the investigated polyamide exhibits energy dissipation (hysteresis) and strain softening. In order to gain insight about the strain hardening, cyclic tests that strain the material beyond the yield point had to be performed. Two such cyclic test protocols were designed, one having equal stress increments while the other having equal strain increments.

Mullins' effect tests with equal stress increment were performed on a Zwick Z250 machine that was programmed in stress control, strains being recorded with an incremental extensometer. The cyclic programme is presented in **Figure 14a**, and the test results are presented in **Figure 14b**. Having more cycles before the yield region, these tests gave insight about the strain decomposition in the elastic domain.

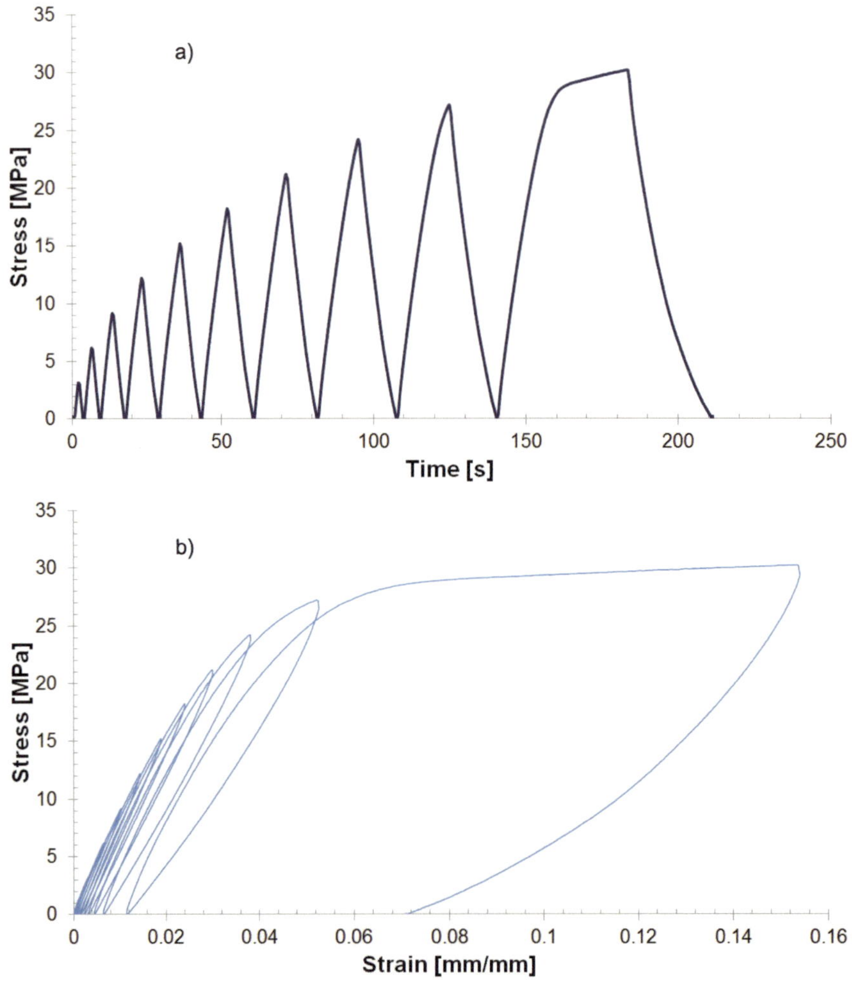

Figure 14. Program set-up (a) and experimental results (b) for the Mullins' effect test with equal stress increment.

Mullins' effect tests with equal strain increment were performed on the same Zwick Z250 machine, the machine being programmed in strain control. The test protocol is presented in **Figure 15a**, and the results are presented in **Figure 15b**. The strains programmed in these experiments went beyond the yield region, showing the material behaviour in the plastic domain. Considering the strain decomposition using cyclic tests, even though the tests deformed the specimens well into the plastic region, the elastic strain did not reach a plateau and kept increasing at a decelerated rate due to the delayed elasticity (**Figure 16**). These viscous effects can be noticed when unloading to zero stress, the material being relaxed and some of the energy consumed in deforming the specimen got recovered (the non-linear unloading path). It was also noticed that, similar to rubbers and elastomers, the studied polyamide exhibits strain hardening as well (the stress-strain response of the specimen tested in cyclic loadings resumes the path described by a virgin specimen, **Figure 17**).

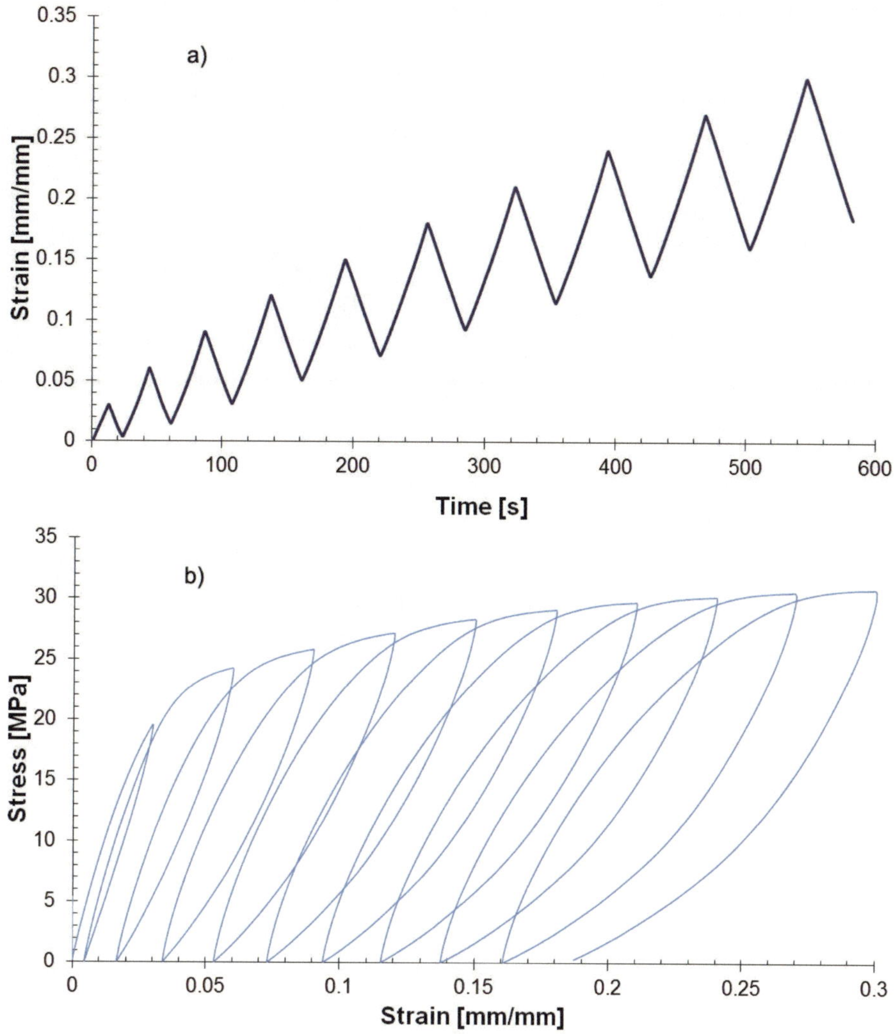

Figure 15. Program set-up (a) and experimental results (b) for the Mullins' effect test with equal stress increment.

2.2.3. *Low-cycle fatigue tests*

A better insight into the behaviour of the investigated polyamide in cyclic loadings was given through low-cycle fatigue tests. The experiments were performed in tension and strain control, with an *R*-ratio of 0.1. A 40-kN Schenk PC63M machine was used and was equipped with an environmental chamber, set at 23°C (to negate the effects of temperature variation) and the strains were recorded with a strain-gauge extensometer.

The influence of three parameters on the fatigue behaviour of the material was investigated: number of cycles, frequency and strain level (each parameter was varied in three steps). The test protocol also implied the investigation of the material recovery and permanent damage, tensile tests being performed on fatigued specimens immediately after tests as well as after a recovery time of 24 h.

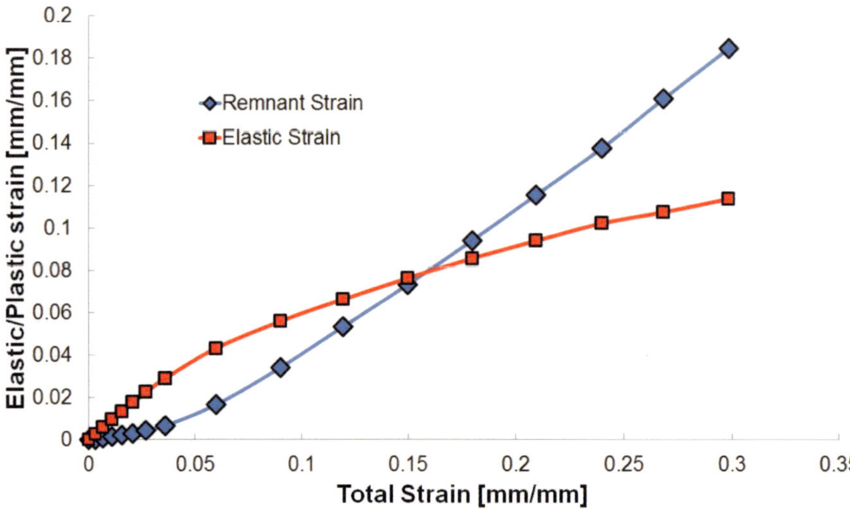

Figure 16. Strain decomposition using Mullins' effect tests.

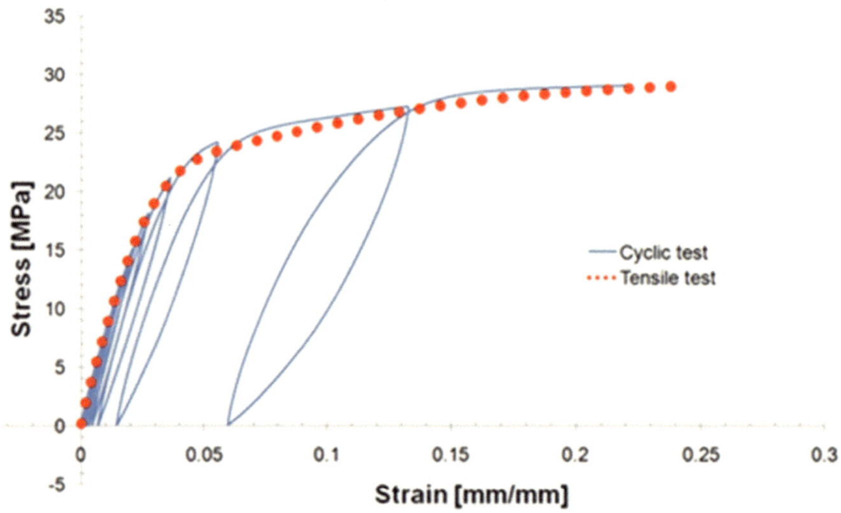

Figure 17. Strain hardening of the PA compound during cyclic loadings.

2.2.3.1. Effect of the number of cycles

Previous cyclic tests at constant amplitude performed below the yield point of the investigated material showed that remnant deformation is accumulated with each cycle, having a logarithmic increase (due to the accumulation of viscous strain, as discussed above). This study aimed to determine after how many loading cycles the polyamide's viscous strain reaches saturation. Three values for the number of cycles were chosen: 5000 cycles (a), 10,000 cycles (b) and 50,000 (c). Each test was divided in five equal steps (blocks of 1000 cycles for (a), 2000 cycles for (b) and 10,000 cycles for (c)), the stress-strain response being recorded after each step. The frequency (5 Hz) and amplitude (0.0075 mm/mm with a pre-strain of 0.025 mm/mm) were maintained constant. **Figure 18** presents the variation in normalized tensile strength (peak stress value recorded after each loading block, divided by the initial value) with the number of cycles.

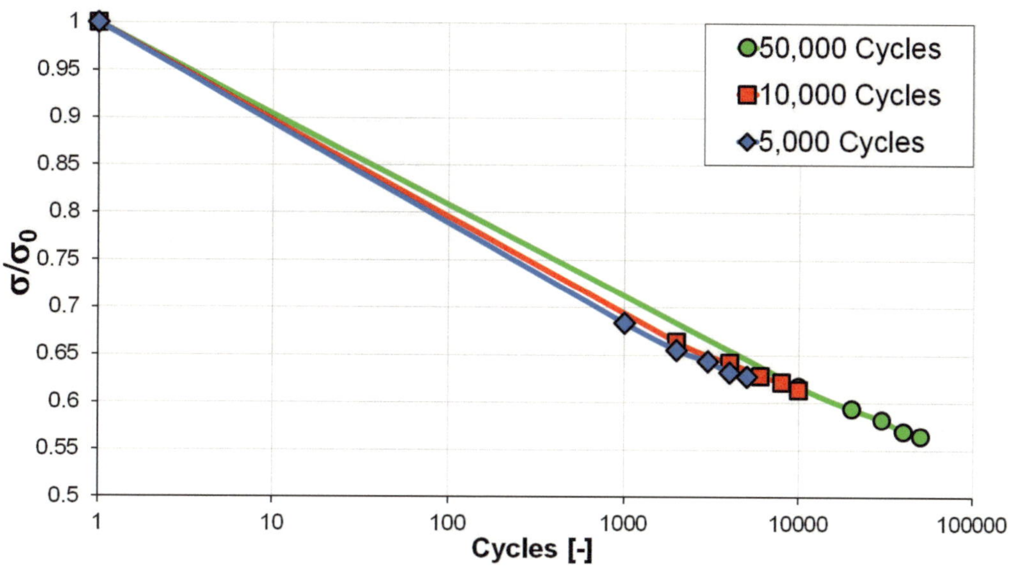

Figure 18. Variation of the normalized stress with the number of cycles.

As with the Mullins' effect tests, the LCF results show a considerable decrease in tensile strength in the beginning of the tests, followed by a steadier decrease as the number of cycles grew, the tensile stress number of cycle curves showing a linear variation on the semi-logarithmical scale.

The results of the tensile tests are presented in **Figure 19**. The behaviour of three specimens was plotted: a specimen subjected to LCF, tested immediately after the tests (denoted 'Instant'), a specimen rested for 24 h after it was subjected to LCF (denoted 'Recovery') and that of a virgin specimen (denoted 'Initial'). A negligible variation in the stress-strain response is observed, consisting of a lower stiffness up to 0.04-mm/mm deformation for the fatigued specimens tested immediately after LCF [32].

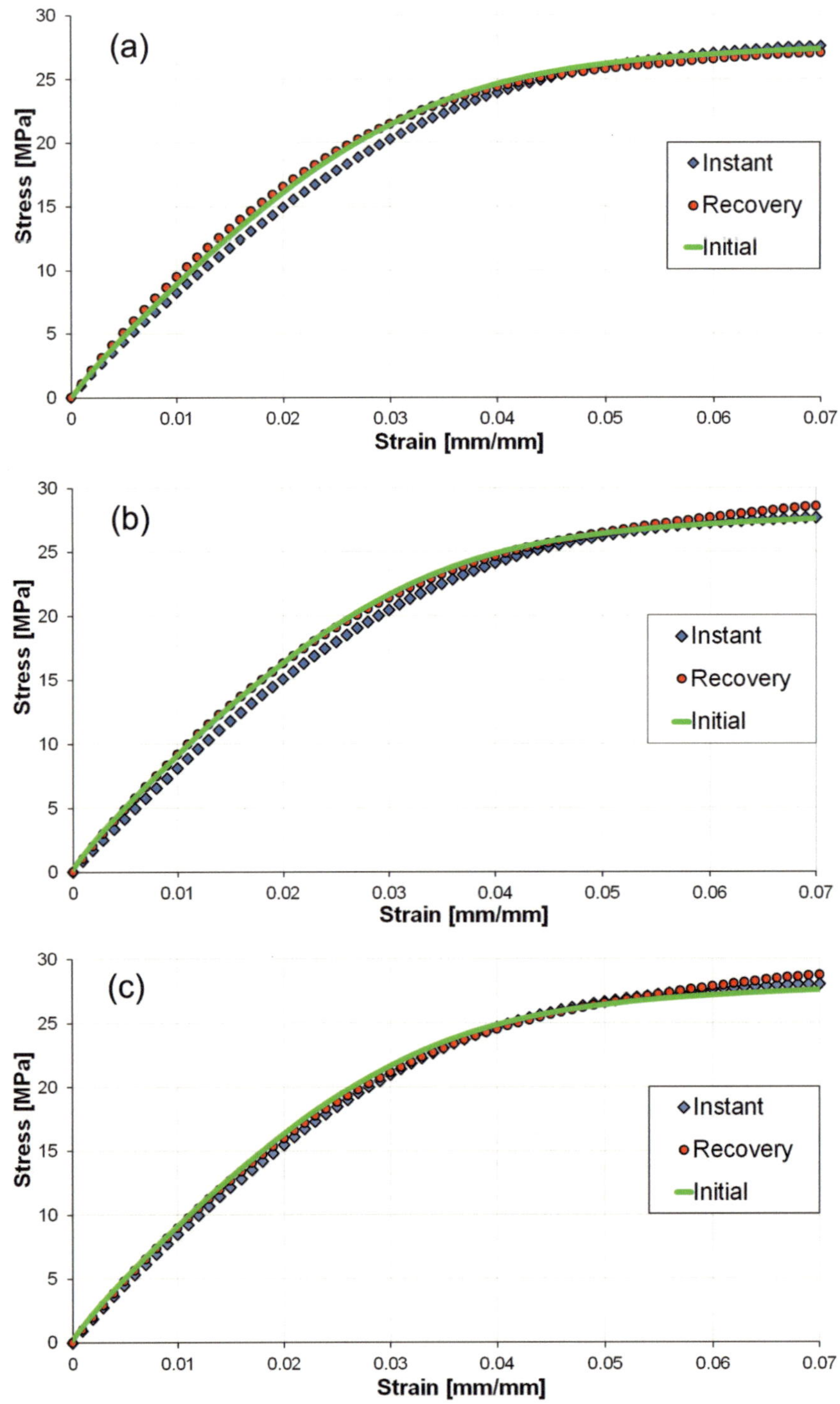

Figure 19. Tensile test comparison for (a) 5000 cycles, (b) 10,000 cycles and (c) 50,000 cycles.

2.2.3.2. Effect of frequency

The effect of strain rate on the mechanical behaviour of the studied polyamide in LCF was studied by the mean of frequency variation. Three values were chosen for this study: 3, 5 and 7 Hz. The strain levels were maintained constant and identical with the previous programme (0.025-mm/mm reference with 0.0075-mm/mm amplitude) and the total number of cycles was 5000. The results for the variation in normalized tensile strength of the cycles with the number of cycles for the three chosen frequencies are presented in **Figure 20**.

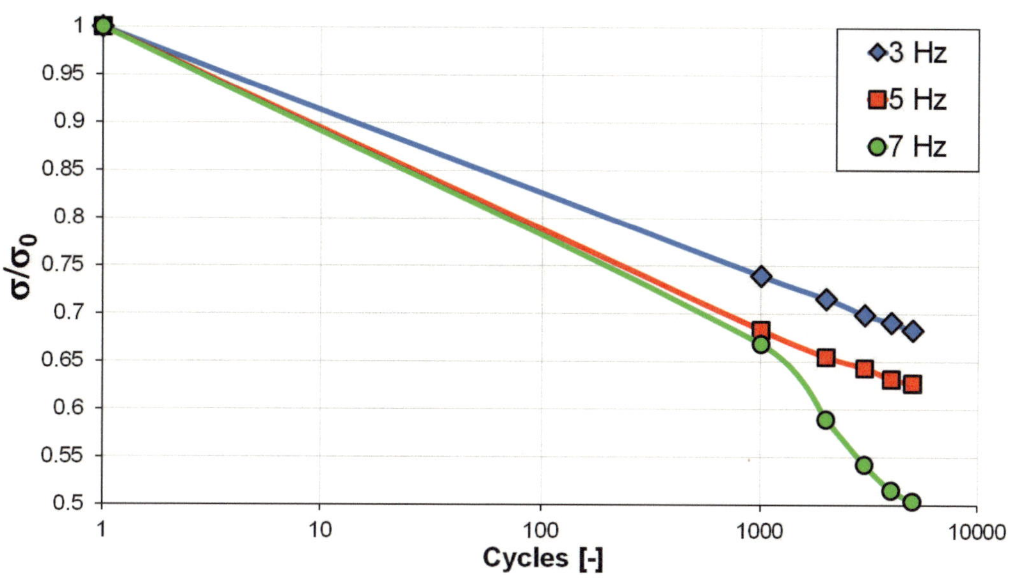

Figure 20. Normalized stress variation with the number of cycles for three different frequencies.

As with previous tests, higher strain rates determine higher reactions. Considering the normalized values, frequency affects the fatigue behaviour of the material: a linear behaviour (on the semi-logarithmic scale) is observed for lower strain rates (3 and 5 Hz), the apparent softening being proportionate with frequency. For the tests performed at 7 Hz, the normalized strength number of cycles behaviour is non-linear in the semi-logarithmic scale as the decrease in strength is more accentuated as the cycle count increases. This is due to the same mechanism responsible for shifting of the yield region observed in strain-rate and temperature-dependency tests (**Figures 3** and **7**): the viscous flow is being hindered and the effects of plasticity are more pronounced.

Tensile tests performed on previously untested specimens, specimens tested immediately after LCF tests and specimens tested 24 h after LCF tests show similar features as the results for the effect of number of cycles in terms of elasticity, but showing higher strengths at 0.07-mm/mm deformation. Results are presented in **Figure 21** [32].

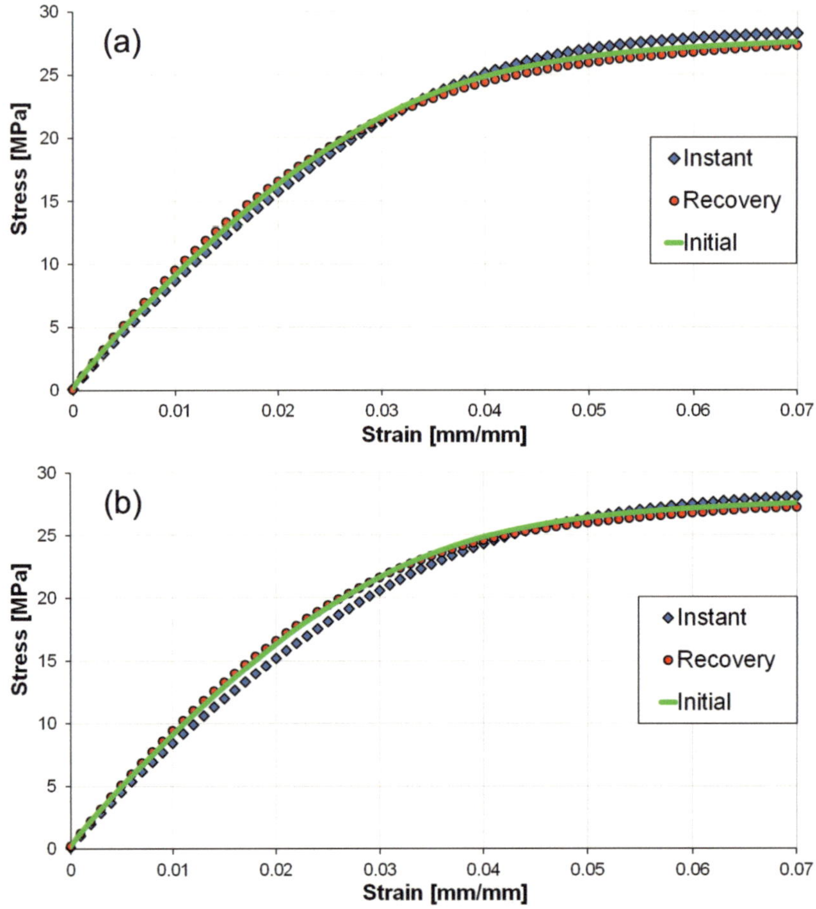

Figure 21. Tensile test comparison: 3 (a) and 7 Hz (b).

2.2.3.3. Effect of strain

As seen with previous tests and depicted in **Figure 12**, the remnant deformation accumulates with the increase in strain. The effect of the straining level in cyclic loadings on the fatigue behaviour of the polyamide was investigated by choosing three pre-strains (0.025, 0.035 and 0.045 mm/mm) and maintaining the same amplitude (0.0075 mm/mm), same frequency (5 Hz) and the same number of cycles (5000). The variation of the normalized strength with the number of cycles is presented in **Figure 22**.

The tensile test results performed on the same three types of specimens are presented in **Figure 23**, showing a slight lower response between 0.01- and 0.05-mm/mm deformation [32].

Figure 22 shows that the reference strain level has a small influence on the normalized stress variation with the number of cycles, showing linear variation on the semi-logarithmic scale and lower values of the normalized stress for higher strains. **Figure 24** explains this phenomenon: as the material is strained to higher values, its remnant deformation increases. At a reference strain of 0.025 mm/mm (denoted with a vertical red line in **Figure 24**), the remnant strain values are around 0.007 mm/mm. At 0.045-mm/mm reference strain (vertical red line in **Figure 24**), the remnant strains reach values of 0.018 mm/mm. The higher remnant strains accumulate and determine a drop in the tensile strength of each cycle.

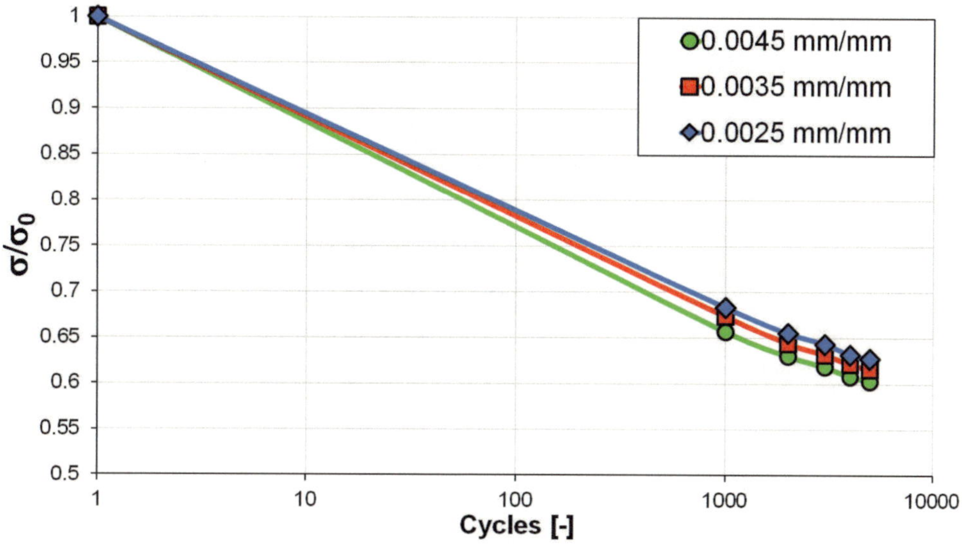

Figure 22. Normalized stress variation with reference strain.

In order to acquire a more in-depth analysis of the LCF behaviour, a study of the energy dissipation was performed, analysing specific stress-strain cycles taken from each recording step, from all tests. The dissipated energy in each cycle was evaluated by calculating the area enclosed by the hysteresis described by the loading-unloading stress-strain curve.

For the tests that investigated the influence of the number of cycles, it was observed that the dissipated energy decreases linearly on the semi-logarithmic graph (**Figure 25a**) until 10,000 cycles, after which the steep of the curve decreases.

A different trend was observed in the case of frequency-influence tests. Experiments performed at 3 and 5 Hz show a linear decrease in dissipated energy with the number of cycles while those performed at 7 Hz show an increase (**Figure 25b**). This is caused by the viscous characteristic of the material. At lower strain rates, viscous flow is more pronounced (as the macromolecules have the time to slip past each other [1, 5, 6]) and more energy is dissipated through slip (for the tests performed at 3 Hz, the values for the dissipated energy are almost triple in comparison to the tests at 5 Hz). At higher strain rates, the material 'freezes' and less energy is dissipated in the beginning of the tests. As the number of cycles grows, at low strain rates, viscous strain reaches saturation (a decrease in dissipated energy) while remnant strain accumulates at high strain rates (due to plastic strain accumulation), hence an increase in dissipated energy.

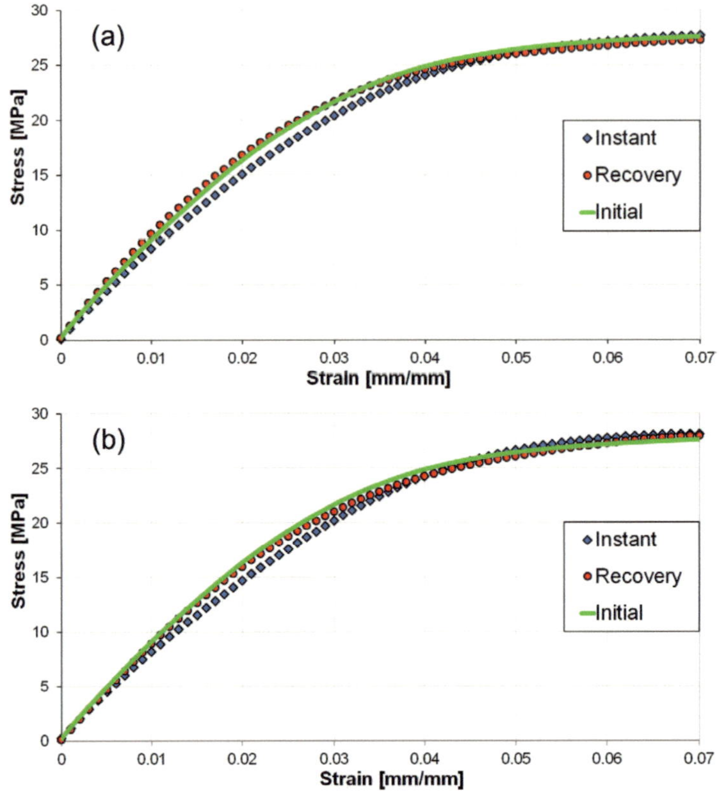

Figure 23. Tensile test comparison for 0.035 (a) and 0.045 mm/mm (b).

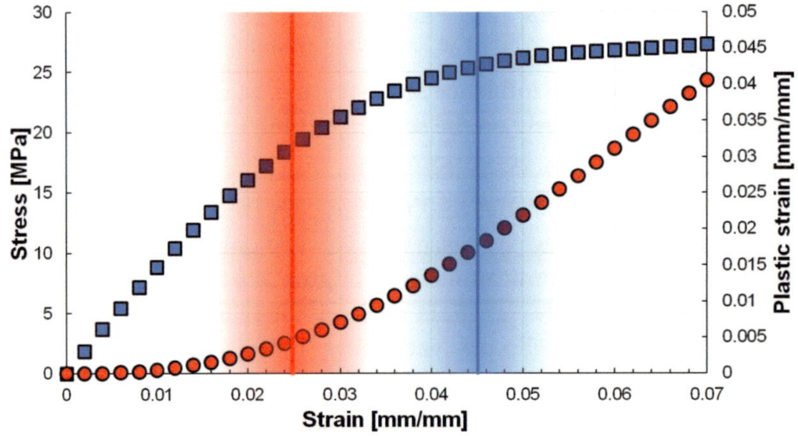

Figure 24. Stresses and plastic strains for reference strain values of 0.025 (vertical red line) and 0.045 mm/mm (vertical blue line) and their correspondent amplitude range (areas with blue gradient and red gradient).

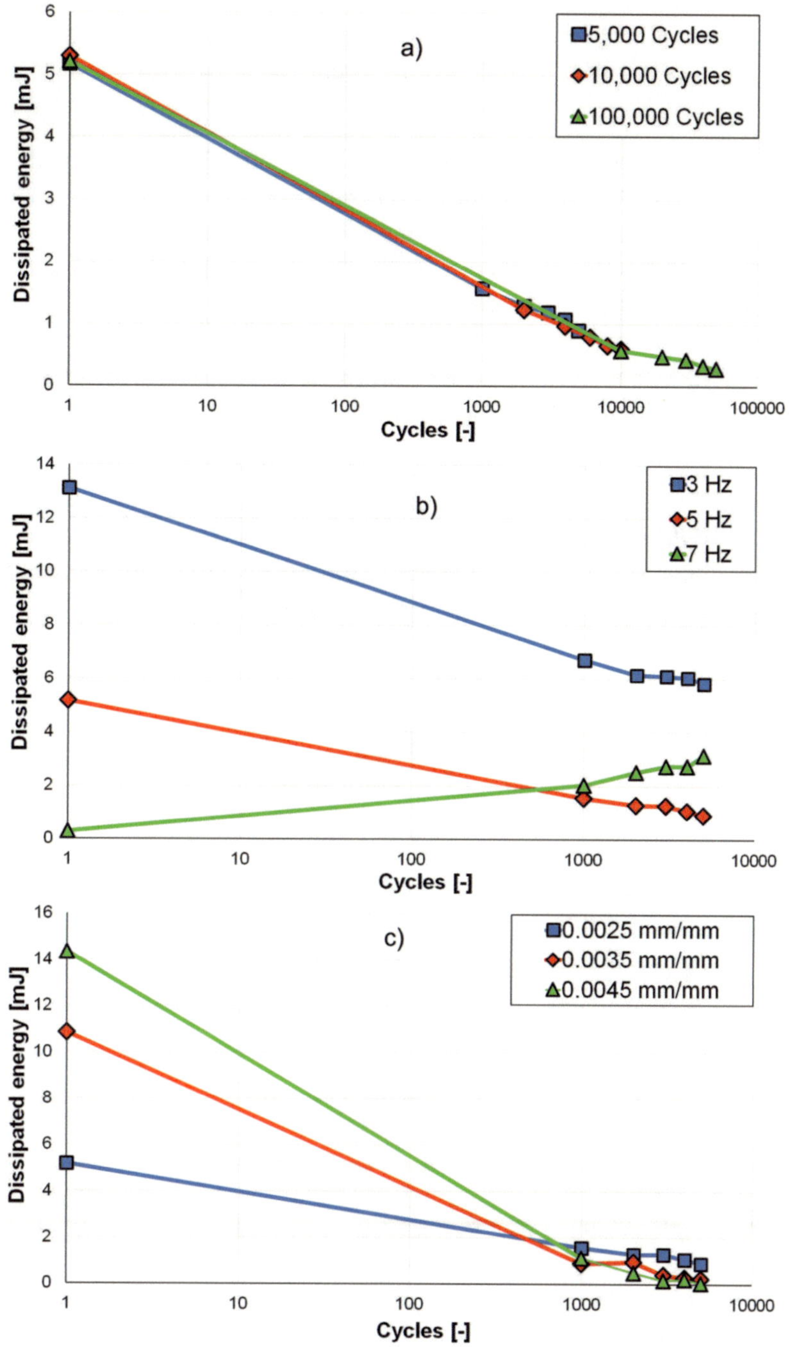

Figure 25. Variation of energy dissipation with cycles for the influence of number of cycles (a), frequency (b) and strain (c).

The same mechanism of deformation determines higher values for the dissipated energy at higher strains, and almost no energy dissipated after 5000 cycles for the tests performed at the highest pre-strain (faster saturation of viscous strain), as seen with the tests performed at varying strains (**Figure 25c**).

Tensile tests show that the LCF test parameters have little influence on the stress-strain response of the material. A slight softening occurred between 0.01- and 0.05-mm/mm deformation in the case of the specimens tested immediately after the fatigue tests, softening that is reverted after the specimen has recovered. It was also observed that, after the LCF tests were completed, when unloading to 0 N, a remnant deformation was observed, ranging from 0.013 to 0.02 mm/mm. Considering that all types of specimens determine similar responses, it can be concluded that no damage was induced to the structure of the material, the apparent softening observed during testing being caused by the accumulation of reversible (viscous) deformations.

This observation is backed by the analysis of the strain variation of the instantaneous elastic moduli variation (which was obtained by dividing the stress increment with the strain increment of the recorded cycles). For all tests, the results are similar, the stiffness of the material increasing as the number of cycles grew (**Figure 26**).

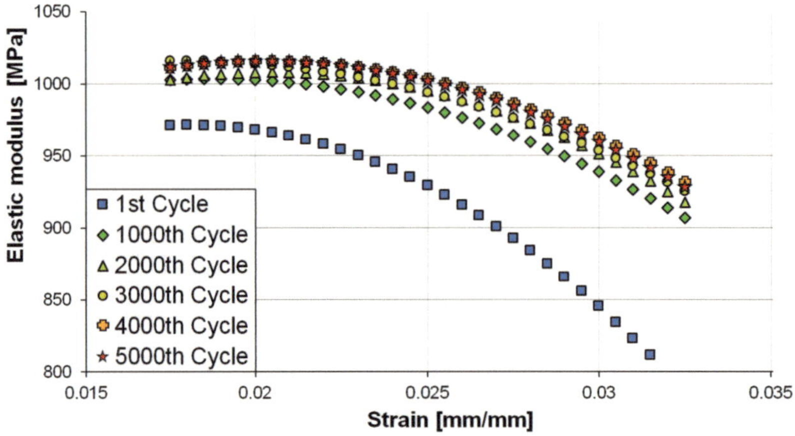

Figure 26. Variation of instant elastic moduli with strain.

To interpret this graph, we need to investigate the stiffness variation with strain, similar to **Figures 5** and **9**. In this case, for the same extensometer opening (LCF tests being performed in strain control), the 5000th cycle determined stiffer responses and lower stresses, compared with the results of the first cycle (**Figure 27**). This happens because, during cyclic loadings, the accumulated viscous strain shifts the loading spectrum towards lower elastic strains. **Figure 28** shows the strain intervals of the first cycle (blue gradient, the blue vertical line being the reference strain) and the equivalent strain interval of the 5000th cycle (red gradient). Overlapped with the stress-strain curve of tests performed on 0.1 s^{-1} (the strain

http://dx.doi.org/10.5772/64563

rate of the 5-Hz tests was 0.075 s^{-1}), it can be observed that values for the stresses and stiffness of both cycles match with the stress and stiffness response of the material [32].

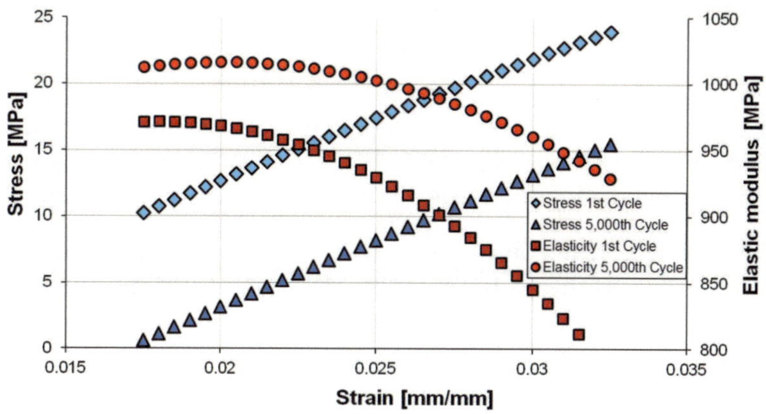

Figure 27. Variation of stress and of the instantaneous elastic modulus with strain for the first cycle and for the 5000th cycle.

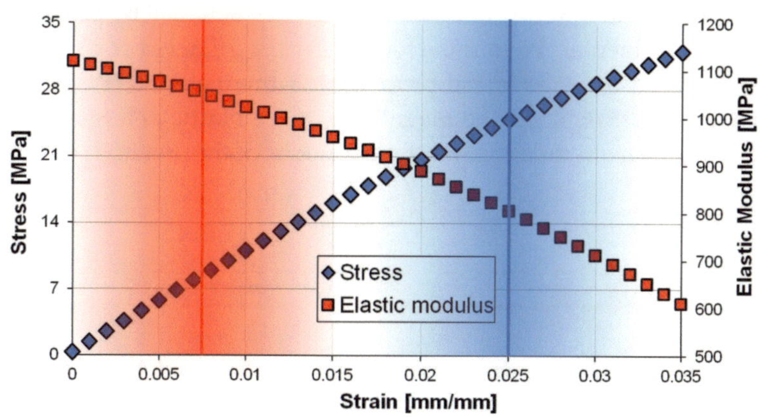

Figure 28. Stress-strain and modulus-strain curves in tension at a 0.1 s-1 strain rate and the deformation interval for the initial cycle (blue gradient) and final cycle (red gradient).

3. Conclusions

This work presents an extensive experimental programme performed on a polyamide compound, in order to determine its non-linear behaviour when subjected to various loading scenarios.

The first set of tests studied the strain-rate dependency of the material with deformation speeds ranging from 0.00028 to 2.8 s^{-1}. Tests were performed in tension and highlighted the strong influence of this parameter on the material's strength and stiffness. Going further than the low-speed tests, creep tests were performed in order to determine the material softening with time. The tensile compliance increased drastically in the first few minutes of the tests and even though it decreased significantly after 7 h, no plateau (saturation of the viscous strain) was observed after 24 h.

Temperature-dependency tests were performed in order to determine the mechanical properties of the polyamide in an envelope of temperatures that might occur during service. A substantial difference in behaviour was observed between the tests performed at low temperatures (−25 and 0°C) and the tests performed at high temperatures (23 and 50 °C).

Considering this aspect, temperature-sweep dynamical mechanical analysis tests were performed at 1 and 10 Hz in order to determine the glass transition temperature of the material. As expected, the GTT is situated near the ambient temperature (around 40°C), which might influence the material's response significantly during service.

Regarding the dynamical properties and keeping in mind the deformation range that products manufactured from this material might be subjected to, DMA tests at high strains (up to ~4% deformation) were performed. The dynamic moduli vary with strain as expected and the damping coefficient reaches a plateau when deformations get close to the yield region.

Further insight about the long-term effects of loadings was obtained by performing low-cycle fatigue tests. The experimental programme focused on the influence of the number of cycles, frequency and strain on the mechanical properties of the material. It was concluded that, considering the given test regimes, the magnitude of the plastic deformation was reduced, viscous flow having the main influence on the behaviour of the material.

Author details

Şerban Dan-Andrei

Address all correspondence to: dan.serban@upt.ro

Research Institute for Renewable Energy, Polytechnic University of Timişoara, Timişoara, Romania

References

[1] Sperling L. An Introduction to Physical Polymer Science. 4th ed. Wiley-Interscience; Hoboken, New Jersey, USA, 2006.

[2] Strobl G. The Physics of Polymers. Springer; Berlin, Germany, 2007.

[3] Kumar A, Gupta R. Fundamentals of Polymer Engineering. 2nd ed. Marcel Dekker Inc.; New York, NY, USA, 2003.

[4] Bower D. An Introduction to Polymer Physics. Cambridge University Press; New York, NY, USA, 2002.

[5] Brinson H, Brinson L. Polymer Engineering Science and Viscoelasticity: An Introduction. Springer Science; New York, NY, USA, 2008.

[6] Shaw M, MacKnight W. Introduction to Polymer Viscoelasticity. Wiley-Interscience; Hoboken, New Jersey, USA, 2005.

[7] Şerban DA, Marşavina L, Silberschmidt VV. Behaviour of semi-crystalline thermoplastic polymers: Experimental studies and simulations. Computational Material Science. 2012;52:139–146.

[8] Şerban DA, Weber G, Marşavina L, Silberschmidt VV, Hufenbach W. Tensile properties of semi-crystalline thermoplastic polymers: Effects of temperature and strain rates. Polymer Testing. 2013;32:413–425.

[9] Bonet J, Wood R. Nonlinear Continuum Mechanics for Finite Element Analysis. Cambridge University Press; New York, NY, USA, 1997.

[10] Yeoh O. Some forms of the strain energy function for rubber. Rubber Chemistry and Technology. 1993;66:754–771.

[11] Ogden R. Large deformation isotropic elasticity—on the correlation of theory and experimental for incompressible rubber like solids. Proceedings of the Royal Society. 1972;326:565–584.

[12] Arruda E, Boyce M. A three-dimensional constitutive model for the large stretch behaviour of rubber elastic materials. Journal of Mechanics & Physics of Solids. 1993;41:389–411.

[13] Marlow R. A general first invariant hyperelastic constitutive model. Constitutive Models for Rubber. 2003; 157–160.

[14] Drozdov A, Christiansen J. Finite viscoplasticity of semicrystalline polymers. Archive of Applied Mechanics. 2003;72:779–803.

[15] Drozdov A, Christiansen J. Cyclic viscoplasticity of carbon black-filled thermoplastic elastomers: Experiments and modelling. Computational Materials Science. 2009;45:398–406.

[16] Drozdov A. Cyclic viscoelastoplasticity and low-cycle fatigue of polymer composites. International Journal of Solids and Structures. 2011;48:2026–2040.

[17] Presidential Commission on the Space Shuttle Challenger Accident. Report to the President of the Presidential Commission on the Space Shuttle Challenger Accident.

National Aeronautics and Space Administration and the Government Printing Office, Washington, D.C., 1986.

[18] ISO 527. Plastics—Determination of Tensile Properties. International Organization for Standardization, Geneva, Switzerland, 1996.

[19] Xiao X. Dynamic tensile testing of plastic materials. Polymer Testing. 2008;27:164–178.

[20] Şerban DA, Marşavina L, Silberschmidt VV. Response of semi-crystalline thermoplastic polymers to dynamic loading: A finite element study. Computational Material Science. 2012;64:116–121.

[21] Şerban DA, Hanson H, Marşavina L, Silberschmidt VV. Viscoelastic properties of semi-crystalline thermoplastic polymers: dynamic analysis and creep. Solid State Phenomena. 2011;188:211–218.

[22] Dunne F, Petrinic N. Introduction to Computational Plasticity. Oxford University Press; New York, NY, USA, 2006.

[23] Rees D. Basic Engineering Plasticity. Elsevier; Oxford, United Kingdom, 2006.

[24] Cantournet S, Desmorat D, Besson J. Mullins Effect and cyclic stress softening of filled elastomers by internal sliding and friction thermodynamics model. International Journal of Solids and Structures. 2009;46:2255–2264.

[25] Drozdov A. Mullins' effect in thermoplastic elastomers: Experiments and modeling. Mechanics Research Communications. 2009;36:437–443.

[26] Diani J, Fayolle B, Gilormini P. A review on the Mullins effect. European Polymer Journal. 2009;45:601–612.

[27] Diani J, Brieu M, Vacherand J. A damage directional constitutive model for Mullins effect with permanent set and induced anisotropy. European Journal of Mechanics A/ Solids. 2006;25:483–496.

[28] Drozdov A. Mullins' effect in semicrystalline polymers. International Journal of Solids and Structures. 2009;46:3336–3345.

[29] Drozdov A, Dusunceli N. Mullins-type phenomena in polypropylene. International Journal of Applied Mathematics and Mechanics. 2012;8:82–98.

[30] Dorfmann A, Ogden R. A constitutive model for the Mullins effect with permanent set in particle-reinforced rubber. International Journal of Solids and Structures. 2004;41:1855–1878.

[31] Şerban DA, Marşavina L, Culea L, Silberschmidt VV. Experimental determination of Mullins effect in semi-crystalline thermoplastic polymers. Acta Technica Napocensis. 2010, Volume 53, pages 317–323.

[32] Şerban DA, Marşavina L, Modler N. Low-cycle fatigue behaviour of polyamides. Fatigue and Fracture of Engineering Materials and Structures. 2015;38:1383–1394.

Chapter 10

Viscoelasticity in Foot-Ground Interaction

Roozbeh Naemi, Sara Behforootan,
Panagiotis Chatzistergos and
Nachiappan Chockalingam

Additional information is available at the end of the chapter

http://dx.doi.org/10.5772/64170

Abstract

Mechanical properties of the plantar soft tissue, which acts as the interface between the skeleton and the ground, play an important role in distributing the force underneath the foot and in influencing the load transfer to the entire body during weight-bearing activities. Hence, understanding the mechanical behaviour of the plantar soft tissue and the mathematical equations that govern such behaviour can have important applications in investigating the effect of disease and injuries on soft tissue function. The plantar soft tissue of the foot shows a viscoelastic behaviour, where the reaction force is not only dependent on the amount of deformation but also influenced by the deformation rate. This chapter provides an insight into the mechanical behaviour of plantar soft tissue during loading with specific emphasis on heel pad, which is the first point of contact during normal gait. Furthermore, the methods of assessing the mechanical behaviour including the in vitro/in situ and in vivo are discussed, and examples of creep, stress relaxation, rate dependency and hysteresis behaviour of the heel pad are shown. In addition, the viscoelastic models that represent the mechanical behaviour of the plantar soft tissue under load along with the equations that govern this behaviour are elaborated and discussed.

Keywords: mechanical behaviour, plantar soft tissue, force-deformation, barefoot, viscoelastic models, finite element analyses, stress-strain, mathematical model, plantar pressure

1. Introduction

The human heel pad is usually the first part of foot that contacts the ground during normal gait. The soft tissue structure, which is located underneath the calcaneus (heel bone) consists of fat pad and skin. While this fat pad, also known as corpus adiposum, works as shock absorber and dampens the ground reaction forces during weight-bearing activities like standing and walking, the skin has another important role to prevent tear and to work as an impermeable barrier to protect the underlying soft tissue [1]. Reported heel pad thickness varies from 12.5 to 24.5mm in different studies [2–8] using different imaging modalities including ultrasound [2–4], magnetic resonance imaging (MRI) [5] and radiography [6, 7].

The plantar soft tissue structure is designed to bear large loads. Similar type of adipose tissue is found in other parts of body that normally need to bear compressive and shear stress, such as fingertips. The structure of the plantar adipocytes consists of a dense network of septa, which prevents free movement of fat cells while allows the lateral movement [9]. The unique structure of the plantar soft tissue enables it to bear large strain in reaction to the ground force. In each heel strike, vertical loading roughly equal to 110% body weight is applied on the heel, whereas 25% of the body weight is applied in anteroposterior and 10% in mediolateral directions [10]. Normalising the loading over an area results in an average pressure around 100–400 kPa for healthy individuals depending on the site of the foot [11]. These plantar pressure values can increase as a result of increase in the stiffness of plantar soft tissue, i.e., due to diabetes that results in a decrease in the contact area [11].

Furthermore, understanding the effect of different pathologies such as diabetes on the mechanical properties of human plantar soft tissue is paramount. While these pathological conditions may not affect the structure (geometry) of the plantar soft tissue they would affect the mechanical properties of the plantar soft tissue [12]. The knowledge of these mechanical properties that determine the behaviour of tissue under load can be utilised for diagnosis of foot pathologies as well as for treatment interventions such as foot orthoses and footwear.

In order to understand the mechanical behaviour of the plantar soft tissue, it is necessary to have an overview of the basic mechanical definitions.

2. Basic mechanics

2.1. Principles of mechanics of materials

Stress is a quantity which expresses the force that neighbouring particles apply on each other. Stress can also be defined as the amount of force, which is required for a unit of cross-section area to compress or extend in the normal direction.

$$\sigma = \frac{F}{A} \tag{1}$$

http://dx.doi.org/10.5772/64170

where σ represents stress, F is the load and A is the cross-section area. The unit of the stress is Nm^{-2}.

Strain is the change in the length of the object in the axial direction which is normal to the surface of applied load. Strain can be defined as the amount of change in the length of the object over the original length as a result of applying load.

$$\epsilon = \frac{dL}{L} \tag{2}$$

where ε is strain, dL is the change in length and L is the original length (**Figure 1**); hence, strain does not have a unit.

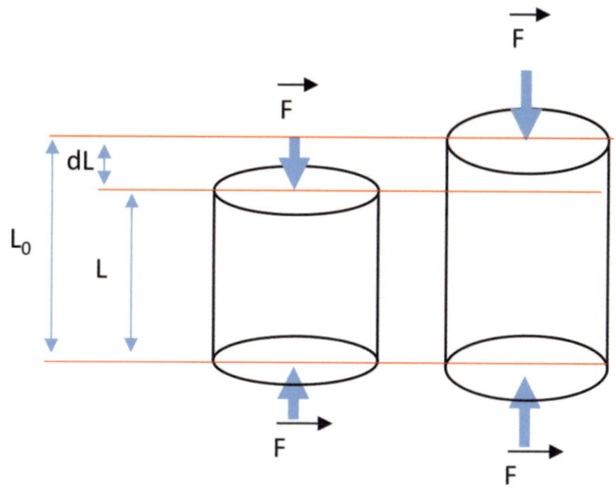

Figure 1. Strain as a result of compressive force applied to a cylindrical specimen.

Shear stress strain is the response of a material to the force, which is applied parallel to the specific surface. This force makes the geometry of the structure to deform but not to stretch/compress.

$$\tau = \frac{F}{A_{Parallel\ surface\ to\ force}} \tag{3}$$

where τ is the shear stress, F is the applied force parallel to the surface and A is the cross-section area (**Figure 2**).

The gradient of the force-deformation graph describes the stiffness of the material, and it is the quantification of the rigidity of the material and is expressed in Nm^{-1}.

$$Stiffness = {dF}/{dL} \tag{4}$$

where dF and dL are changes in load and displacement, respectively.

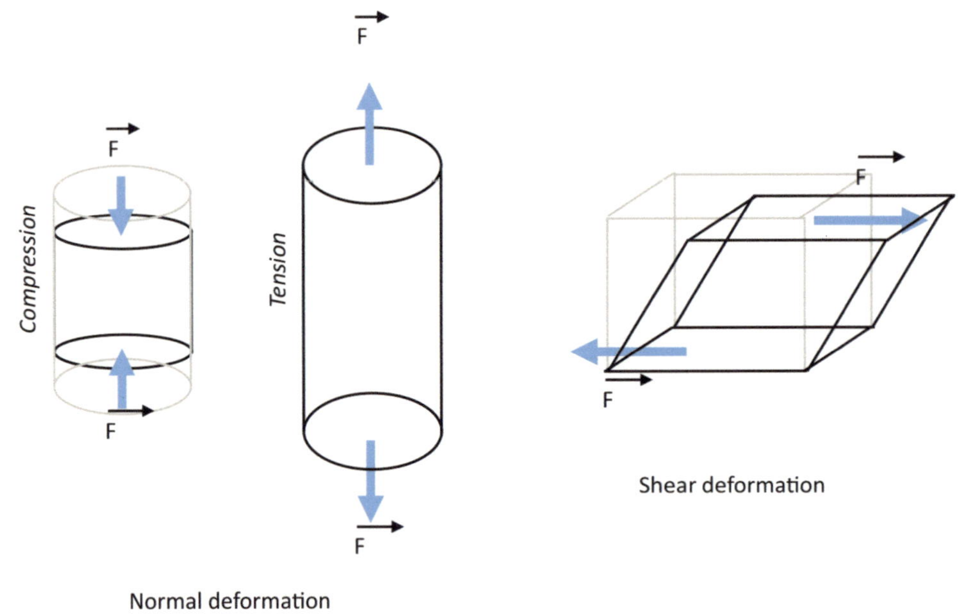

Figure 2. Normal deformation as a result of tensile and compressive forces applied to cylindrical specimens along with the shear deformation as a result of shear force applied to a cubic specimen.

2.2. Principles of elastic solid material

Elasticity is the ability of a material to resist force and return to its original shape when the force is removed. Elastic solid materials are divided into two main groups: Hookean and non-Hookean [13].

In the Hookean material where stress increases linearly with increase in strain, the slope of the stress-strain graph is defined as Young's modulus or modulus of elasticity (E) expressed in Pa.

$$E = \frac{\sigma}{\varepsilon} \tag{5}$$

where σ is stress and ε is the strain.

However, in the non-Hookean material stress is not linearly proportional to strain. The relationship between stress and strain changes during different stages of loading.

2.3. Principles of viscous fluid materials

A fluid material is defined as a material that bears shear deformation and consists of liquids and solids [14]. Liquids consist of atoms with interatomic connections and molecules with weak intermolecular connections. The shear force can be applied to break the weak intermolecular bonds to allow the material to flow [15].

In continuum mechanics, a Newtonian fluid is a fluid in which the arising viscous stresses are linearly proportional to the local strain rate [16]. In other words, the shear stress is proportional to the rate of change of the fluid's velocity vector.

$$\tau = \mu . \frac{dv}{dy} \tag{6}$$

where τ represents the shear stress, μ represents viscosity and dv/dy represents the velocity gradient in the direction perpendicular to the velocity.

The simplest mathematical models that take viscosity into account can be applied for Newtonian fluids. Although there is no real fluid that fits this definition properly, many common liquids and gases, such as water and air, are assumed to be Newtonian under ordinary conditions.

On the other hand, a non-Newtonian fluid is the one with different properties from a Newtonian fluid. To be more specific, the viscosity, which is the quantity of a fluid's ability to resist gradual deformation by shear or tensile stresses, depends on shear rate or shear rate history [16].

2.4. Viscoelastic behaviour

A viscoelastic material combines properties of the elastic solid with viscous fluid. The elastic solid can be Hookean or non-Hookean and the viscous fluid can be Newtonian or non-Newtonian [17]. There is a variety of behaviour within different viscoelastic materials which ranges from completely elastic solid behaviour to completely viscous fluid behaviour [17]. In addition, the viscoelastic material has the distinctive characteristics which show both viscous and elastic behaviour when exposed to loading. The soft tissue exhibits a viscoelastic behaviour under compression which means that the force-deformation behaviour depends on the amount of deformation and deformation rate [18]. Viscoelastic materials behave in different ways under various types of loading exhibiting differences in deformation/force rate dependency, creep, stress relaxation and hysteresis [18]. For example, a viscoelastic material under cyclic loading behaves differently during loading and unloading. In a sense, the stiffness of the material decreases during unloading when compared to the stiffness that material shows during loading. The area between force-deformation graph in loading and unloading is called hysteresis that represents dissipated energy [18] (**Figure 3**).

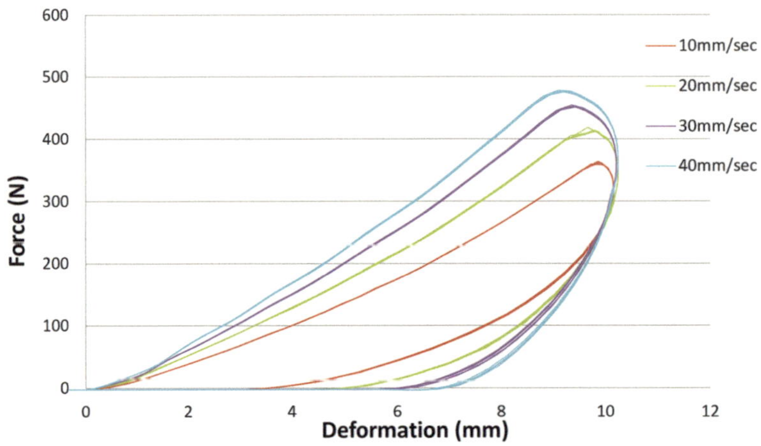

Figure 3. The load deformation graph of a heel pad tested using ultrasound indentation technique [19]. The different colours represent data gathered at different deformation rates as presented in the legend. The area surrounded between the loading and unloading curves represents hysteresis.

The force-deformation behaviour of the plantar soft tissue like other biological materials is influenced by the loading velocity [18] (**Figure 3**). Stress relaxation characteristics indicate that when a viscoelastic material is deformed suddenly and the deformation is kept constant for a specific time, the force decreases with time [18] (**Figure 4**). The creep characteristics indicate that when a specific amount of load is suddenly applied to the viscoelastic material and kept constant for a specific time, there would be an increase in deformation over time [18] (**Figure 5**).

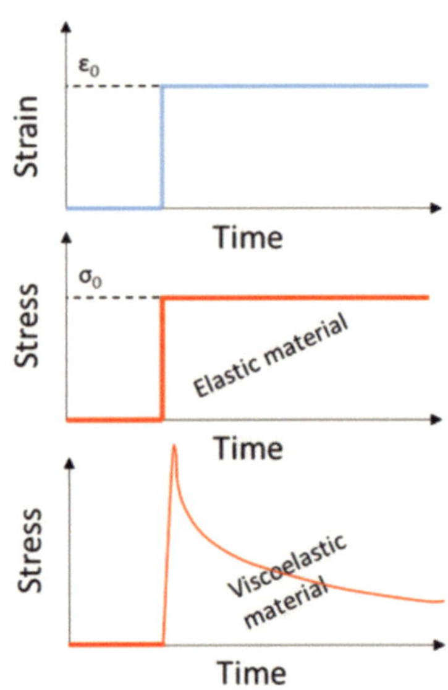

Figure 4. Schematic representations of the stress/time and strain/time graphs for a stress relaxation test for an elastic and a viscoelastic material.

http://dx.doi.org/10.5772/64170

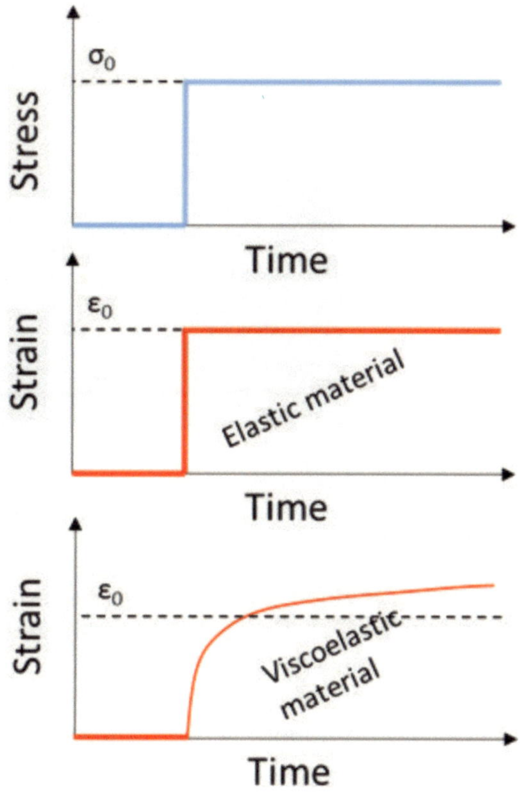

Figure 5. Schematic representations of the stress/time and strain/time graphs for a creep test for an elastic and a viscoelastic material.

3. Mechanics of the heel pad

Because of the liquid content of the heel pad tissue along with the arrangement of solid components which has an influence on regulating the fluid movement, the heel pad behaviour is mostly assumed as that of a nearly incompressible material [20].

3.1. The structure

The fat pad consists of two layers: microchambers and macrochambers. Microchambers shape the plantar layer of heel pad, which protects the fat pad from excessive bulging during loading [12]. The deeper layer is composed of sparser, fibro-adipose structure called a macrochamber [21]. The microchambers is the thinner layer of the two and contains mainly elastic fibres, but the thicker layer (macrochambers) contains roughly equal amount of elastic fibres and collagen [22]. Therefore, a different behaviour between two layers is expected [12], which is discussed under the mechanical behaviour section (**Figure 6**).

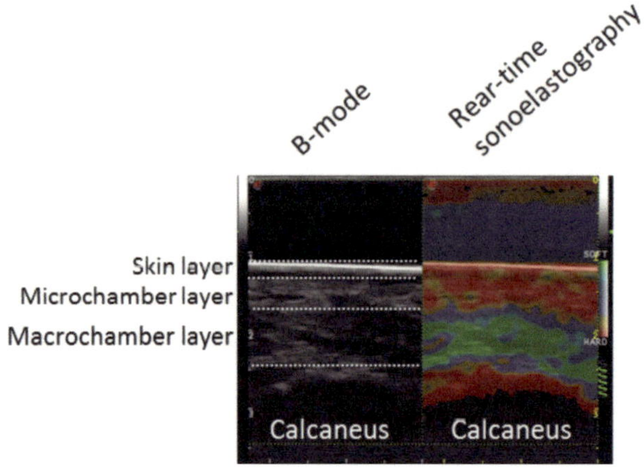

Figure 6. Ultrasound image of heel pad in B mode (left) along with the ultrasound elastography image superimposed on the B mode (right) showing the micro and macrochambers colour coded by the stiffness with the spectrum showing tissue in red colour as hard tissue and tissue in green colour as soft tissue.

The heel pad atrophy is a clinical condition, which is usually linked with diabetes, collagen disorders and peripheral neuropathy [22]. The results of histological studies have revealed that the adipocytes in the subcutaneous layers closer to the foot surface of the fat pad were 25% smaller in the mean cell area and 10% smaller in mean maximum diameter in an atrophic heel compared to a normal heel [22]. Additionally, the adipocytes in the deep subcutaneous layers are 45% smaller in the mean cell area and 25% smaller in mean maximum diameter in an atrophic heel compared to a normal heel [22]. Septa in both deep and superficial subcutaneous layers in the atrophic heel are 25% thicker than in a normal heel and include more percentage of elastic tissue which seems to be uneven in some cases [23].

Furthermore, in atrophic feet, collagen septa are found to be thicker, and also the adipose cells are smaller than healthy feet [22, 24]. The amount of internal stress and strain depends on the structure of the heel (geometry) as well as on the material properties of the tissue and also the value and direction of force. Hence, the above-mentioned changes in the structure of the heel pad can increase the internal stress and strain that are claimed to be the main causes of tissue injuries [25–27]. In addition, the interface between the soft tissue structure and underlying bony prominence can be the area of high stress concentration and it is claimed that tissue damage starts in deep tissue close to the bony prominences and then develops up to the skin surface [25, 26, 28, 29].

3.2. The function

The foot supports stabilising the whole body during standing and is the interface between the body and the ground during walking. The heel pad as the first part of the foot which has contact with the ground during locomotion acts as a shock absorber and shock reducer to protect the foot from local stress [2]. The strain and pressure applied to the heel during gait can be withstood by the honeycomb structure of the heel pad. The heel pad as a structure shows viscoelastic behaviour and provides cushioning during heel strike and absorbs shocks by dissipating energy [2]. This mechanical energy dissipates as heat just after heel strike and decreases the possibility of mechanical trauma to the foot [2].

The heel pad can absorb the impact shock during heel strike by carrying out deformation under loading and by distributing the force over a wider area of the skin to prevent stress concentration [30]. The chambered structure helps to spread the compressive force over the whole area of plantar surface of the bone and prevents any injuries on the calcaneus bone during heel strike [2]. Under compression, the heel pad expands easily as a result of low stiffness, but later the tension on the collagen fibres of the fat pad and skin limits the movement of the tissue and increases the stiffness of the heel pad gradually in the loading direction [31]. This justifies the nonlinear force-deformation behaviour of the heel pad and causes the strain stiffening inherent in the heel pad's mechanical behaviour.

Additionally, it was reported that the mechanical behaviour of microchambers and macrochambers is different under compression [12]. The microchamber layer experiences less strain compared to the macrochamber layer [12]. Hsu et al. [12] indicated that the stiffness of the microchambers is 10 times greater than macrochambers, and concluded that the observed difference plays an important role in the heel pad mechanical behaviour. It appears that the macrochamber layer is responsible for large deformation and cushioning behaviour of the heel pad during gait; however, the microchamber layer is responsible in preventing the heel pad from excessive bulging [12].

The main role of the heel pad is to decrease the impact shock during heel strike and to distribute pressure during the foot-ground contact through undergoing deformation. The deformability of the heel pad may reduce through tearing the fibrous septa or atrophy of heel pad due to a trauma or ageing [32, 33]. After severe injuries such as tearing or breaking of the honeycomb structure, the heel pad does not have the ability to remodel itself [32, 33]. As the fat pad is a semi-liquid structure with hydrostatic properties of fluids [33], a decrease in water content of the heel as well as a decrease in the elastic fibrous tissue and loss of collagen are the main reasons for the gradual weakening of the tissue due to ageing [32]. Overall, it can be concluded that the loss of soft tissue substances due to ageing, atrophy or any previous injury prevents the tissue from responding to load in an optimal way [33].

4. Mechanical behaviour of heel pad

Plantar soft tissue like other biological tissues has a nonlinear elastic behaviour. Initially by applying small amount of load, the tissue deforms easily (low stiffness) and by increasing the loads the stiffness increases gradually [34]. The heel pad expands easily as a result of low stiffness but afterwards the tension on the collagen fibres of the fat pad and skin limits the movement of the tissue and increases the stiffness of the heel pad in the loading direction [31].

Pathological changes in foot may not be detectable from the structure of the foot but normally correlate with the alteration in the mechanical behaviour of the tissue [12]. Therefore, quantifying the mechanical properties of the plantar soft tissue is important to assess the risk of mechanical trauma to the foot.

4.1. Method of assessing mechanical behaviour of heel pad

Several methods have been used to extract the mechanical behaviour of the heel pad that can be divided into three main groups: in vitro, in situ and in vivo.

4.1.1. In vitro tests

In some studies, the heel pad behaviour was characterised using an in vitro or in situ method to quantify the material properties of the plantar soft tissue [32, 35–40]. Miller-Young et al. [35] performed a series of tests on the cylindrical samples of plantar soft tissue extracted from cadaveric feet. Three series of tests that were performed on the samples included the quasi-static test to obtain the hyperelastic mechanical properties of the soft tissue; the stress-relaxation tests to calculate the viscoelastic time constants; and the dynamic compression test to extract the viscoelastic relaxation material coefficients. The reported results clearly showed the time dependency and viscoelastic behaviour of the heel pad [35].

Some other studies [39, 41] used 2 × 2 cm samples from different sites of plantar soft tissue of cadaveric feet to calculate the elastic and viscoelastic coefficients of a mathematical model for plantar fat pad to compare the properties of the plantar soft tissue between a healthy and a diabetic foot. The results showed that stiffness and energy dissipation increase with loading frequency [39, 41]. This was completely in contrast with Bennet and Ker's results, which indicated no changes in energy dissipation and stiffness with different testing frequencies [36]. The results also showed that frequency dependency is higher in younger subjects and this was attributed to the difference in the heel pad's water content in younger versus older tissue specimen [32, 37]. The water content of the soft tissue can be considered as the main reason for viscoelastic behaviour of the soft tissue. Therefore, a decrease in water content of the soft tissue in older subjects can lead to a decrease in the viscoelastic characteristic of the soft tissue such as loading frequency dependency.

While there was no significant difference in the values of viscoelastic coefficients between diabetic and non-diabetic feet, it was claimed that changes in plantar soft tissue were mainly recognised at the structural level, and not reflected effectively at the material level [38, 39].

4.1.2. In situ tests

Aerts et al. [40] compared the results of in vitro tests with the in situ tests in which the heel bone was fixed to the wall and a pendulum was used to impact the heel region.

In this study, differences between the test results in two conditions were observed and they were attributed to the differences in soft tissue behaviour at structural and material level [40]. In another study, Bennet and Ker [36] compared the results of in vitro versus in situ tests. The researchers performed two series of tests in which one group of specimens was tested while attached to the calcaneus and the surrounding tissues, while the other group of specimens were tested after complete removal from the calcaneus [36]. In this study, the specimens were tested using a dynamic loading machine and load-displacement data were recorded during the test. The energy dissipation ratios and stiffness were higher in isolated heel pads compared to the heel pads attached to the calcaneus [36]. Therefore, it was concluded that the results of the in vitro test show the properties of the isolated heel pad material, while the in situ test results represent the behaviour of the heel pad structure [40]. The fact that the mechanical properties of heel pad extracted from the in vitro and in situ results are different can be explained by the indications that structural factors such as heel pad skin and geometry of the calcaneus have an influence on the heel pad behaviour.

Although the in vitro test can provide more repeatable data due to eliminating geometrical complexities of the plantar soft tissue, it cannot provide a realistic assessment of the mechanical behaviour of the plantar soft tissue during weight-bearing activities [38, 39]. Furthermore, the in situ tests cannot be used to assess the mechanical behaviour of the heel pad in different individuals.

4.1.3. In vivo assessments

In a number of studies, the human heel pad was characterised using the force-deformation data extracted from in vivo experiments.

Investigating the mechanical behaviour of the heel pad during walking can enhance our knowledge about the heel pad in realistic loading conditions.

In one study, radiographic fluoroscopy was utilised to measure the thickness of the heel pad during walking [42] while the plantar pressure was also measured using optical display method. Their results showed that maximum strain of the heel pad during walking was 40% and the absorbed energy ratio was estimated as 17.8% (SD 0.8) for different velocities (0.5–0.9 m/s) [43], which is considerably lower than 35% ratio that was reported for in vitro studies of the heel pad [40, 41].

While the radiographic method allows measuring the deformation of the heel pad during actual gait, the control on the direction of loading affects the results and this may lead to variation in the observed results.

To overcome these limitations, a number of studies utilised ultrasound indentation technique to develop an accurate and repeatable dynamic loading condition [44–51].

The device commonly consists of linear array ultrasound probe, load cell, motor and mechanical body, which is composed of a foot place perpendicular to the axial of loading (same axis to the probe head). The ultrasound probe is necessary to measure the deformation of the soft tissue. The force that is applied to the foot to compress/indent the foot can be measured by a load cell. It can be mounted at the back of the probe to measure the axial force. The mechanical part should be mounted in such a way that the probe can be compressed/indented to different feet sizes. The motor can generate uniform movement of the probe. A number of studies which conducted experiments with ultrasound indentation device [52, 53] used a custom loading device that consists of a linear array ultrasound probe which was in series with a dynamometer (load cell) and mounted on a rigid frame (**Figure 7**).

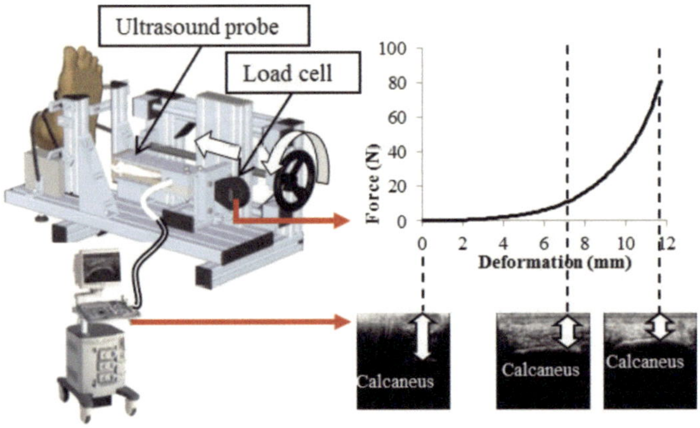

Figure 7. The ultrasound indentation device and a schematic representation of the procedure followed to create the tissue's force/deformation curve [52].

In addition to the use of indentation device in experimental analyses of the plantar soft tissue, this device has been commonly used to quantify the mathematical and finite element (FE) models that govern the behaviour of soft tissue during loading. This is discussed in the next section under mechanical behaviour model. Ultrasound strain elastography has been used for assessment of plantar soft tissue stiffness in patients with diabetic neuropathy and was recognised to have potentials for diagnosing tissue mechanical malfunction in clinical setting [54].

4.1.4. Plantar soft tissue stiffness and measurement methods

The mechanical properties of the plantar soft tissue show a high level of dependency on the measurement method [40, 55]. For example, it was reported that the stiffness of the heel pad of humans using in vitro method is almost six times higher compared to the stiffness measured during in vivo tests, while the absorbed energy ratio is about three times lower using the in vitro method [40]. This can be explained by the indications that structural factors such as heel pad thickness and geometry of the calcaneus have a significant influence on the heel pad mechanical behaviour [3, 23, 56]. Aerts et al. [40] compared the energy dissipation and stiffness of the soft tissue in an amputated leg and in an isolated heel pad. They showed that the whole lower leg which is involved during in vivo test affects the test results in different ways and influences stiffness and energy dissipation in in situ test. The presence of lower leg makes a difference in terms of limiting expansion in some directions and dissipating the energy [40].

Furthermore, while indentation seems to provide more realistic and reliable assessment of the mechanical behaviour of the plantar soft tissue during in vivo conditions, the effects of indenter's shape (i.e. probe's head geometry) together with the effects of calcaneus bone geometry need to be taken into account in analysing the results from these type of studies.

4.2. Changes to mechanical behaviour

The mechanical behaviour of the plantar soft tissue can be changed by ageing, heel pain and other pathologies; some of which are described below.

4.2.1. Effect of ageing on the mechanical behaviour

Hsu et al. [57] compared the average unloaded heel pad thickness in a young group and an elderly group of participants. The average unloaded heel pad thickness was 20.1 (±2.4) mm in the elderly group and 17.6 (±2.0) mm in the young group. These results are in line with the findings by Kwan et al. [58] who showed that the thickness of the soft tissue was higher in the elderly group and that the stiffness increases with age [58]. The stiffening of the soft tissue by ageing may reduce the adaptability of the tissue in responding to the stress, which may lead to foot diseases in elderly people [59].

The shock absorption of the heel pad was determined in two different age groups for two different impact velocities by Kinoshita et al. [60]. It was reported that the absorbed energy in younger adults is significantly higher than in elderly [60]. The energy absorbed density depends on the viscoelastic properties of the plantar soft tissue and can be calculated by subtracting the energy return density from the energy input density [53]. The energy dissipation of soft tissue is directly linked to its viscoelastic characteristic. Therefore, it can be concluded that the viscoelastic properties of the soft tissue alter with age.

4.2.2. Effect of heel pain on mechanical behaviour

Physical activity and repetitive high-impact loading during sports activity can cause some micro-damage in the heel pad and consequently lead to heel pain. The sports that involve running or jumping apply repetitive impact loading on the heel pad and may cause collapse of the heel pad and finally cause a greater force impact on the calcaneus.

The inflammatory oedema may be a sign of changes in the structure of the heel pad that can lead to decrease in shock absorption capability during heel strike [61]. To compare the mechanical behaviour of the heel pad, a parameter known as heel pad compressibility index (HPCI) was defined as the ratio of the loaded tissue thickness to unloaded tissue thickness in percentage [46]. This parameter was utilised by Prichasuk et al. [62] to compare between normal and painful heel and revealed that the compressibility index increases in the patients with heel pain.

Tong et al. [50] compared the plantar tissue thickness and HPCI in normal, plantar heel pain and participants with diabetes. The results showed that compressibility in patients with diabetes and heel pain was less than the corresponding value in healthy volunteers. While the findings of Tong et al. [50] on the effect of ageing on HPCI contradict Prichasuk et al.'s findings [62], in another study Ozdemir et al. [63] reported the increase in heel pad thickness and HPCI with ageing and increase in body weight. This was attributed to the gradual loss of collagen, water content and elastic fibrous tissue of the heel pad as result of ageing [60], all of which can lead to a change in the viscoelastic behaviour of the heel pad.

4.2.3. Effect of other pathologies on mechanical behaviour

The structural changes in the soft tissue of diabetic patients will cause some changes in the macroscopic and microscopic behaviour of plantar soft tissue and make it more vulnerable to mechanical trauma which can lead to ulceration [54]. Less elastic tissue and impaired ability in distributing pressure are the other changes in diabetic tissues which can lead to a weakened cushioning effect [54]. Increase in energy dissipation during weight bearing is the other factor which can increase the risk of ulceration [54].

A tissue with increased viscosity and decreased elasticity may provide the similar amount of stiffness during loading, but during the swing phase of gait and as the tissue unloads the increased viscosity may not allow the tissue to completely return to its original thickness. This may cause the next loading cycle to start with a tissue that is partially deformed, which may increase the likelihood of tissue bottoming out.

In addition, generally the stiffening and thinning of the fat pad would make the tissue more fragile that makes it more likely to damage compared to a healthy tissue [64]. This also applies to the behaviour of skin in a diabetic foot, which can be less flexible and more brittle when it becomes drier [50]. It was also found that the deformability of the heel pad is less in participants with diabetes compared to their healthy counterparts [50]. However Hsu et al. [46] measured the elastic modulus, compressibility index and unloaded tissue thickness in people with diabetes and found no difference with the respective tissue characteristics in healthy participants. Hsu et al. [48] compared the strain and elastic modulus in macrochambers and microchambers of the heel pad in people with diabetes and healthy subjects. Strain in microchambers in people with diabetes was significantly greater than that of the healthy subjects [48]. In healthy subjects, macrochambers' strain was significantly greater than that of the people with diabetes, which can be a result of uneven distribution of the collagen fibrils in a diabetic heel [65].

Furthermore, Hsu and co-workers concluded that the heel pad tissue properties are altered heterogeneously in people with diabetes, indicated by an increased stiffness in macrochambers but a decreased stiffness in microchambers, which were attributed to a diminished cushioning capacity in diabetic heels [48]. The energy dissipation or hysteresis (area between loading and unloading in force-deformation graph) was also shown to be significantly higher in people with diabetes [38, 65]. Furthermore, in people with diabetes the ability of recovering the shape of the heel pad after unloading reduces which can be linked to the increase in the amount of energy dissipation [62, 65].

Chatzistergos et al. [66] investigated the correlation between the mechanical behaviour of the heel pad in type-2 diabetes and blood biochemical parameters such as triglycerides and fasting blood sugar (FBS). A medium strength positive correlation was found between the stiffness of the heel pad and FBS and a strong correlation was found between the triglycerides and stiffness. In addition, a strong negative correlation was found between the triglycerides and energy absorption during loading [66]. Chatzistergos and co-workers [66] concluded that people with type-2 diabetes and high levels of triglycerides and FBS are more likely to have stiffer heel pads that may hinder the tissue's ability to evenly distribute loads that makes the tissue more vulnerable to trauma and ulceration.

5. Quantifying mechanical behaviour

As mentioned before, heel pad provides a cushioning interface between calcaneus bone and the ground and has a natural function of shock absorption. The mechanical properties of the heel pad govern the force-deformation behaviour of the heel pad during heel strike and therefore these properties affect the loading on musculoskeletal system [67]. In order to investigate the force-deformation behaviour of the heel pad in in vitro, in situ and in vivo conditions, several mathematical models have been developed and utilised [36, 40, 42, 53, 65, 68–74]. Additionally, a number of FE analyses were used to quantify the mechanical behaviour of the heel pad [45, 52, 75–80].

5.1. Mathematical models

A number of studies developed mathematical models to quantify the mechanical behaviour of the heel pad in vitro [35, 41], while others utilised mathematical models to describe the in vivo mechanical behaviour of the heel pad [42, 44, 49, 51, 53, 65, 70, 81, 82]. Most of the studies measured the elastic behaviour of the heel pad and just a few of them represented the viscoelastic behaviour of the heel during dynamic loading.

Ledoux and Belvis [41] performed in vitro test on a freshly frozen cadaver's plantar soft tissue and used the following equation to characterise the mechanical behaviour of the plantar soft tissue:

$$\sigma = A.(e^{B.\varepsilon} - 1) \tag{7}$$

The elastic modulus and absorbed energy of different plantar sites were calculated in which the subcalcaneal heel pad showed higher elastic modulus and absorbed energy compared to other sites of the plantar surface. This can be partly related to the fact that the subcalcaneal fat pad is designed to absorb energy during heel strike, while the fat pad underneath other plantar sites is mainly adapted to provide functions such as even pressure distribution.

In another study by Pai and Ledoux [39], quasilinear viscoelastic (QLV) model was used in two approaches: traditional frequency-sensitive and indirect frequency-sensitive:

$$\sigma = \int_t^0 G(t-\tau)\frac{\partial\sigma^e(\varepsilon)}{\partial\varepsilon}\frac{\partial\varepsilon}{\partial\tau}\partial\tau \tag{8}$$

where G is the time-dependent function, τ is the relaxation time, σ^e is the elastic stress, ε is the strain, $\frac{\partial \sigma^e(\varepsilon)}{\partial \varepsilon}$ is the derivative of elastic strain over strain and $\frac{\partial \varepsilon}{\partial \tau}$ is the derivative of strain over time. QLV theory normally assumes that the elastic and time-dependent properties can be separated and a linear combination of the elastic behaviour and viscous behaviour can describe the mechanical behaviour of the system.

The coefficients of stress-relaxation response for diabetic and non-diabetic subjects were compared and significant differences were found in the value of B between diabetes versus healthy subjects. However, no significant differences were found in viscous coefficients between diabetic and non-diabetic specimens. The lack of differences between diabetic and non-diabetic tissue was attributed to the changes at the structural level that have not been reflected effectively at the material level [38, 39].

A number of studies characterise the mechanical behaviour of the heel pad during in situ tests [40, 83, 84]. Ker [83] employed a nonlinear equation to characterise the force-deformation behaviour of the heel pad. However, the stiffness values determined from this equation were dependent on the stage of loading cycle.

Although a number of studies utilised the in vitro and in situ data for mathematical modelling of the plantar soft tissue, in vivo assessment of the biomechanical behaviour of the plantar soft tissue has been a more common method for obtaining data for mathematical model.

In order to quantify the in vivo mechanical properties of the heel pad, the following function was utilised to represent the force-deformation data from in vivo test on the heel pad [49]:

$$F = a^{b.x} \tag{9}$$

where a and b are the constants that are calculated from fitting the function to the in vivo data. The parameters were extracted from two groups of subjects with and without heel pain. Although the value of b was significantly lower in the group with heel pain compared to the group without heel pain, there was no significant difference in the value of a between the two groups [49]. While the a value that represents the slope of the force-deformation graph is more related to the stiffness, the value of b relates to the rate of the changes in the stiffness by loading and represents the curvature of the force-deformation graph.

Challis et al. [70] used the same formula as proposed by Ledoux and Belvis [41] for modelling the in vivo force deformation relationship of the heel pad during indentation. They compared the thickness, strain, energy loss and stiffness of the heel pad in cyclists versus runners. Although the heel pad stiffness was found to be significantly less in runners compared to cyclists, the percentage of absorbed energy was not found to be significantly different.

A general formula with separate terms for geometry and material parameters was utilised in a number of other studies in order to introduce coefficients which reflect the material characteristic of the heel pad.

http://dx.doi.org/10.5772/64170

$$F = E.\frac{2a\kappa(\nu, a/h)}{1-\nu^2}.x \tag{10}$$

where E is the elastic modulus of the heel pad, a is the radius of the indenter, ν is the Poisson ratio, h is the thickness of the soft tissue and K is the function of ratio of the radius of the indenter to the thickness of the tissue. Zheng et al. [51] calculated Young's modulus for different regions of the plantar soft tissue which was different from 40 to 50 kPa in healthy subjects, while the values were 160% more in average for the same sites of the plantar soft tissue in diabetic feet [51]. While there was no indication about how the K value is different between the two groups, limiting the maximum deformation to 10% of the initial thickness of soft tissue may have an influence on the calculated coefficients. As the heel pad shows nonlinear elastic behaviour, the coefficients may be different with higher deformation.

Chao et al. [44] used the same formulation (Eq. 10) to compare the elastic modulus between two different age groups. The air-jet indentation system was used along with non-contact optical coherence tomography in four loading-unloading cycles with 0.4 mm/s deformation rate. It was found that the modulus of elasticity under the second metatarsal head is significantly higher in older group compared to their younger counterparts.

Most studies concentrated on representing the force-deformation behaviour during loading, while one of the characteristics of viscoelastic behaviour is having a different force-deformation behaviour during loading and unloading. Hsu et al. [65] utilised a formula that considers different coefficients for loading and unloading:

$$\sigma = \sigma_{\max} \cdot \left(\frac{\varepsilon}{\varepsilon_{\max}} \right)^{\alpha} \tag{11}$$

where σ is stress, ε represents the strain, σ_{max} is the maximum stress, ε_{max} represents the maximum strain and α is the curvature constant that is different for loading and unloading graphs. Hsu et al. [65] compared in vivo data from diabetic and healthy subjects in which the curvature constant was significantly higher in diabetics compared to healthy participants during unloading; however, there was no significant difference in the α value during loading.

In addition to the importance of elasticity in the mechanical role of the heel pad, viscosity also plays an important role in dissipating energy and hysteresis. In order to identify the dissipating energy ratio, a number of studies added viscous term to the force-deformation formulas.

One of the in vivo models representing both elasticity and viscosity of the heel pad was proposed by Gefen et al. [42]. They proposed Kelvin-Voigt model in which elasticity behaviour was characterised by a linear spring and viscous behaviour was represented by a nonlinear damper.

$$\sigma = E.\varepsilon + \eta.\varepsilon.\varepsilon^{\bullet} \tag{12}$$

where σ represents stress, ε and $\varepsilon^{\bullet}$ represents the strain and strain rate respectively, E is the elastic modulus and η is the viscosity parameter of the soft tissue. Digital fluoroscopy along with pressure plate was utilised to measure the pressure and deformation during walking.

The proposed model improved the predicted force-deformation behaviour of the soft tissue significantly by adding viscosity; however, the assumption of linear spring is an oversimplification of the behaviour of the soft tissue due to the fact that the quasi-static behaviour of the soft tissue still shows nonlinear force-deformation behaviour [20, 34].

In a more comprehensive approach, Natali et al. [20] developed a constitutive mathematical model for the mechanical behaviour of the heel pad based on in vitro and in vivo tests. They considered hyperelastic and viscoelastic behaviour of the plantar soft tissue and developed stress-strain curve based on the second Piola-Kirchhoff stress tensor and used Miller-Young et al. [35] and Ledoux and Belvins's [41] in vitro data and also Zheng et al. [85] and Erdemir et al.'s [45] in vivo data in order to adapt the formula for the plantar soft tissue:

$$S(C,q^{i}) = S^{\infty}(C,K_{v},r,C_{1},\alpha_{1}) + \sum_{i=1}^{n} q^{i}(t,\gamma^{i},\gamma^{\infty},\tau^{i}) \tag{13}$$

where S was the second Piola-Kirchhoff stress tensor, S^{∞} was the elastic stress when viscous condition is completely relaxed, q^{i} was the viscous stress, C was right Cauchy-Green strain tensor, k_{v} and C_{1} relate to initial bulk modulus and shear stiffness, respectively, whereas r and α are hyperelastic coefficients of the soft tissue. γ^{∞} and γ^{i} relate to stress-strain history and τ^{i} represents the relaxation time.

On the other hand, Sciume et al. [86] used a mathematical model, which was based on thermodynamically constrained averaging theory (TCAT) [87]. The soft tissue was modelled as a porous medium filled by an interstitial fluid.

Ultrasound indentation has also been frequently used to inverse engineer the material coefficients of heel pad and also to measure heel pad stiffness and energy dissipation [20, 45, 88]. However, there have been only a few studies that have used mathematical modelling to characterise the elastic and viscous behaviour of the plantar soft tissue. Naemi et al. [53] developed a mathematical model, which considered both elastic and viscous behaviour of the heel pad. They also modelled the nonlinear behaviour of the heel pad using nonlinear spring and nonlinear damper. They claimed that during quasi-static tests in which only the elastic component of the heel pad plays role in resistance to compression, strain-stiffening can be observed. Therefore, they used power function for depicting the elasticity component proposed by Scott and Winter [89]. This study proposed a method to quantify the force-deformation behaviour of the heel pad under compression in cyclic loading and took into account the nonlinear viscoelastic behaviour of the heel pad.

Energy input and energy return were derived as follows:

$$E_i(\varepsilon) = \frac{a}{b+1}.\varepsilon^{b+1} + \frac{c}{d+1}.\varepsilon^{d+1}.\varepsilon^{\bullet} \tag{14}$$

By fitting the elastic and viscous energy densities to the data, Naemi et al. [53] showed that elastic energy density was much higher than absorbed energy density and elastic stress was significantly more than viscous stress. Elastic scaling factor and exponent were also 1.9 times and 14 times higher than viscous scaling factor, respectively. Despite the findings, they also reported that the deformation rates at which the tests were performed were much lower than the realistic deformations of the heel pad achieved during walking and they recommended testing the heel pad at more realistic strain rates to achieve realistic coefficients.

Although a significant number of studies utilised mathematical modelling in order to model elasticity and viscosity behaviour of the soft tissue at the heel, there has been a paucity of studies in which mathematical models were utilised for quantifying the viscoelastic characteristics of the plantar soft tissue at the forefoot. Although Hsu et al. [12, 48, 65] found the differences between macrochamber and microchamber behaviour of the soft tissue at the heel between healthy and diabetic subjects, no study has investigated the model parameters of different layers of this soft tissue.

5.2. FE models

As shown earlier, in pathological conditions such as diabetes there is an increased interest in investigating the mechanical characteristics of the human plantar soft tissue. In vivo observations indicating that plantar soft tissue properties change as a result of tissue damage [49] or diabetes [12, 46, 65, 90, 91] have highlighted the clinical relevance of plantar soft tissue biomechanics.

In this context, inverse FE analysis can be employed for the deterministic assessment of plantar soft tissue mechanical properties. FE analysis is a powerful numerical method that enables solving problems with complicated geometry, material properties or loading that cannot be approached using analytical solutions. In its direct application, FE analysis enables the calculation of the mechanical response to loading (e.g. internal stresses/strains) of deformable bodies that have known geometry and material properties. However, in inverse FE analysis the values of the mechanical properties that minimise the difference between the in vivo (experimentally measured) response to loading and the simulated (FE) one are calculated.

Two main in vivo material testing techniques have been used to inform inverse FE analyses: indentation [45, 52, 75–78] and compression [79, 80]. In both cases, the plantar soft tissue is compressed between a rigid loading surface and a bony prominence but in the case of indentation, the loading surface is significantly smaller than the plantar surface of the foot. In both cases, the applied force is measured using a load sensor and tissue deformation is either inferred from the displacement of the loading surface [75, 92] or directly measured using medical imaging techniques such as ultrasound [45, 52] or MRI [80]. These measurements enable the calculation of an indentation/compression force-deformation graph that describes the macroscopic mechanical response to loading of the tissue.

As a next step, a FE model simulating the same loading scenario is designed and used to numerically calculate the same force/deformation graph. At first, the material coefficients of the tissue are assigned random values (within predefined range of values) and the difference between the experimental in vivo graph and the numerical one is calculated. An optimisation algorithm is used to update the material coefficients of the plantar soft tissue in search of those values that minimise the difference between experiment and FE simulation.

In this context, Erdemir et al. [45] combined ultrasound indentation with FE modelling to calculate the material properties of plantar soft tissue on a subject-specific basis using an axisymmetric model of the heel. This technique was used to find the hyperelastic (Ogden 1st order) coefficients of the heel pads of two age-matched groups of diabetic and non-diabetic volunteers without however revealing any statistically significant difference between them. In order to improve the level of subject specificity of the model, Chatzistergos et al. [52] developed a 2D plane stress model of the heel from frontal ultrasound images that took into account the subject-specific geometry of heel pad. This modelling technique was later enhanced with a method for the automated generation of 3D models of the heel pad from ultrasound images [19].

The aforementioned studies [19, 52] assumed a bulk plantar soft tissue; however, in a more comprehensive approach Petre et al. [80] differentiated between three layers of soft tissue: skin, fat pad and muscle. For this purpose, a 3D model of the forefoot was developed based on load-bearing MRI and an optimisation-based method was used to inverse engineer the material coefficients for all three different layers. The case of a single layer of soft tissue (i.e. bulk plantar soft tissue) was also considered indicating that simulating different layers affects the value of the calculated peak pressures; however, it was noted that the location where they appear does not get affected.

At this point, it has to be emphasised that there is a limit to the number of coefficients that can be calculated from inverse FE based on indentation or compression alone. To solve this limitation, Fontanella et al. [79] combined in vitro data with the information gathered from in vivo tests. They utilised the in vitro tests to estimate the values of all twelve coefficients for their visco-hyperelastic material model and then used in vivo compression tests to modify six of the coefficients on subject-specific basis.

Even though FE analyses have shed new light on plantar soft tissue biomechanics [35, 45, 79, 93–97], their actual contribution for the improvement of the diagnosis and treatment of the diabetic foot or other foot-related pathologies is limited [98]. This is mainly attributed to the difficulty of using FE analysis outside the research domain and particularly within the context of clinical practice [98]. Developing reliable subject-specific FE modelling techniques that are easy to use and not computationally demanding remains the key barrier for clinically applicable FE modelling.

http://dx.doi.org/10.5772/64170

6. Concluding remarks

The mechanical behaviour of plantar soft tissue shows viscoelasticity characterised by the reaction force being affected by the amount of deformation and deformation rate. The behaviour of the plantar soft tissue is highly nonlinear and shows the strain stiffening effect, with the stress-strain graph showing difference between loading and unloading. In a sense, viscosity causes the reaction force at the same deformation to be less during loading compared to the force during unloading and the difference that is caused by hysteresis increases with an increase in deformation rate. As a complex structure, the plantar soft tissue's mechanical behaviour shows a high degree of dependency on the method of testing evidenced by the fact that the mechanical properties of heel pad extracted from the in vivo and in situ results were observed to be different. This, for example, can be explained by the indication that structural factors such as heel pad thickness and geometry of the calcaneus have an influence on the heel pad behaviour. The models that represent the viscoelastic behaviour of heel pad are scarce and there are few that can fully justify the features of the mechanical behaviour of the plantar soft tissue in different testing scenarios. Despite these limitations, the model parameters that show the viscoelastic behaviour of tissue under load have the potential to be used to assess the mechanical behaviour of soft tissue under load with implication in identifying its malfunction as a result of disease or injury.

Acknowledgements

Support from *DiaBSmart* that was funded by the European Commission under Industry Academia Partnerships and Pathways—November 2011–October 2015—is acknowledged (FP7-PEOPLE-2011-IAPP, Grant Agreement Number 285985).

Author details

Roozbeh Naemi*, Sara Behforootan, Panagiotis Chatzistergos and Nachiappan Chockalingam

*Address all correspondence to: r.naemi@staffs.ac.uk

Faculty of Health Sciences, Staffordshire University, Stoke on Trent, UK

References

[1] Hagisawa, S., and Shimada, T. Skin Morphology and Its Mechanical Properties Associated with Loading, Pressure Ulcer Research: Current and Future Perspectives, D.L. Bader, C.V.C. Bouten, D. Colin, and C.W.J. Oomens, eds., Springer Berlin Heidelberg, Berlin, Heidelberg. 2005;161–185.

[2] Rome K, Campbell RS, Flint AA, Haslock I. Ultrasonic heel pad thickness measurements: a preliminary study. Br J Radiol. 1998;71(851):1149–52.

[3] Cavanagh PR, Morag E, Boulton AJM, Young MJ, Deffner KT, Pammer SE. The relationship of static foot structure to dynamic foot function. J Biomech. 1997;30(3):243–50.

[4] Gooding GA, Stress RM, Graf PM, Grunfeld C. Heel pad thickness: determination by high-resolution ultrasonography. J Ultrasound Med. 1985;4:173–4.

[5] Levy AS, Berkowitz R, Franklin P, Corbett M, Whitelaw GP. Magnetic resonance imaging evaluation of calcaneal fat pads in patients with os calcis fractures. Foot Ankle. 1992;13(2):57–62.

[6] Steinbach HL, Russell W. Measurement of the heel-pad as an aid to diagnosis of acromegaly. Radiography. 1964;82:418–23.

[7] Greene E. Plantar fasciitis and the plantar heel fat pad. Aust Pod. 1995;12:89–93.

[8] Uzel M, Cetinus E, Ekerbicer HC, Karaoguz A. Heel pad thickness and athletic activity in healthy young adults: a sonographic study. J Clin Ultrasound. 2006;34:231–6.

[9] Ledoux, W. R. The Biomechanics of Aging and Diabetic Plantar Soft Tissue. In: Mechanical Properties of Aging Soft Tissues. Springer International Publishing. 2015;pp. 187–206.

[10] Perry J. Gait analysis: normal and pathological function. Slack, New Jersy. 1992.

[11] Cavanagh P, Ulbrecht JS, Capulo G. The biomechanics of the foot in diabetes mellitus. In: Bokwer JH, Pfeifer MA (eds) The diabetic foot. Mosby, St. Louis. 2001.

[12] Hsu C-C, Tsai W-C, Wang C-L, Pao S-H, Shau Y-W, Chuan Y-S. Microchambers and macrochambers in heel pads: are they functionally different? J Appl Physiol [Internet]. 2007;102(6):2227–31. Available from: http://jap.physiology.org/content/102/6/2227.short [cited 2016 Mar 8].

[13] Vincent J. Basic elasticity and viscoelasticity. In: Structural Biomaterials. Princeton University Press, Princeton, New jersey Oxford. 2012;1–28.

[14] Vogel S. Life in moving fluids: the physical biology of flow. Princeton University Press. Princeton University Press, Princeton, New jersey. 1994.

[15] Frost H. Introduction to biomechanics. Charles C Thomas Pub Ltd., Springfield, IL. 1971.

[16] Batchelor GK. An introduction to fluid dynamics. Cambridge University Press, Cambridge UK. 1967.

[17] Lockett FJ. Nonlinear viscoelastic solids. Academic Press, London. 1972.

[18] Fung YC. Biomechanics: mechanical properties of living tissues. Springer-Verlag Inc, New York, USA. 1993.

[19] Behforootan S, Chatzistergos PE, Naemi R, Chockalingam N. Ultrasound based subject specific modelling of the heel. Abstract Book of the XXV Congress of the International Society of Biomechanics, Isb 2015, 12-16 July 2015, International Society of Biomechanics, ISB, Glasgow, Uk, 2015;787–788.

[20] Natali N, Fontanella CG, Carniel EL. Constitutive formulation and analysis of heel pad tissues mechanics. Med Eng Phys. 2010;32(5):516–22.

[21] Bletchschmidth E. The structure of the calcaneal padding. Foot Ankle. 1982;2:260–83.

[22] Buschmann WR, Jahss MH, Kummer F, Desai P, Gee RO, Ricci JL. Histology and histomorphometric analysis of the normal and atrophic heel fat pad. Foot Ankle Int [Internet]. 1995;16(5):254–8. Available from: http://fai.sagepub.com/content/16/5/254.full [cited 2016 Mar 8].

[23] Jahss MH, Michelson JD, Desai P, Kaye R, Kummer F, Buschman W, et al. Investigations into the fat pads of the sole of the foot: anatomy and histology. Foot Ankle Int [Internet]. 1992;13(5):233–42. Available from: http://fai.sagepub.com/content/13/5/233.full [cited 2016 Mar 8].

[24] Resnick RB, Hudgins LC, Buschmann WR, Kummer FJ, Jahss MH. Analysis of the heel pad fat in rheumatoid arthritis. Foot Ankle. 1999;20(8):481–4.

[25] Berlowitz DR, Brienza DM. Are all pressure ulcers the result of deep tissue injury? A review of the literaturet. J Ostomy Wound Manag. 2007;53(10):34–8.

[26] Stekelenburg A, Gawlitta D, Bader DL, Oomens C. Deep tissue injury: how deep is our understanding? J Phys Med Rehabil. 2008;89(7):1410–3.

[27] Gefen A. The biomechanics of heel ulcers. J Tissue Viability. 2010;19(4):124–31.

[28] Le KM, Madsen BL, Barth PW, Ksander GA, Angell JB VL. An in-depth look at pressure sores using monolithic silicon pressure sensors. Plast Reconstr Surg. 1984;76(6):745–54.

[29] Brand PW. The pathomechanics of the diabetic neurotrophic ulcer and its conservative treatment. Acta Orthop Scand. 1982;53(2):309–9.

[30] Brinckmann P. Musculoskeletal biomechanics. Georg Thieme Verlag, Stuttgart, Germany, New York. 2002.

[31] Natali AN, Fontanella CG, Carniel EL. A numerical model for investigating the mechanics of calcaneal fat pad region. J Mech Behav Biomed Mater. 2012;5(1):216–23.

[32] Kuhns JG. Changes in elastic adipose tissue. J Bone Jt Surg-Am. 1949;31(A-3):541–7.

[33] Tietze A. Concerning the architectural structure of the connective tissue in the human sole. Foot Ankle. 1982;2:252–9.

[34] Rome K. Mechanical properties of the heel pad: current theory and review of the literature. Foot. 1998;8:179–85.

[35] Miller-Young JE, Duncan NA, Baroud G. Material properties of the human calcaneal fat pad in compression: experiment and theory. J Biomech [Internet]. 2002;35(12):1523–31. Available from: http://www.ncbi.nlm.nih.gov/pubmed/12445605 [cited 2014 Mar 31].

[36] Bennett MB, Ker RF. The mechanical properties of the human subcalcaneal fat pad in compression. J Anat. 1990;171:131–8.

[37] Alcover EA. Differences between machine and human testing of shock absorbing system. In: Health, biological and enviromental sciences. Middlesex University. PhD thesis, Middlesex University, London, UK. 2000;194.

[38] Pai S, Ledoux WR. The compressive mechanical properties of diabetic and non-diabetic plantar soft tissue. J Biomech [Internet]. 2010;43(9):1754–60. Available from: http://www.sciencedirect.com/science/article/pii/S0021929010001041 [cited 2016 Mar 8].

[39] Pai S, Ledoux WR. The quasi-linear viscoelastic properties of diabetic and non-diabetic plantar soft tissue. Ann Biomed Eng. 2011;39(5):1517–27.

[40] Aerts P, Ker RF, de Clercq D, Ilsley DW. The effects of isolation on the mechanics of the human heel pad. J Anat. 1996;188(Pt 2):417–23.

[41] Ledoux WR, Blevins JJ. The compressive material properties of the plantar soft tissue. J Biomech. 2007;40(13):2975–81.

[42] Gefen A, Megido-Ravid M, Itzchak Y. In vivo biomechanical behavior of the human heel pad during the stance phase of gait. J Biomech. 2001;34(12):1661–5.

[43] Cavanagh PR, et al. Biological aspects of modelling shoe/foot interaction during running. In: Sport shoes play surfaces. 1984;24–46.

[44] Chao CYL, Zheng Y-P, Huang Y-P, Cheing GL-Y. Biomechanical properties of the forefoot plantar soft tissue as measured by an optical coherence tomography-based air-jet indentation system and tissue ultrasound palpation system. Clin Biomech (Bristol, Avon). 2010;25(6):594–600.

[45] Erdemir A, Viveiros ML, Ulbrecht JS, Cavanagh PR. An inverse finite-element model of heel-pad indentation. J Biomech [Internet]. 2006;39(7):1279–86. Available from: http://www.ncbi.nlm.nih.gov/pubmed/15907330 [cited 2014 Mar 31].

[46] Hsu TC, Wang CL, Shau YW, Tang FT, Li KL, Chen CY. Altered heel-pad mechanical properties in patients with type 2 diabetes mellitus. Diabet Med [Internet]. 2000;17(12): 854–9. Available from: http://www.ncbi.nlm.nih.gov/pubmed/11168328 [cited 2014 Mar 31].

[47] Hsu C-C, Tsai W-C, Shau Y-W, Lee K-L, Hu C-F. Altered energy dissipation ratio of the plantar soft tissues under the metatarsal heads in patients with type 2 diabetes

mellitus: a pilot study. Clin Biomech (Bristol, Avon) [Internet]. 2007;22(1):67–73. Available from: http://www.ncbi.nlm.nih.gov/pubmed/17011684 [cited 2014 Mar 31].

[48] Hsu C-C, Tsai W-C, Hsiao T-Y, Tseng F-Y, Shau Y-W, Wang C-L, et al. Diabetic effects on microchambers and macrochambers tissue properties in human heel pads. Clin Biomech (Bristol, Avon) [Internet]. 2009;24(8):682–6. Available from: http://www.ncbi.nlm.nih.gov/pubmed/19619918 [cited 2014 Mar 31].

[49] Rome K, Webb P, Unsworth A, Haslock I. Heel pad stiffness in runners with plantar heel pain. Clin Biomech (Bristol, Avon). 2001;16(10):901–5.

[50] Tong J, Lim C, Goh O. Technique to study the biomechanical properties of the human calcaneal heel pad. Foot [Internet]. 2003;13(2):83–91. Available from: http://www.thefootjournal.com/article/S0958-2592(02)00149-9/abstract [cited 2014 Mar 31].

[51] Zheng YP, Choi YK, Wong K, Chan S, Mak AF. Biomechanical assessment of plantar foot tissue in diabetic patients using an ultrasound indentation system. Ultrasound Med Biol. 2000;26(3):451–6.

[52] Chatzistergos, P. E., Naemi, R., and Chockalingam, N. A method for subject-specific modelling and optimisation of the cushioning properties of insole materials used in diabetic footwear. Medical engineering and physics. 2015;37(6):531–538.

[53] Naemi, R., Chatzistergos, P. E., and Chockalingam, N. A mathematical method for quantifying in vivo mechanical behaviour of heel pad under dynamic load. Medical and biological engineering and computing. 2016;54(2-3):341–350.

[54] Naemi, R., Chatzistergos, P., Sundar, L., Chockalingam, N., and Ramachandran, A. Differences in the mechanical characteristics of plantar soft tissue between ulcerated and non-ulcerated foot. Journal of Diabetes and its Complications. 2016.

[55] Pain MTG, Challis JH. The role of the heel pad and shank soft tissue during impacts: a further resolution of a paradox. J Biomech. 2001;34(3):327–33.

[56] Spears IR, Miller-Young JE. The effect of heel-pad thickness and loading protocol on measured heel-pad stiffness and a standardized protocol for inter-subject comparability. Clin Biomech. 2006;21(2):204–12.

[57] Hsu TC, Wang CL, Tsai WC, Kuo JK, Tang FT. Comparison of the mechanical properties of the heel pad between young and elderly adults. Arch Phys Med Rehabil. 1998;79:1101–4.

[58] Kwan RLC, Zheng YP. Cheing GL. The effect of aging on the biomechanical properties of plantar soft tissues. Clin Biomech. 2010;25(6):601–5.

[59] Hsu CC, Tsai WC, Chen CPC, Shau YW, Wang CL, Chen MJL, et al. Effects of aging on the plantar soft tissue properties under the metatarsal heads at different impact velocities. Ultrasound Med Biol. 2005;31:1423–9.

[60] Kinoshita H, Francio PR, Murase T, Kawai S, Ogawa T. The mechanical properties of the heel pad in elderly adults. J Appl Physiol. 1996;73(5):404–9.

[61] Rome K, Campbell R, Flint A, Haslock I. Heel pad thickness – a contributing factor associated with plantar heel pain. Foot Ankle. 2002;23(2):142–7.

[62] Prichasuk S, Mulpruek P, Siriwongpairat P. The heel pad compressibility. Clin Orthop Relat Res. 1994;300:197–200.

[63] Özdermir H, Söyüncü Y, Özgörgen M, Dabak K. Effects of changes in heel fat pad thickness and elasticity on heel pain. J Am Podiatr Med Assoc. 2004;94(1):47–52.

[64] Kao PF, Davis BL, Hardy PA. Characterization of the calcaneal fat pad in diabetic and non-diabetic patients using magnetic resonance imaging. Magn Reson Imaging [Internet]. 1999;17(6):851–7. Available from: http://www.sciencedirect.com/science/article/pii/S0730725X99000193 [cited 2016 Jan 22].

[65] Hsu T, Lee Y, Shau Y. Biomechanics of the heel pad for type 2 diabetic patients. Clin Biomech. 2002;17:291–6.

[66] Chatzistergos PE, Naemi R, Sundar L, Ramachandran A, Chockalingam N. The relationship between the mechanical properties of heel-pad and common clinical measures associated with foot ulcers in patients with diabetes. J Diabetes Complications http://www.sciencedirect.com/science/article/pii/S105687271400066X. 2014;28(4):488–93. [cited 2016 Feb 28].

[67] Jorgensen U, Larsen E, Varmarken JE. The Hpc-device – a method to quantify the heel pad shock absorbency. Foot Ankle. 1989;10(2):93–8.

[68] Aerts P, De Clercq D. Deformation characteristics of the heel region of the shod foot during a simulated heel strike: the effect of varying midsole hardness. J Sports Sci. 1993;11(5):449–61.

[69] Aerts P, Ker RF, De Clercq D, Ilsley DW, Alexander RM. The mechanical properties of the human heel pad: a paradox resolved. J Biomech. 1995;28(11):1299–308.

[70] Challis JH, Murdoch C, Winter SL. Mechanical properties of the human heel pad: a comparison between populations. J Appl Biomech. 2008;24(4):377–81.

[71] Gefen A. Plantar soft tissue loading under the medial metatarsals in the standing diabetic foot. Med Eng Phys. 2003;25(6):491–9.

[72] Gilchrist LA, Winter DA. A two-part, viscoelastic foot model for use in gait simulations. J Biomech. 1996;29(6):795–8.

[73] Gilchrist LA, Winter DA. A multisegment computer simulation of normal human gait. IEEE Trans Rehabil Eng. 1997;5(4):290–9.

[74] Miller RH, Hamill J. Computer simulation of the effects of shoe cushioning on internal and external loading during running impacts. Comput Methods Biomech Biomed Engin. 2009;12(4):481–90.

[75] Gu Y, Li J, Ren X, Lake MJ, Zeng Y. Heel skin stiffness effect on the hind foot biomechanics during heel strike. Ski Res Technol. 2010;16(3):291–6.

[76] Lin, S. C., Chen, C. P. C., Tang, S. F. T., Chen, C. W., Wang, J. J., Hsu, C. C., ... and Chen, W. P. Stress distribution within the plantar aponeurosis during walking—a dynamic finite element analysis. Journal of Mechanics in Medicine and Biology. 2014;14(04): 1450053.

[77] Lemmon D, Shiang TY, Hashmi A, Ulbrecht JS, Cavanagh PR. The effect of insoles in therapeutic footwear – a finite element approach. J Biomech. 1997;30(6):615–20.

[78] Luo G, Houston VL, Garbarini MA, Beattie AC, Thongpop C. Finite element analysis of heel pad with insoles. J Biomech. 2011;44(8):1559–65.

[79] Fontanella CG, Matteoli S, Carniel EL, Wilhjelm JE, Virga A., Corvi A, et al. Investigation on the load-displacement curves of a human healthy heel pad: in vivo compression data compared to numerical results. Med Eng Phys. 2012;34(9):1253–9.

[80] Petre M, Erdemir A, Panoskaltsis VP, Spirka TA, Cavanagh PR. Optimization of nonlinear hyperelastic coefficients for foot tissues using a magnetic resonance imaging deformation experiment. J Biomech Eng. 2013;135(6):61001–61012. doi: 10.1115/1.4023695.

[81] Cole GK, Nigg BM, van Den Bogert AJ, Gerritsen KGM. The clinical biomechanics award paper 1995 Lower extremity joint loading during impact in running. Clin Biomech (Bristol, Avon). 1996;11(4):181–93.

[82] Spears IR, Miller-young JE, Waters M, Rome K. The effect of loading conditions on stress in the barefooted heel pad. Med Sci Sports Exerc. 2005;37(7):1030–6.

[83] Kerr MF. The time-dependent mechanical properties of the human heel pad in the context locomotion. J Exp Biol. 1996;199:1501–1508.

[84] Noe DA, Voto SJ, Hoffmann MS, Askew MJ, Gradisar I. Role of the calcaneal heel pad and polymeric shock absorbers in attenuation of the heel strike impact. J Biomed Eng. 1993;15:23–6.

[85] Zheng YP, Choi YK, Wong K, Chan S, Mak AFT. Biomechanical assessment of plantar foot tissue in diabetic patients using an ultrasound indentation system. Ultrasound Med Biol. 2000;26:1–20.

[86] Sciumè G, Boso DPP, Gray WGG, Cobelli C, Schrefler BAA. A two-phase model of plantar tissue: a step toward prediction of diabetic foot ulceration. Int J Numer Method Biomed Eng. 2014;30(11):1153–69.

[87] Gray WG, Miller CT. Thermodynamically constrained averaging theory approach for modelling flow and transport phenomena in porous medium systems: 1. Motivation and overview. Adv Water Resour. 2005;28:161–80.

[88] Natali N, Fontanella CG, Carniel EL, Young JM. Biomechanical behaviour of heel pad tissue experimental testing, constitutive formulation, and numerical modelling. Proc Inst Mech Eng Part H J Eng Med. 2011;225(5):449–59.

[89] Scott SH, Winter DA. Biomechanical model of the human foot: kinematics and kinetics during the stance phase of walking. J Biomech [Internet]. 1993;26(9):1091–104. Available from: http://www.ncbi.nlm.nih.gov/pubmed/8408091 [cited 2014 Mar 31].

[90] Klaesner JW, Hastings MK, Zou D, Lewis C, Mueller MJ. Plantar tissue stiffness in patients with diabetes mellitus and peripheral neuropathy. Arch Phys Med Rehabil [Internet]. 2002;83(12):1796–801. Available from: http://www.sciencedirect.com/science/article/pii/S0003999302006068 [cited 2016 Mar 8].

[91] Chao CYL, Zheng Y-P, Cheing GLY. Epidermal thickness and biomechanical properties of plantar tissues in diabetic foot. Ultrasound Med Biol [Internet]. 2011;37(7):1029–38. Available from: http://www.sciencedirect.com/science/article/pii/S0301562911002018 [cited 2016 Mar 8].

[92] Thomas VJ, Patil KM, Radhakrishnan S. Three-dimensional stress analysis for the mechanics of plantar ulcers in diabetic neuropathy. Med Biol Eng Comput. 2004;42:230–5.

[93] Agić A, Nikolić V, Mijović B, Reischl U. Biomechanical model of the diabetic foot. Coll Antropol. 2008;32(2):881–6.

[94] Chen W-PP, Ju C-WW, Tang F-TT. Effects of total contact insoles on the plantar stress redistribution: a finite element analysis. Clin Biomech. 2003;18(6):S17–24.

[95] Budhabhatti SP, Erdemir A, Petre M, Sferra J, Donley B, Cavanagh PR. Finite element modeling of the first ray of the foot: a tool for the design of interventions. J Biomech Eng. 2007;129(5):750–6.

[96] Erdemir A, Saucerman JJ, Lemmon D, Loppnow B, Turso B, Ulbrecht JS, et al. Local plantar pressure relief in therapeutic footwear: design guidelines from finite element models. J Biomech. 2005;38(9):1798–806.

[97] Petre M, Erdemir A, Cavanagh PR. An MRI-compatible foot-loading device for assessment of internal tissue deformation. J Biomech. 2008;41(2):470–4.

[98] Telfer S, Erdemir A, Woodburn J, Cavanagh PR. What has finite element analysis taught us about diabetic foot disease and its management? A systematic review. PLoS One. 2014;9(10):e109994.

Chapter 11

Dynamic Transient Response of Viscoelastic Structures

Jon García-Barruetabeña and
Fernando Cortés Martínez

Additional information is available at the end of the chapter

http://dx.doi.org/10.5772/64253

Abstract

This chapter focuses on investigating the dynamic transient response of viscoelastic structures. First, the influence of nonviscous modes on the vibrational response of exponentially damped systems has been studied on lumped parameter systems where the analytical solution has been derived by modal superposition and by means of Laplace transformation. Then, the analytical solution is obtained by modal superposition and compared to two numerical solutions derived for continuous systems by finite element formulations. These numerical solutions have been solved by modal superposition and by direct integration applying through a particularly built method together with the Newmark method. Finally, an experimental procedure for studying the influence that geometrical properties of viscoelastic joints have on the vibrational response of a metallic beam doubly supported on viscoelastic adhesive joints has been developed.

Keywords: viscoelastic adhesives, relaxation functions, noise and vibration reduction, numerical simulation, experimental test

1. Introduction

This chapter focuses on investigating the dynamic transient response of viscoelastic structures. Mechanical behavior of viscoelastic materials is not only related to the instantaneous stress, but it is also a consequence of the past history of the stress. If dissipative forces in a structural system arise from viscoelastic materials, the classical governing equations of structural dynamics is not reached and direct methods or internal variables should be employed. Nevertheless, direct methods do not provide any information about the contribution of each vibration mode, information of main importance for engineering applications.

For the point of view of the practical application, viscoelastic adhesive joints are used in structural noise control due to its capability to introduce effective modal damping below 1 kHz. In structures under dynamic loads, the transmission of noise and vibration is governed by joint behavior.

Hence, an experimental procedure is presented to analyze the influence of geometry of viscoelastic joint over the dynamic response of the low-order flexural modes of adhesively bonded beams.

2. Analysis of exponentially damped systems

This chapter is aimed at investigating the influence of nonviscous modes on vibrational response of viscoelastic systems. Thus, exponential damping models are considered.

2.1. Lumped parameters systems

In short, the main objective of the present section is to study the influence of nonviscous modes on the vibrational response of exponentially damped systems. The analysis is carried out in time domain over a single degree of freedom (dof). The conditions for the equivalence between Zener and exponential damping models are stated. Next, the time response for free vibration is obtained. On the one hand, the need for solving internal variables is avoided through the Laplace transform properties. On the other hand, internal variables are used, and modal superposition is applied.

Figure 1. Single dof system: (a) using a Zener model, (b) using relaxation functions.

2.1.1. Analysis via a single degree-of-freedom system

The equations of motion for the free vibration of a single dof system using a Zener and an exponential model are obtained in this section. The conditions for the full equivalence between both models are also deduced. The analytical response is reached by two ways: first, the internal variables are eliminated and the transient response is computed by means of Laplace transformation; and second, the internal variables are kept, allowing to apply traditional modal superposition.

2.1.1.1. Solution without solving internal variables

Next, the single dof systems represented in **Figure 1** are analyzed, whose dissipative forces are modeled by means of Zener and exponential damping models, respectively.

The equation of motion for the lumped mass m results in

$$m\ddot{u}_z + c_1(\dot{u}_z - \dot{y}) + k_0 u_z = 0 \tag{1}$$

where c_1 denotes the damping coefficient, k_0 the stiffness of the parallel spring, u_z represents the displacement of the mass, and y is an internal variable needed to solve the problem. This internal variable represents the displacement of the connection point between the dashpot and the in-series spring. In fact, it is necessary to write the force equilibrium equation for the internal variable y, yielding

$$k_1 y = c_1(\dot{u}_z - \dot{y}), \tag{2}$$

where k_1 denotes the stiffness of the spring in series with the damper. Combining Eqs. (1) and (2) the linear differential equation for the displacement u_z is obtained as

$$\frac{c_1}{k_1} m\dddot{u}_z + m\ddot{u}_z + c_1\left(1 + \frac{k_0}{k_1}\right)\dot{u}_z + k_0 u_z = 0. \tag{3}$$

To solve this third-order differential equation, the initial displacement $u_z(0)=u_0$, the initial velocity $\dot{u}(x,0) = \dot{u}_0(x)$, and the initial acceleration $\ddot{u}_z(0)=\ddot{u}_0$ are needed. (In structural dynamics just initial displacement and the initial velocity are needed.) However, this third condition (initial acceleration $\ddot{u}_z(0)=\ddot{u}_0$) can be obtained as a function of the initial displacement.

One way to solve Eq. (3) is making use of the Laplace transform properties, the response $\bar{u}_z(s)$ being

$$\bar{u}_z(s) = \frac{\frac{c_1}{k_1} m u_0 s^2 + m\left(\frac{c_1}{k_1}\dot{u}_0 + u_0\right)s + m\dot{u}_0 + c_1 u_0 + \frac{c_1}{k_1}\left(m\ddot{u}_0 + k_0\dot{u}_0\right)}{\frac{c_1}{k_1} m s^3 + m s^2 + c_1\left(1 + \frac{k_0}{k_1}\right)s + k_0} \tag{4}$$

where $\bar{u}_z(s)$ represents the Laplace transform for the time response $u_z(t)$. Consequently, this time response $u_z(t)$ can be obtained by means of the inverse Laplace transform of Eq. (4).

However, the conditions for the equivalence between both considered damping models are deduced first. Regarding the exponential model, the motion equation [1] for the 1 dof system yields

$$m\ddot{u}_e(t)+\int_0^t c\mu e^{-\mu(t-\tau)}\dot{u}_e(\tau)d\tau+k_0u_e(t)=0 \tag{5}$$

where u_e represents the displacement for the exponential formulation and c the damping coefficient. In contrast to the previous case, to solve Eq. (5) only two initial conditions are needed, the initial displacement $u_e(0)=u_0$ and the initial velocity $\dot{u}_e(0)=\dot{u}_0$.

Accordingly to the Zener formulation, by transforming Eq. (5) into the Laplace domain, the Laplace transform of the displacement $\bar{u}_e(s)$ satisfies

$$\bar{u}_e(s)=\frac{\frac{1}{\mu}mu_0s^2+m\left(\frac{1}{\mu}\dot{u}_0+u_0\right)s+m\dot{u}_0+cu_0}{\frac{1}{\mu}ms^3+ms^2+c\left(1+\frac{k_0}{c\mu}\right)s+k_0} \tag{6}$$

At this point, by comparing Eqs. (4) and (6), a useful observation can be made. These two considered damping models are fully equivalents if three conditions are fulfilled. The first condition to be hold is that the damping coefficients of both models must be the same, $c_1=c$. The second condition is that the relaxation parameter μ of the exponential model and the parameters k_1 and c_1 of the Zener one are related according to $\mu=k_1/c_1$.

And finally, the third condition for the fully equivalence between both formulations results from the relationship between initial displacement and acceleration

$$\ddot{u}_0=-\frac{k_0}{m}u_0, \tag{7}$$

implying that in reality, only two independent initial conditions must be taken into account. Having verified that both formulations are equivalent, from now on, a unique response $u(t)=u_z(t)=u_e(t)$ is considered. Thus, the transient response $u(t)$ can be obtained by means of the inverse Laplace transform of Eq. (4) or (6), either, resulting in

$$u(t)=\sum_{i=1}^{3}\left\{\frac{u_0(s_i^2m+c\mu)+m(\dot{u}_0\mu+u_0s_i\mu+\dot{u}_0s_i)}{3ms_i^2+2sm\mu+c\mu+k_0}e^{s_it}\right\} \tag{8}$$

where s_i denotes the roots of the third-order characteristic equation

$$s^3 + \mu s^2 + \frac{c\mu + k_0}{m} s + \frac{k_0 \mu}{m} = 0. \tag{9}$$

These three roots may be solved by means of the Cardan method [2], yielding

$$s_{1,2}^* = -\frac{c}{3} - \frac{1}{2}(u+v) \pm \mathrm{i}\frac{\sqrt[3]{3}}{2}(u-v) \tag{10}$$

and

$$s_3 = -\frac{c}{3} + (u+v) \tag{11}$$

where $u=\sqrt[3]{q+\sqrt{p}} v=\sqrt[3]{q-\sqrt{p}}$, the parameters p and q being $p=\frac{3B-C^2}{9}$ and $q=\frac{9CB-27A-2C^3}{54}$, respectively, with $A = \mu$, $B=\frac{1}{m}(c\mu + k_0)$, and $C=\frac{k_0\mu}{m}$.

It should be pointed out that, in contrast to a viscous 1 dof system, three roots have been found instead of two, the third extra root being always real, involving an overdamped vibration mode. This fact implies that the system could not oscillate even if the roots $s_{1,2}^*$ are complex (this question was extensively studied by Muller [3] and Adhikari [4]).

2.1.1.2. *Solution solving internal variables*

Next, the time response for the system modeled in **Figure 1(a)** is obtained by means of classical modal superposition, using the internal variable $y(t)$. Rewriting Eqs. (1) and (2) in matrix form, a classical second-order differential equation for free vibration

$$\mathbf{M}\ddot{\mathbf{z}}(t) + \mathbf{C}\dot{\mathbf{z}}(t) + \mathbf{K}\mathbf{z}(t) = \mathbf{0} \tag{12}$$

is reached, where **M**, **C**, and **K** are the mass, damping, and stiffness matrices, given by

$$\mathbf{M} = \begin{bmatrix} m & 0 \\ 0 & 0 \end{bmatrix}; \quad \mathbf{C} = \begin{bmatrix} c_1 & -c_1 \\ -c_1 & c_1 \end{bmatrix}; \quad \mathbf{K} = \begin{bmatrix} k_0 & 0 \\ 0 & k_1 \end{bmatrix} \tag{13}$$

where $z(t)$ vector satisfies $\mathbf{z}(t)=\{u(t) \quad y(t)\}^{\mathrm{T}}$.

Solving the eigenproblem for Eq. (12), the eigenvalues are obtained from the characteristic equation

$$s^3 + \frac{k_1}{c_1} s^2 + \frac{1}{m}(k_1 + k_0)s + \frac{k_1 k_0}{c_1 m} = 0. \tag{14}$$

As it was expected, Eqs. (9) and (14) are equivalent, and the same three eigenvalues s_1^*, s_2^*, and s_3 given by Eqs. (9) and (10) are obtained. Associated to the ith eigenvalue s_i, the ith eigenvector $\mathbf{Z}_i$ satisfies

$$\mathbf{Z}_i = \begin{Bmatrix} U_i \\ Y_i \end{Bmatrix} = \begin{Bmatrix} 1 \\ \dfrac{1}{1 + \dfrac{k_1}{s_i c_1}} \end{Bmatrix} \tag{15}$$

Thus, applying modal superposition, the time response $\mathbf{z}(t)$ can be written as

$$\mathbf{z}(t) = q_1 \mathbf{Z}_1 e^{s_1^* t} + q_2 \mathbf{Z}_2 e^{s_2^* t} + q_3 \mathbf{Z}_3 e^{s_3 t} \tag{16}$$

where q_1, q_2, and q_3 denote the modal participation factors. To solve them, initial conditions $\mathbf{z}(0)$ and $\dot{\mathbf{z}}(0)$ have to be employed. Nevertheless, to establish the initial value for the internal variable $y(0)$ and its derivative $\dot{y}(0)$, some physical assumptions can be made. In order to respect the force equilibrium at the initial instant, the value of the internal variable $y(0)$ must be zero and its time derivative $\dot{y}(0)$ must be the same as the initial velocity $\dot{u}_0$. Hence, the internal variable does not introduce extra energy into the system: the spring of stiffness k_1 is not deformed and the dashpot presents a rigid movement dissipating no energy. Therefore, the initial conditions are given by $\mathbf{z}(0) = \mathbf{z}_0 = \{u(0) \quad y(0)\}^{\mathrm{T}} = \{u_0 \quad 0\}^{\mathrm{T}}$ and $\dot{\mathbf{z}}(0) = \dot{\mathbf{z}}_0 = \{\dot{u}(0) \quad \dot{y}(0)\}^{\mathrm{T}} = \{\dot{u}_0 \quad \dot{u}_0\}^{\mathrm{T}}$.

However, applying these initial conditions, a four equation system with three unknowns is reached. Nevertheless, the system has a unique solution because the rank of the resulting system matrix is 3. This fact is due to the linear combination among the internal variable $y(t)$, its time derivative $\dot{y}(t)$, and the velocity $\dot{u}(t)$ according to Eq. (2). Thus, the three equations needed to solve the modal participation factors yield

$$\begin{Bmatrix} u_0 \\ \dot{u}_0 \\ 0 \end{Bmatrix} = \begin{bmatrix} 1 & 1 & 1 \\ s_1^* & s_2^* & s_3 \\ Y_1 & Y_2 & Y_3 \end{bmatrix} \begin{Bmatrix} q_1 \\ q_2 \\ q_3 \end{Bmatrix} \tag{17}$$

Therefore, once the modal participation factors are solved, the transient response can be achieved using Eq. (16) by modal superposition.

In short, the conditions for the equivalence between the Zener and the exponential models have been established. Using these nonviscous damping models, a 1 dof system has been analyzed and three roots have been obtained. The analytical response has been reached without solving and solving the internal variable. For the former, Laplace transform properties have been employed. For the latter, using the initial value of the internal variable, modal superposition has been applied.

2.2. Continuous systems

Next, different procedures for solving the dynamics of an exponentially damped rod are described. First, the analytical solution is derived by means of modal superposition, solving and without solving internal variables. Then, two different finite element formulations are proposed; one is derived for direct integration methods, and the other to apply modal superposition procedures. The latter allows analyzing the impact of nonviscous modes into the global response.

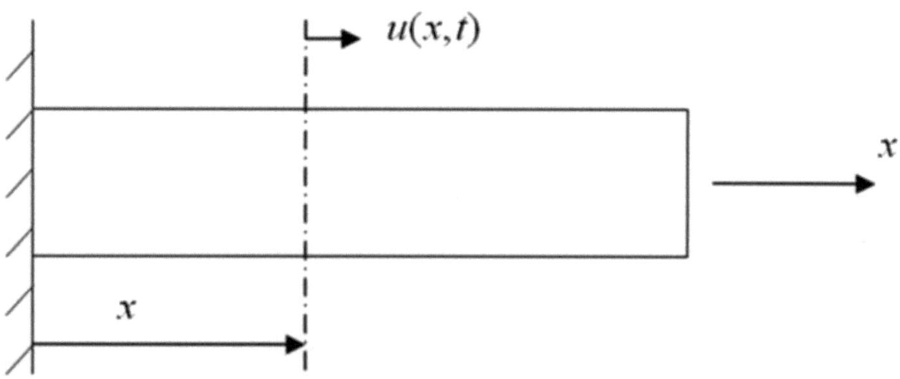

Figure 2. Representation of a continuous and homogeneous rod.

2.2.1. Analytical solution

Regarding the analytical solution, two different procedures are presented. The first developed procedure solves internal variables and the solution is obtained by modal superposition. In the second one, internal variables are avoided and a third-order time derivative field equation is found. The solution conditions are obtained and applied reducing the system equation to that reached using internal variables and enabling the application of modal superposition (**Figure 2**).

2.2.1.1. Solution using internal variables

To solve the displacement field $u(x, t)$ of the rod, the material behavior law

$$\sigma(t)+\frac{\bar{c}_1}{E_1}\frac{\mathrm{d}\sigma(t)}{\mathrm{d}t}=E_0\,\varepsilon(t)+\bar{c}_1\left(1+\frac{E_0}{E_1}\right)\frac{\mathrm{d}\varepsilon(t)}{\mathrm{d}t} \tag{18}$$

has to be taken into account, where $\varepsilon(x,\ t)=\frac{\partial u(x,\ t)}{\partial x}$ represents the strain field and where σ denotes the stress, ε the strain, $\bar{c}_1$ represents the damping coefficient, t is the current time, and E_1 and E_0 are stiffness parameters. The governing equations are obtained using an internal variable field $y(x, t)$. Its gradient $\varepsilon_y(x,\ t)=\frac{\partial y(x,\ t)}{\partial x}$ is related with its time derivative $\dot{\varepsilon}_y(x,t)$ and the strain rate $\dot{\varepsilon}(x,\ t)\dot{\varepsilon}(x,t)$ as

$$E_1\,\varepsilon_y(x,t)=c_1\left[\dot{\varepsilon}(x,t)-\dot{\varepsilon}_y(x,t)\right] \tag{19}$$

Besides, the force equilibrium equation satisfies

$$\sigma(x,t)-c_1\left[\dot{\varepsilon}(x,t)-\dot{\varepsilon}_y(x,t)\right]-E_0\,\varepsilon(x,t)=0 \tag{20}$$

Thus, the field equation can be written in matrix form yielding

$$\begin{bmatrix}\rho S & 0\\ 0 & 0\end{bmatrix}\begin{Bmatrix}\ddot{u}(x,t)\\ \ddot{y}(x,t)\end{Bmatrix}-\begin{bmatrix}c_1 S & -c_1 S\\ -c_1 S & c_1 S\end{bmatrix}\frac{\partial^2}{\partial x^2}\begin{Bmatrix}\dot{u}(x,t)\\ \dot{y}(x,t)\end{Bmatrix}-\begin{bmatrix}E_0 S & 0\\ 0 & E_1 S\end{bmatrix}\frac{\partial^2}{\partial x^2}\begin{Bmatrix}u(x,t)\\ y(x,t)\end{Bmatrix}=\begin{Bmatrix}0\\ 0\end{Bmatrix} \tag{21}$$

where ρ is material density and S denotes cross-sectional area. It should be remarked the correspondence between Eqs. (19) and (21), to those relating the 1 dof case (Eqs. (1)–(3)). Thus, applying variable separation for $u(x, t)$ and $y(x, t)$ as

$$\mathbf{z}(x,t)=\begin{Bmatrix}u(x,t)\\ y(x,t)\end{Bmatrix}=\begin{Bmatrix}U(x)\\ Y(x)\end{Bmatrix}\mathrm{e}^{st} \tag{22}$$

$U(x)$ being the spatial component, $Y(x)$ the component for the internal variable y, and s a complex variable, the time dependence can be eliminated yielding

$$s^2\begin{bmatrix}\rho & 0\\ 0 & 0\end{bmatrix}\begin{Bmatrix}U(x)\\ Y(x)\end{Bmatrix}+\begin{bmatrix}-sc_1-E_0 & sc_1\\ -sc_1 & sc_1+E_1\end{bmatrix}\frac{\mathrm{d}^2}{\mathrm{d}x^2}\begin{Bmatrix}U(x)\\ Y(x)\end{Bmatrix}=\begin{Bmatrix}0\\ 0\end{Bmatrix} \tag{23}$$

The solution of the eigenproblem (23) provides the eigenfunctions $\{U_j(x) \quad Y_j(x)\}^{\mathrm{T}}$ and the eigenvalues λ_j given by

$$\begin{Bmatrix}U_j(x)\\ Y_j(x)\end{Bmatrix}=q_j\begin{Bmatrix}1\\ A_j\end{Bmatrix}\sin(\lambda_j x) \tag{24}$$

and

$$\lambda_j=\frac{\pi}{2\ell}(2j-1) \tag{25}$$

respectively, j being the mode number, q_j the modal participation factor, A_j the amplitude relationship, and ℓ the rod length. As it can be verified through Eq. (25), all eigenvalues λ_j are real. Actually, proportional damping can be considered based on the homogeneity of the material. Therefore, the system presents normal modes, the eigenfunctions $U_j(x)$ being those of the undamped one (see, e.g., Muller [3], for the solution of an undamped rod).

Using eigenfunctions $\{U_j(x) \quad Y_j(x)\}^{\mathrm{T}}$ and the eigenvalues λ_j, the characteristic equation for Eq. (23) yields

$$s^3+s^2\frac{E_1}{c_1}+s\frac{E_1+E_0}{\rho}\lambda_j^2+\frac{E_0E_1}{c_1\rho}\lambda_j^2=0, \tag{26}$$

which is a third-order equation, analogous to that of 1 dof case. Consequently, the solution is configured by infinite groups of three roots $s^*_{j,1}$, $s^*_{j,2}$, and $s_{j,3}$, associated to each eigenfunction $\{U_j(x) \quad Y_j(x)\}^{\mathrm{T}}$ and eigenvalue λ_j. Thus, in each vibration mode, two elastic and one nonviscous components are involved.

Thus, the time response by means of modal superposition can be written as

$$\mathbf{z}(x,t)=\sum_{j=1}^{\infty}\left(q_{j,1}\mathbf{Z}_{j,1}\mathrm{e}^{s^*_{j,1}t}+q_{j,2}\mathbf{Z}_{j,2}\mathrm{e}^{s^*_{j,2}t}+q_{j,3}\mathbf{Z}_{j,3}\mathrm{e}^{s_{j,3}t}\right)\sin(\lambda_j x) \tag{27}$$

where $q_{j,k}$ ($k = 1, 2, 3$) are the modal participation factors and $\mathbf{Z}_{j,k}$ are eigenvectors given by

$$\mathbf{Z}_{j,k} = \begin{Bmatrix} 1 \\ A_{j,k} \end{Bmatrix} = \begin{Bmatrix} 1 \\ \dfrac{1}{1 + \dfrac{E_1}{s_{j,k} c_1}} \end{Bmatrix} \tag{28}$$

It should be noted that the resulting expression for the eigenvectors of the continuous system hold in Eq. (28) is analogous to the one corresponding to the 1 dof case (see Eq. (15)). Therefore, the time response can be obtained by modal superposition by solving, for each jth mode the modal participation factors $q_{j,k}$. However, from the equations provided by the initial conditions $\mathbf{z}_0(x)$ and $\dot{\mathbf{z}}_0(x)$,

$$\mathbf{z}_0(x) = \sum_{j=1}^{\infty} \left(q_{j,1}\mathbf{Z}_{j,1} + q_{j,2}\mathbf{Z}_{j,2} + q_{j,3}\mathbf{Z}_{j,3} \right) \sin(\lambda_j x) \tag{29}$$

and

$$\dot{\mathbf{z}}_0(x) = \sum_{j=1}^{\infty} \left(s_{j,1}^* q_{j,1}\mathbf{Z}_{j,1} + s_{j,2}^* q_{j,2}\mathbf{Z}_{j,2} + s_{j,3} q_{j,3}\mathbf{Z}_{j,3} \right) \sin(\lambda_j x) \tag{30}$$

these $q_{j,k}$ cannot be directly solved because infinite unknowns are present. To overcome this situation, the consideration of proportional damping can be recalled, concluding that the eigenfunctions $U_j(x)$ have to be orthogonal with respect to the mass and stiffness operators. Hence, the modes can be decoupled according to (see, e.g., Ref. [5] for details)

$$\int_{\ell} U_i(x) \rho S U_j(x) \mathrm{d}x = 0, \quad \text{for } i \neq j \tag{31}$$

Accordingly, applying Eqs. (29)–(31), it is obtained

$$\int_{\ell} \mathbf{z}_0(x) \sin(\lambda_j x) \mathrm{d}x = \left(q_{j,1}\mathbf{Z}_{j,1} + q_{j,2}\mathbf{Z}_{j,2} + q_{j,3}\mathbf{Z}_{j,3} \right) \int_{\ell} \sin^2(\lambda_j x) \mathrm{d}x \tag{32}$$

and

$$\int_{\ell} \dot{\mathbf{z}}_0(x) \sin(\lambda_j x) \mathrm{d}x = \left(s_{j,1}^* q_{j,1}\mathbf{Z}_{j,1} + s_{j,2}^* q_{j,2}\mathbf{Z}_{j,2} + s_{j,3} q_{j,3}\mathbf{Z}_{j,3} \right) \int_{\ell} \sin^2(\lambda_j x) \mathrm{d}x \tag{33}$$

which constitute a four-equation and three unknown system. Here, it should be pointed out (see Eq. (19)) that the internal variable $y(x, t)$ and its time derivative $\dot{y}(x,t)$ are linearly combined together with the displacement $u(x, t)$. Consequently, the rank of the equation system is 3. Therefore, only the initial conditions are needed, namely $u_0(x)$ and $\dot{u}_0(x)$, together with the initial value for the internal variable field $y_0(x)$ or together with its time derivative $\dot{y}_0(x)$. The initial condition for $y_0(x)$ and its time derivative $\dot{y}_0(x)$ have to satisfy $y_0(x)=0$ and $\dot{y}_0(x)=\dot{u}_0(x)$, respectively. Hence, the equation to resolve each group $q_{j,1}$, $q_{j,2}$, and $q_{j,3}$ of modal participation factors is given by

$$\frac{2}{\ell}\int_{\ell}\begin{Bmatrix} u_0(x) \\ \dot{u}_0(x) \\ 0 \end{Bmatrix}\sin(\lambda_j x)\,\mathrm{d}x = \begin{bmatrix} 1 & 1 & 1 \\ s^*_{j,1} & s^*_{j,2} & s_{j,3} \\ A_{j,1} & A_{j,2} & A_{j,3} \end{bmatrix}\begin{Bmatrix} q_{j,1} \\ q_{j,2} \\ q_{j,3} \end{Bmatrix} \tag{34}$$

Thus, once the modal participation factors are solved for each mode, the transient response by modal superposition can be achieved, giving

$$u(x,t) = \sum_{j=1}^{\infty}\left(q_{j,1}\mathrm{e}^{s^*_{j,1}t} + q_{j,2}\mathrm{e}^{s^*_{j,2}t} + q_{j,3}\mathrm{e}^{s_{j,3}t}\right)\sin(\lambda_j x) \tag{35}$$

2.2.1.2. Solution without using internal variables

Next, the analytical solution for the displacement field $u(x, t)$ of the rod is reached without introducing internal variables. Thus, recalling the material behavior law (Eq. (18)), the field equation can be written as

$$\frac{c_1}{E_1}\rho S\dddot{u}(x,t) + \rho S\ddot{u}(x,t) - c_1 S\left(1+\frac{E_0}{E_1}\right)\frac{\partial^2 \dot{u}(x,t)}{\partial x^2} - E_0 S\frac{\partial^2 u(x,t)}{\partial x^2} = 0 \tag{36}$$

Eq. (36) is a third-order equation in time, analogous to the one relating the 1 dof case (Eq. (3)). Applying variable separation for $u(x, t)$ as $u(x, t) = U(x)\,\mathrm{e}^{st}$ it yields

$$-\left[s^3\rho\frac{c_1}{E_1} + s^2\rho\right]U(x) + \left[sc_1\left(1+\frac{E_0}{E_1}\right) + E_0\right]\frac{\partial^2 U(x)}{\partial x^2} = 0 \tag{37}$$

The solution of the eigenproblem (37) provides the eigenfunctions $U_j(x)$, given by $U_j(x) = B_j \sin(\lambda_j x)$ where the eigenvalues λ_j are those of Eq. (25), j being the mode number and B_j the amplitude. Therefore, the characteristic Eq. (26) is reached also.

Hence, if the solution is going to be derived without using internal variables, it is formed by infinite groups of three roots $s^*_{j,1}$, $s^*_{j,2}$, and $s_{j,3}$, associated to each eigenfunction $U_j(x)$ and eigenvalue λ_j. Although, three initial conditions are needed to solve Eq. (36). These are the initial displacement $u(x, 0)=u_0(x)$, the initial velocity $\dot{u}(x,0)=\dot{u}_0(x)$, and the initial acceleration $\ddot{u}(x, 0)=\ddot{u}_0(x)$, but, it can be proved that this initial acceleration $\ddot{u}_0(x)$ can be written as a function of the initial displacement $u_0(x)$.

In fact, using the Boltzmann superposition principle [6], the memory of a viscoelastic material can be properly modeled using hereditary models. The stress can be evaluated using relaxation functions $R(t)$ through convolution integrals given by

$$\sigma(t)=\int_0^t E(t-\tau)\dot{\varepsilon}(\tau)\mathrm{d}\tau = E_0\varepsilon(t)+\int_0^t R(t-\tau)\dot{\varepsilon}(\tau)\mathrm{d}\tau \tag{38}$$

where $E(t)$ is the relaxation modulus of the material, τ corresponds to the retardation time, and $(\bullet)$ represents time derivative. A relaxation function $R(t)$ widely used in the literature [7–9] is the exponential model $R(t) = c\mu\, \mathrm{e}^{-\mu t}$, where μ is the relaxation parameter and c represents the damping coefficient. Hence, using Eq. (38) and writing the force equilibrium for the continuous rod, it is obtained

$$\rho\ddot{u}(x,t)=E_0\frac{\partial^2 u(x,t)}{\partial x^2}+\int_0^t R(t-\tau)\frac{\partial^2 u(x,t)}{\partial x^2}\mathrm{d}\tau. \tag{39}$$

Assuming causality, the material presents no memory prior to initial conditions and therefore, the initial acceleration $\ddot{u}_0(x)$ can be written as

$$\ddot{u}_0(x)=\frac{E_0}{\rho}\frac{\partial^2 u(x,0)}{\partial x^2}. \tag{40}$$

Hence, recalling Eq. (37), from Eq. (40) the initial acceleration yields

$$\ddot{u}_0(x)=-\frac{E_0}{\rho}\lambda^2 u_0(x). \tag{41}$$

Thus, the time response can be written also as

$$u(x,t)=\sum_{j=1}^{\infty}\left(B_{j,1}\mathrm{e}^{s^*_{j,1}t}+B_{j,2}\mathrm{e}^{s^*_{j,2}t}+B_{j,3}\mathrm{e}^{s_{j,3}t}\right)\sin(\lambda_j x) \tag{42}$$

where $B_{j,k}$ ($k = 1, 2, 3$) are the modal participation factors. Thus, these modal participation factors $B_{j,k}$ must be solved for each jth mode in order to solve the time response by means of modal superposition. However, from the equations derived from the initial conditions $u_0(x)$, $\dot{u}_0(x)$, and $\ddot{u}_0(x)$

$$u_0(x) = \sum_{j=1}^{\infty}\left(B_{j,1} + B_{j,2} + B_{j,3}\right)\sin(\lambda_j x) \tag{43}$$

$$\dot{u}_0(x) = \sum_{j=1}^{\infty}\left(s_{j,1}^{*}B_{j,1} + s_{j,2}^{*}B_{j,2} + s_{j,3}B_{j,3}\right)\sin(\lambda_j x) \tag{44}$$

and

$$\ddot{u}_0(x) = \sum_{j=1}^{\infty}\left[\left(s_{j,1}^{*}\right)^2 B_{j,1} + \left(s_{j,2}^{*}\right)^2 B_{j,2} + \left(s_{j,3}\right)^2 B_{j,3}\right]\sin(\lambda_j x) \tag{45}$$

these $B_{j,k}$ cannot be directly solved because infinite unknowns are present. Hence, applying the orthogonality conditions, the modes can be decoupled according to Eq. (31) where variable separation together with Eq. (25) is needed. Consequently, it is obtained

$$\int_{\ell} u_0(x)\sin(\lambda_j x)\,\mathrm{d}x = \left(B_{j,1} + B_{j,2} + B_{j,3}\right)\int_{\ell}\sin^2(\lambda_j x)\mathrm{d}x \tag{46}$$

$$\int_{\ell} \dot{u}_0(x)\sin(\lambda_j x)\mathrm{d}x = \left(s_{j,1}^{*}B_{j,1} + s_{j,2}^{*}B_{j,2} + s_{j,3}B_{j,3}\right)\int_{\ell}\sin^2(\lambda_j x)\mathrm{d}x \tag{47}$$

and

$$\int_{\ell} \ddot{u}_0(x)\sin(\lambda_j x)\mathrm{d}x = \left[\left(s_{j,1}^{*}\right)^2 B_{j,1} + \left(s_{j,2}^{*}\right)^2 B_{j,2} + \left(s_{j,3}\right)^2 B_{j,3}\right]\int_{\ell}\sin^2(\lambda_j x)\mathrm{d}x \tag{48}$$

which constitute a 3-equation and 3-unknown system. Hence, the equation to solve each group $B_{j,1}$, $B_{j,2}$, and $B_{j,3}$ of modal participation factors is given by

$$\frac{2}{\ell}\int_{\ell}\begin{Bmatrix} u_0(x) \\ \dot{u}_0(x) \\ \ddot{u}_0(x) \end{Bmatrix}\sin(\lambda_j x)\mathrm{d}x = \begin{bmatrix} 1 & 1 & 1 \\ s_{j,1}^* & s_{j,2}^* & s_{j,3} \\ \left(s_{j,1}^*\right)^2 & \left(s_{j,2}^*\right)^2 & \left(s_{j,3}\right)^2 \end{bmatrix}\begin{Bmatrix} B_{j,1} \\ B_{j,2} \\ B_{j,3} \end{Bmatrix} \tag{49}$$

or taking into account Eq. (41), it can be also written as

$$\frac{2}{\ell}\int_{\ell}\begin{Bmatrix} u_0(x) \\ \dot{u}_0(x) \\ -\dfrac{E_0}{\rho}\lambda_j^2 u_0(x) \end{Bmatrix}\sin(\lambda_j x)\mathrm{d}x = \begin{bmatrix} 1 & 1 & 1 \\ s_{j,1}^* & s_{j,2}^* & s_{j,3} \\ \left(s_{j,1}^*\right)^2 & \left(s_{j,2}^*\right)^2 & \left(s_{j,3}\right)^2 \end{bmatrix}\begin{Bmatrix} B_{j,1} \\ B_{j,2} \\ B_{j,3} \end{Bmatrix} \tag{50}$$

It should be remarked that the system equations provided in Eq. (34) to solve the modal participation factors $q_{j,1}$, $q_{j,2}$, and $q_{j,3}$ are equivalent to that provided in Eq. (50) to solve those of the system without internal variables $B_{j,1}$, $B_{j,2}$, and $B_{j,3}$. In fact, it can be verified that the third row of Eq. (50) is a linear combination of the third and first rows of Eq. (34) as

$$\mathrm{row}_{3,39} = \alpha_j \times \mathrm{row}_{1,22} + \beta_j \times \mathrm{row}_{3,22} \tag{51}$$

where $\alpha_j = -\frac{E_0}{\rho}\lambda_j^2$ and $\beta_j = -\frac{E_1}{\rho}\lambda_j^2$.

As a conclusion it should be highlighted that if internal variables are avoided, the system equation reached (Eq. (49)) is equivalent to the one obtained using internal variables (Eq. (34)). Accordingly, the response of Eq. (34) can be derived by modal superposition through Eq. (35).

As a conclusion, the governing equations for a nonviscously damped rod have been obtained, using and without using internal variables. First, considering internal variables, the analytical response has been reached applying modal superposition. Then, without considering internal variables a third-order equation system in time is found where three initial conditions needed (see solve Eq. (36)). However, it has been proved that the initial acceleration $\ddot{u}_0(x)$ is a function of the initial displacement $u_0(x)$. Also, the equality between the equation systems (49) and (34) has been validated. Consequently, the response has been solved by modal superposition.

2.2.2. Finite element formulation

In this section, two finite element formulations are developed. These formulations allow to compute the motion (see Eq. (21) or (36)) of continuous rods with exponential damping. One is conceived to apply modal superposition and the other for direct integration methods.

The one for modal superposition allows to compute the time response (Eq. (21)) by solving two internal variables per finite element. The one created for direct integration, solves Eq. (36) by direct methods. This solution is accurate and efficient (no internal variables are solved).

2.2.2.1. Finite element formulation for modal superposition

Next, a finite element formulation to solve Eq. (21) is presented. It is a two-node formulation, leading to a four dof nodal displacement vector $\mathbf{z}_e(t)=\{u_1(t)\quad y_1(t)\quad u_2(t)\quad y_2(t)\}^T$ considering the nodal displacements $u_1(t)$ and $u_2(t)$ and the corresponding internal variables $y_1(t)$ and $y_2(t)$. The approximate displacement $\mathbf{z}_e(x, t)$ can be written as

$$\mathbf{z}_e(x,t)=\mathbf{N}(x)\mathbf{z}_e(t)=\begin{bmatrix} N_1(x) & 0 & N_2(x) & 0 \\ 0 & N_1(x) & 0 & N_2(x)\end{bmatrix}\begin{Bmatrix} u_1(t) \\ y_1(t) \\ u_2(t) \\ y_2(t)\end{Bmatrix} \tag{52}$$

where $\mathbf{N}(x)$ is the matrix of interpolation functions, in which linear interpolation ones are employed (see, e.g., Refs. [5, 10] for details about the finite element method). Thus, the weighted residual technique from the Galerkin point of view gives

$$\begin{aligned}&\int_{\ell_e}\mathbf{N}^T(x)\begin{bmatrix}\rho S & 0\\ 0 & 0\end{bmatrix}\ddot{\mathbf{z}}_e(x,t)\mathrm{d}x-\int_{\ell_e}\mathbf{N}^T(x)\begin{bmatrix}c_1S & -c_1S\\ -c_1S & c_1S\end{bmatrix}\frac{\partial^2\dot{\mathbf{z}}_e(x,t)}{\partial x^2}\mathrm{d}x-\\ &\int_{\ell_e}\mathbf{N}^T(x)\begin{bmatrix}E_0S & 0\\ 0 & E_1S\end{bmatrix}\frac{\partial^2\mathbf{z}_e(x,t)}{\partial x^2}\mathrm{d}x=\mathbf{0}.\end{aligned} \tag{53}$$

Solving by parts the second and the third integrals, Eq. (53) yields

$$\begin{aligned}&\int_{\ell_e}\mathbf{N}^T(x)\begin{bmatrix}\rho S & 0\\ 0 & 0\end{bmatrix}\mathbf{N}(x)\mathrm{d}x\,\ddot{\mathbf{z}}_e(t)+\int_{\ell_e}\frac{\mathrm{d}\mathbf{N}^T(x)}{\mathrm{d}x}\begin{bmatrix}c_1S & -c_1S\\ -c_1S & c_1S\end{bmatrix}\frac{\mathrm{d}\mathbf{N}(x)}{\mathrm{d}x}\mathrm{d}x\,\dot{\mathbf{z}}_e(t)+\\ &\int_{\ell_e}\frac{\mathrm{d}\mathbf{N}^T(x)}{\mathrm{d}x}\begin{bmatrix}E_0S & 0\\ 0 & E_1S\end{bmatrix}\frac{\mathrm{d}\mathbf{N}(x)}{\mathrm{d}x}\mathrm{d}x\,\mathbf{z}_e(t)=\mathbf{F}_e(t).\end{aligned} \tag{54}$$

Therefore, a classical second-order equation with constant coefficients

$$\mathbf{M}_e\ddot{\mathbf{z}}_e(t)+\mathbf{C}_e\dot{\mathbf{z}}_e(t)+\mathbf{K}_e\dot{\mathbf{z}}_e(t)=\mathbf{F}_e(t) \tag{55}$$

is reached. Consequently, the mass, damping and stiffness elementary matrices yield

$$\mathbf{M}_e=\frac{\rho S\ell_e}{6}\begin{bmatrix}2&0&1&0\\0&0&0&0\\1&0&2&0\\0&0&0&0\end{bmatrix}, \tag{56}$$

$$\mathbf{C}_e=\frac{c_1 S}{\ell_e}\begin{bmatrix}1&-1&-1&1\\-1&1&1&-1\\-1&1&1&-1\\1&-1&-1&1\end{bmatrix} \tag{57}$$

and

$$\mathbf{K}_e=\frac{S}{\ell_e}\begin{bmatrix}E_0&0&-E_0&0\\0&E_1&0&-E_1\\-E_0&0&E_0&0\\0&-E_1&0&E_1\end{bmatrix} \tag{58}$$

Assembling the elementary matrices, the state-space equation

$$\begin{bmatrix}\mathbf{C}&\mathbf{M}\\\mathbf{M}&\mathbf{0}\end{bmatrix}\begin{Bmatrix}\dot{\mathbf{z}}(t)\\\ddot{\mathbf{z}}(t)\end{Bmatrix}=\begin{Bmatrix}\mathbf{F}(t)\\\mathbf{0}\end{Bmatrix}+\begin{bmatrix}-\mathbf{K}&\mathbf{0}\\\mathbf{0}&\mathbf{M}\end{bmatrix}\begin{Bmatrix}\mathbf{z}(t)\\\dot{\mathbf{z}}(t)\end{Bmatrix} \tag{59}$$

is reached. It should be remarked that the size of the eigenproblem obtained from Eq. (59) becomes $4N$. However, reminding that the extra degrees of freedom have no associated mass, the rank of the matrix system is $3N$. Hence, the response yields

$$\mathbf{z}(t)=\sum_{j=1}^{3N} q_j\mathbf{Z}_j e^{s_j t} \tag{60}$$

where q_j represents the modal participation factors, $\mathbf{Z}_j$ the system eigenvectors, and s_j the system roots. To compute the $3N$ modal participation factors, $2N$ equations can be written from the initial conditions $\mathbf{u}_0=\mathbf{u}(0)$ and $\dot{\mathbf{u}}_0=\dot{\mathbf{u}}(0)$, where the N remaining equations can be derived from the initial values for the internal variables, satisfying $\dot{\mathbf{y}}_0 = \dot{\mathbf{y}}(0) = \dot{\mathbf{u}}_0$. Therefore, once the modal participation factors q_j are solved, the response can be computed by modal superposition through Eq. (60).

2.2.2.2. Finite element formulation for direct integration

Next, a finite element formulation to solve Eq. (36) is presented. For that, the rod is discretized in two-node finite elements of length ℓ_e with linear interpolation functions. The weighted residual technique from the Galerkin point of view gives

$$\begin{aligned}&\int_{\ell_e} \mathbf{N}^{\mathrm{T}}(x)\frac{c_1}{E_1}\rho S\mathbf{N}(x)\mathrm{d}x\,\dddot{\mathbf{u}}_e(t) + \int_{\ell_e} \mathbf{N}^{\mathrm{T}}(x)\rho S\mathbf{N}(x)\mathrm{d}x\,\ddot{\mathbf{u}}_e(t) - \\ &\int_{\ell_e} \mathbf{N}^{\mathrm{T}}(x)c_1 S\left(1+\frac{E_0}{E_1}\right)\frac{\partial^2 \dot{\mathbf{u}}_e(x,t)}{\partial x^2}\mathrm{d}x - \\ &\int_{\ell_e} \mathbf{N}^{\mathrm{T}}(x)E_0 S\frac{\partial^2 \mathbf{u}_e(x,t)}{\partial x^2}\mathrm{d}x = \mathbf{0},\end{aligned} \tag{61}$$

where $(\bullet)^{\mathrm{T}}$ denotes the transposition operator, $\mathbf{N}(x)$ is the matrix of interpolation functions, and $\mathbf{u}_e(t) = \{u_1(t) \;\; u_2(t)\}^{\mathrm{T}}$ is the nodal displacement vector. Solving by parts the third and the fourth integrals of Eq. (61), it gives

$$\begin{aligned}&\int_{\ell_e} \mathbf{N}^{\mathrm{T}}(x)\frac{c_1}{E_1}\rho S\mathbf{N}(x)\mathrm{d}x\,\dddot{\mathbf{u}}_e(t) + \int_{\ell_e} \mathbf{N}^{\mathrm{T}}(x)\rho S\mathbf{N}(x)\mathrm{d}x\,\ddot{\mathbf{u}}_e(t) + \\ &\int_{\ell_e} \frac{\mathrm{d}\mathbf{N}^{\mathrm{T}}(x)}{\mathrm{d}x}c_1 S\left(1+\frac{E_0}{E_1}\right)\frac{\mathrm{d}\mathbf{N}(x)}{\mathrm{d}x}\mathrm{d}x\,\dot{\mathbf{u}}_e(t) + \\ &\int_{\ell_e} \frac{\mathrm{d}\mathbf{N}^{\mathrm{T}}(x)}{\mathrm{d}x}E_0 S\frac{\mathrm{d}\mathbf{N}(x)}{\mathrm{d}x}\mathrm{d}x\,\mathbf{u}_e(t) = \mathbf{F}_e(t),\end{aligned} \tag{62}$$

where $\mathbf{F}_e(t)$ represents the external forces nodal vector. Thus, the third order in time matrix system

$$\mathbf{J}_e\dddot{\mathbf{u}}_e(t) + \mathbf{M}_e\ddot{\mathbf{u}}_e(t) + \mathbf{C}_e\dot{\mathbf{u}}_e(t) + \mathbf{K}_e\mathbf{u}_e(t) = \mathbf{F}_e(t) \tag{63}$$

is reached, in which the elementary matrices to be assembled are

$$\mathbf{J}_e = \frac{\rho S\ell_e}{6}\frac{c_1}{E_1}\begin{bmatrix}2 & 1\\ 1 & 2\end{bmatrix} \tag{64}$$

$$\mathbf{M}_{\mathrm{e}} = \frac{\rho S \ell_{\mathrm{e}}}{6} \begin{bmatrix} 2 & 1 \\ 1 & 2 \end{bmatrix} \tag{65}$$

$$\mathbf{C}_{\mathrm{e}} = \frac{c_1 S}{\ell_{\mathrm{e}}} \left(1 + \frac{E_0}{E_1} \right) \begin{bmatrix} 1 & -1 \\ -1 & 1 \end{bmatrix} \tag{66}$$

and

$$\mathbf{K}_{\mathrm{e}} = \frac{E_0 S}{\ell_{\mathrm{e}}} \begin{bmatrix} 1 & -1 \\ -1 & 1 \end{bmatrix} \tag{67}$$

Assembling all the finite element matrices, Eq. (63) yields

$$\mathbf{J}\ddot{\mathbf{u}}(t) + \mathbf{M}\ddot{\mathbf{u}}(t) + \mathbf{C}\dot{\mathbf{u}}(t) + \mathbf{K}\mathbf{u}(t) = \mathbf{F}(t) \tag{68}$$

To solve Eq. (68) by means of traditional direct methods of structural dynamics, the procedure employed in the Cortes Mateos Elejabarrieta (CME) method [11] is used next. For that, the backward definition of the first derivative $\dot{\mathbf{u}}(t)$ of the displacement $\mathbf{u}(t)$ at the time t_{n+1} can be approximated as $\dot{\mathbf{u}}_{n+1} \simeq \Delta t^{-1}\left(\mathbf{u}_{n+1} - \mathbf{u}_n\right)$, where $\Delta t = t_{n+1} - t_n$ is a finite time step. Hence, Eq. (68) yields

$$\mathbf{M}_{\mathrm{eq}}\ddot{\mathbf{u}}_{n+1} + \mathbf{C}\dot{\mathbf{u}}_{n+1} + \mathbf{K}\mathbf{u}_{n+1} = \mathbf{F}_{\mathrm{eq},n+1} \tag{69}$$

with $\mathbf{M}_{\mathrm{eq}} = \mathbf{M} + \Delta t^{-1}\mathbf{J}$ and $\mathbf{F}_{\mathrm{eq},n+1} = \mathbf{F}_{n+1} + \Delta t^{-1}\mathbf{J}\ddot{\mathbf{u}}_n$ representing an equivalent second-order forced system. The system response can be computed through direct integration methods as, for example, the Newmark method.

However, assuming causality, there is no memory prior the initial conditions are applied and therefore, the initial acceleration $\ddot{\mathbf{u}}_0 = \ddot{\mathbf{u}}(0)$ is computed from

$$\mathbf{M}\ddot{\mathbf{u}}_0 + \mathbf{K}\mathbf{u}_0 = \mathbf{F}_0 \tag{70}$$

http://dx.doi.org/10.5772/64253

3. Experimental analysis of the vibrational response of an adhesively bonded beam

In this section, an experimental procedure for studying the influence that geometrical properties of adhesive joints have on the vibrational response of a metallic beam doubly supported on viscoelastic adhesive joints is presented. A test bench has been specifically constructed for the experimental program. At a first step, the modal shapes have been experimentally identified. Regarding the experiments, the beams are seismically excited and the influence of joint thickness and overlapping length on the beam motion is studied by computing the root-mean-square (rms) value of 21 transmissibility functions obtained along the length of the beam. The analysis is carried out on resonance frequencies, peak amplitudes, and modal loss factors.

3.1. Experimental program

The objective of the experiments is to obtain sets of frequency f dependent transmissibility functions $T(f)$ to study the influence that overlapping length ℓ_0 and joint thickness h have on the beam vibrational response.

Item	Use
PC 1	Controls the dSPACE card
dSPACE (+charge amplifier)	Activates the piezoelectric actuator
Piezoelectric actuator PPA40M	Imposes the motion to the adhesive base
Tri-axial accelerometer ICP PCB 356A16	Measures the adhesive base acceleration
PULSE B&K acquisition system	Samples and processes the transducer time signal
PC 2	Controls the PULSE B&K acquisition system
Laser interferometer (+signal amplifier)	Measures the beam velocity response

Table 1. Measurement equipment.

3.1.1. Equipment

Figure 1 shows a schematic diagram of the experimental setup, whereas **Table 1** gives the details about measuring equipment. **Figure 1** shows three groups of components: excitation of the system, data acquisition and processing, and the adhesively bonded beam itself (**Figure 3**).

The base motion is imposed by a PPA40M piezoelectric actuator [12] controlled by a DS1104 dSPACE real-time control card [13]. This card governs the piezoelectric actuator by a control program developed in Simulink® that generates a digital control signal. This is converted into an analogical one by the dSPACE card. Finally, it is amplified and sent to the piezoelectric actuator.

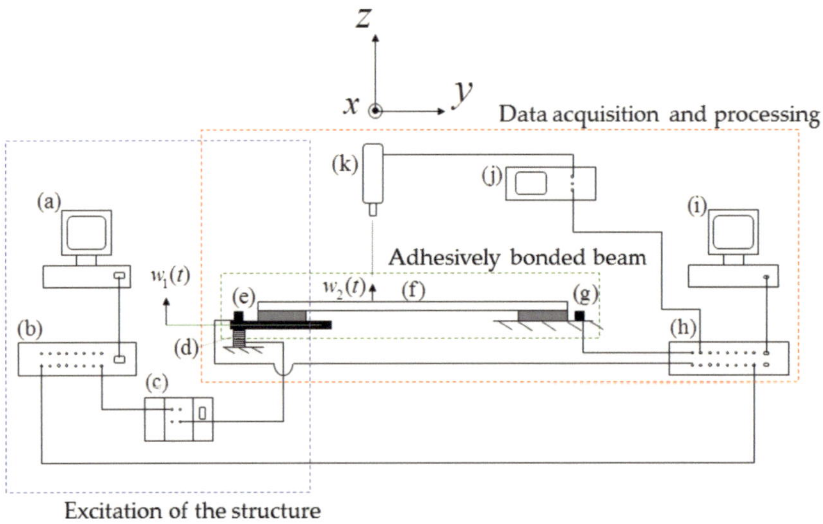

Figure 3. Experimental setup: (a) computer 1, (b) real-time dSPACE card, (c) signal conditioner, (d) piezoelectric device, (e) triaxial accelerometer (1), (f) beam specimen, (g) triaxial accelerometer (2), (h) PULSE acquisition system, (i) computer 2, (j) laser signal amplifier, (k) laser interferometer OFV-505.

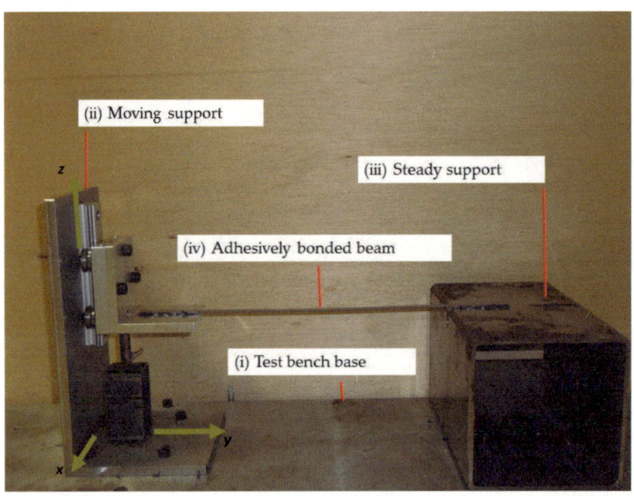

Figure 4. Test bench: general view.

The data acquisition is done by a B&K PULSE acquisition system [14]. Two response sensors are used: a triaxial accelerometer ICP 356A16 of PCB electronics [15] for measuring the adhesive base motion and a laser interferometer OFV-505 of POLYTEC [16] for measuring the beam response. Hence, to validate the excitation, the acceleration of the adhesive base is measured in three directions, whereas the beam motion is measured just transversally. This velocity signal $\dot{w}_2(t)$ is derived to obtain the corresponding acceleration $\ddot{w}_2(t)$.

The designed test bench ensures a repetitive test procedure. Mainly, it is composed by a moving and a steady support assembled on a rigid base. **Figure 4** illustrates a general view of the test bench while **Figure 5** shows the moving support.

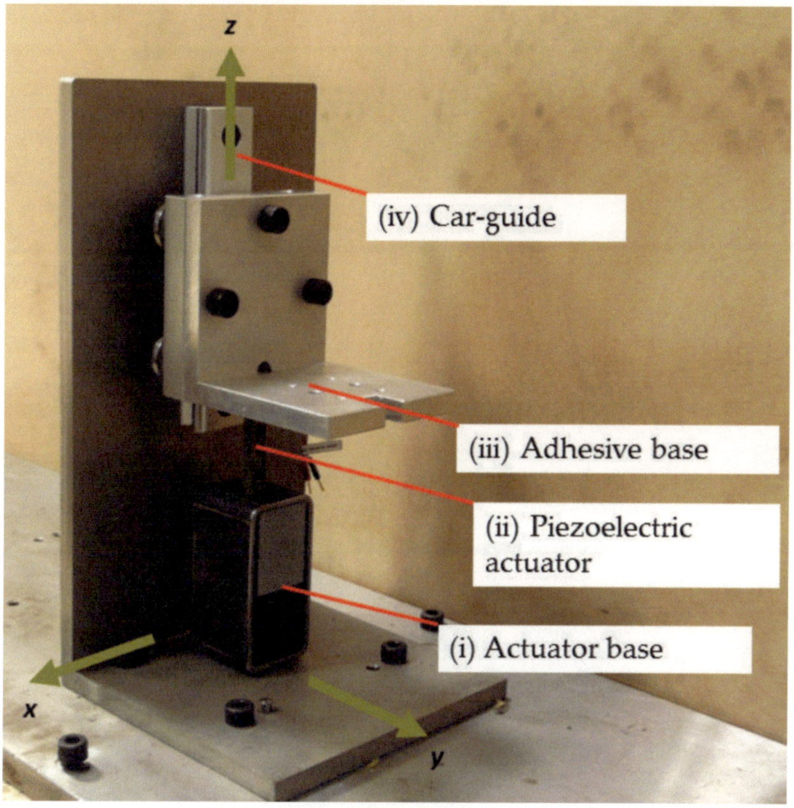

Figure 5. Test bench: detailed view of the excitation system.

Figure 4 shows the test bench and **Figure 5** shows the excitation system. Particularly, **Figure 4** shows (i) the test bench base, (ii) where the excitation support, (iii) the steady support, and (iv) an adhesively bonded beam can be appreciated. **Figure 5** shows the components of the excitation support; these are: (i) the actuator base, (ii) the piezoelectric device used as excitatory (PPA40M of CEDRAT technologies), (iii) the base of the adhesive, and (iv) the car-guide set. It should be noted that the seismic motion is directly imposed to the adhesive base by the piezoelectric actuator.

Two groups of metallic pieces are used to ensure the overlapping length ℓ_0. **Figure 6** shows these pieces. It should be noted that the evenly spaced slots are used to ensure five different overlapping lengths. The procedure is analogous for both sides and therefore left and right overlapping lengths are always nominally identical. Experimental results are presented for three different overlapping lengths, ℓ_0=10, ℓ_0=30, and ℓ_0=50 mm, respectively.

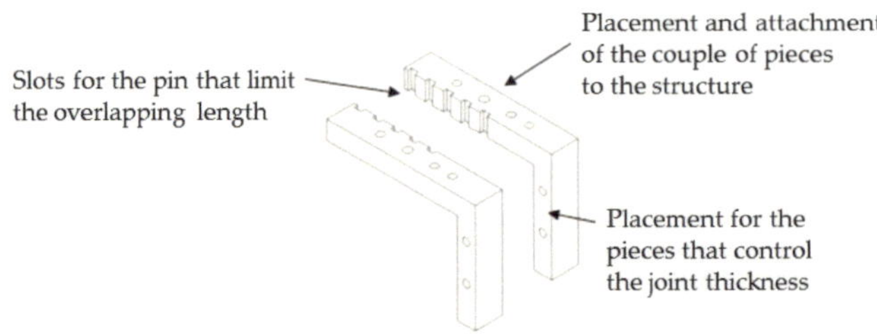

Figure 6. Components for controlling the overlapping length ℓ_0.

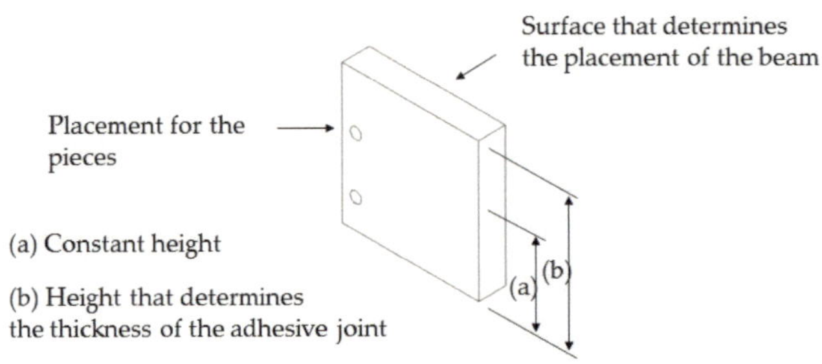

Figure 7. Components for controlling the thickness h.

Figure 8. Detailed view of the hollow for the adhesive joint.

The adhesive joint thickness h is determined by a set of pieces that are mounted onto the ones used for determining the overlapping length. Analogously, it is repeated exactly and simultaneously in both supports ensuring that both adhesive joints are nominally identical. These devices support the beam during the adhesive curing. **Figure 7** shows one of them.

Figure 7 shows placement strategy where five sets have been manufactured. Each set has a particular height related to a particular joint thickness.

The components determining overlapping length and joint thickness are assembled together onto the adhesive bases before the adhesive material is applied. Once they are mounted, the hollow for the adhesive joint is formed among them and the beam itself. This can be appreciated in **Figure 8**.

These components are removed prior to the measurements.

3.1.2. *Experiments and data processing*

The experiment consists on exciting the adhesive base and measuring the beam vibration response. The response is sequentially measured at 21 beam locations using the laser vibrometer.

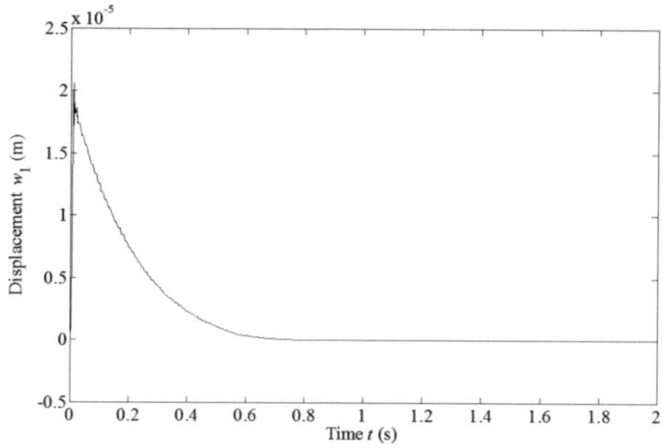

Figure 9. Imposed displacement to the adhesive base: time domain response $w_1(t)$.

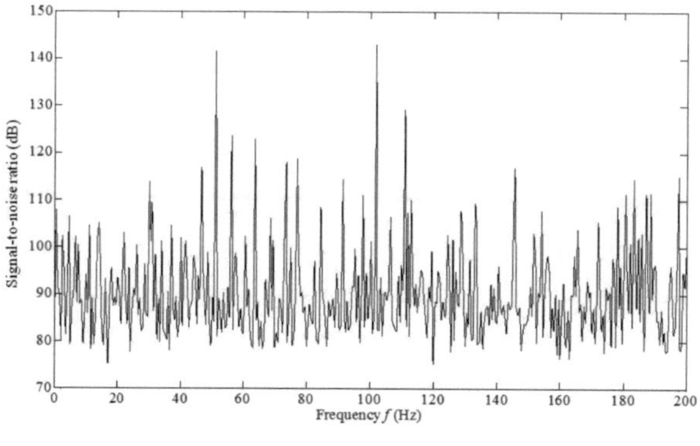

Figure 10. Imposed displacement to the adhesive base: signal-to-noise ratio for autospectrum $W_{11}(f)$.

The piezoelectric actuator receives a 0.8 V step signal that induces the displacement of the adhesive base shown in **Figure 9**. The signal-to-noise ratio for the autospectrum of $W_{11}(f)$ represented in **Figure 10**. This signal-to-noise ratio is calculated as the ratio for the signal at the adhesive base while the beam is vibrating, to the signal while the beam is at rest. The adhesive base displacement was computed by numerical integration from the acceleration measured signal. The measure was carried out for a sample period T_s=2 s, the time resolution being Δt=0.0025 s, involving 800 data lines.

From the experimental data, transmissibility functions $T_j(f)$ between adhesive base and each jth point beam are obtained by means of the $H_1(f)$ definition [14]

$$H_1(f) = \frac{W_{12}(f)}{W_{11}(f)} \tag{71}$$

where $W_{12}(f)$ represents the cross-spectrum between input $w_1(t)$ and output $w_2(t)$ and $W_{11}(f)$ the input autospectrum. The motion $w_2(t)$ is computed by numerical integration from the velocity $\dot{w}_2(t)$ measured by means of the laser vibrometer [16] for each of the 21 points of the beam. The study is performed in the 0–200 Hz frequency range, the frequency resolution being Δf =0.5 Hz.

Prior to the experiments and aimed at understanding the flexural behavior of the vibrating beam, the first three flexural modal shapes are obtained from these transmissibility functions [17]. This is done just for one of the specimens.

For the experimental analysis, the rms value (for the set of 21 points) is computed as

$$\text{rms}(T(f)) = \sqrt{\frac{1}{N}\sum_{j=1}^{N}\left(\left|T_j(f)\right|^2\right)} \tag{72}$$

with N =21. From the rms($T(f)$), the ith resonance frequency f_i, peak amplitude A_i, and modal loss factor η_i of each specimen are studied.

3.1.3. Materials and specimens

The considered adhesive is a flexible one, concretely a Bostik™ modified silane commercially named ISR 70-03. Its mechanical behavior was analyzed in earlier works [18, 19], where the relaxation modulus and the complex modulus were determined. It was done under tensile strain where the strain level imposed was of 0.5%. The test specimens were obtained from plates of the cured adhesive that were manufactured using casts of 50 mm × 70 mm × h, where h represents the nominal thickness. Three casts were manufactured with different thickness h of 0.5, 1, and 1.5 mm using Teflon™, which ensures that after the curing of the adhesive, a plate of solid material can be easily demolded without degradation. For the present case, the curing time was 48 h for all plates; the curing process was carried out at room temperature and no specific equipment was employed [20].

Figure 11 represents the complex modulus $E^*(f)$ for the adhesive in the form of storage modulus $E'(f)$ and loss factor $\eta(f)$.

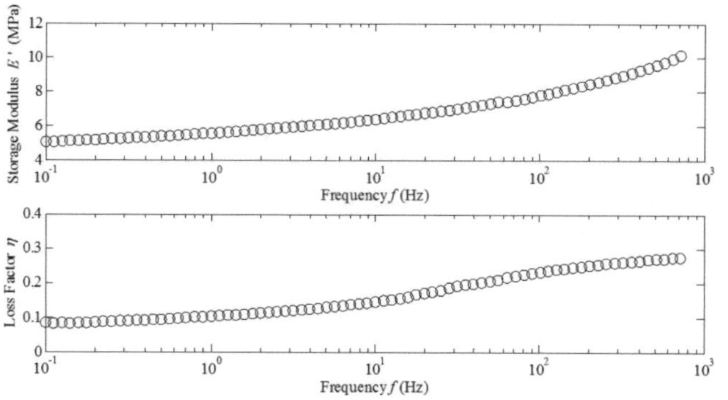

Figure 11. Tested adhesive material complex modulus $E^*(f)$ in the form of storage modulus $E'(f)$ and loss factor $\eta(f)$.

The metallic beam specimens were manufactured from the same stainless steel sheet by water-jet cutting in order to ensure homogeneous properties and undeformed specimens. The nominal beam dimensions are 400 mm × 10 mm × 1 mm , representing length, width, and thickness, respectively; the experimental values being 397 mm × 10 mm × 1 mm.

To prepare the adhesively bonded beam specimens an adhesive curing time of 72 h was established. This curing process was carried out at normal conditions (room temperature and atmospheric pressure). The test sample preparation procedure is outlined as follows:

- At a first step, the devices determining overlapping length and joint thickness h (see **Figures 6** and **7**) are mounted onto the supports (see **Figure 8**).
- Then, the adhesive is applied. It should be emphasized that this is done in a single motion ensuring that the nozzle does not get in touch with the adhesion surface. Otherwise, void creation is promoted.
- Immediately, the beam is placed forcing the adhesive to fill the gap among the components and the beam itself. The spare adhesive material is removed once it is cured.

Five different configurations were tested with three specimens for each configuration. **Table 2** shows, for each configuration, the nominal joint overlapping length and thickness together with the particular dimensions for each specimen. Also, the mean value is reported (in parenthesis). Three different overlapping lengths (BS1, BS2, and BS3) and three joint thicknesses (BS1, BS4, and BS5) were tested using the ISR 70–03 adhesive. It should be highlighted that constant length beam specimens are employed. This means that the higher the overlapping length, the lower the vibrating length.

Specimen group	Overlapping length ℓ_0(mm)			Joint thickness h (mm)		
	Nominal	Excitation side individual (mean)	Steady side individual (mean)	Nominal	Excitation side individual (mean)	Steady side individual (mean)
BS1	55	55.2/55.0/55.0 (55.1)	55.1/55.1/55.2 (55.1)	0.5	0.6/0.6/0.5 (0.6)	0.5/0.5/0.6 (0.5)
BS2	35	35.2/35.1/35.2 (35.2)	35.1/35.2/35.0 (35.1)	0.5	0.6/0.5/0.6 (0.6)	0.6/0.5/0.5 (0.5)
BS3	15	15.2/15.2/15.1 (15.2)	15.1/15.1/15.2 (15.1)	0.5	0.5/0.5/0.6 (0.5)	0.6/0.6/0.6 (0.6)
BS4	55	55.1/55.1/55.2 (55.1)	55.0/55.3/55.1 (55.2)	1	1.2/1.0/1.2 (1.1)	1.1/1.1/1.0 (1.1)
BS5	55	55.1/55.1/55.0 (55.1)	55.1/55.2/55.1 (55.1)	2.5	2.6/2.5/2.6 (2.6)	2.5/2.5/2.6 (2.6)

Table 2. Joint dimensions for the beam specimens: individual and mean values.

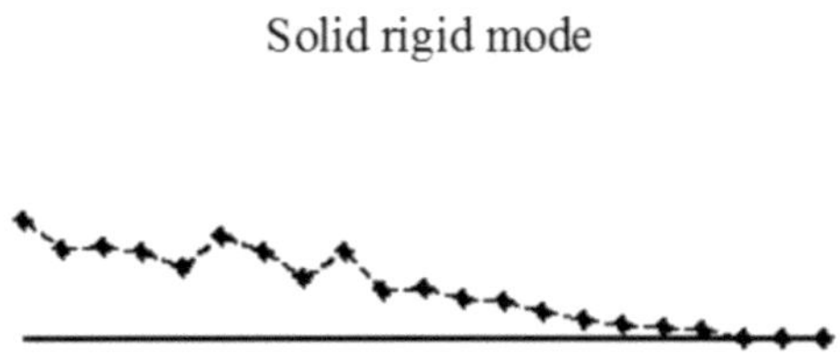

Figure 12. Modal shapes below 200 Hz: rigid body mode.

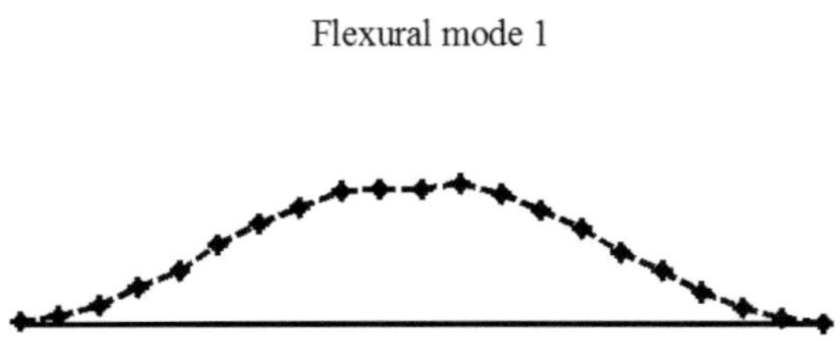

Figure 13. Modal shapes below 200 Hz: first bending mode.

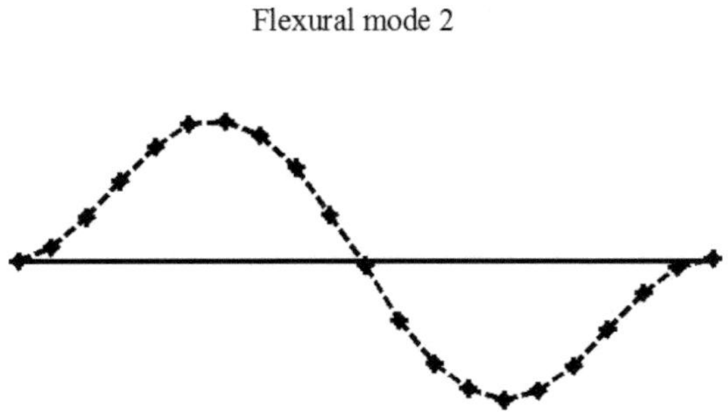

Figure 14. Modal shapes below 200 Hz: second bending mode.

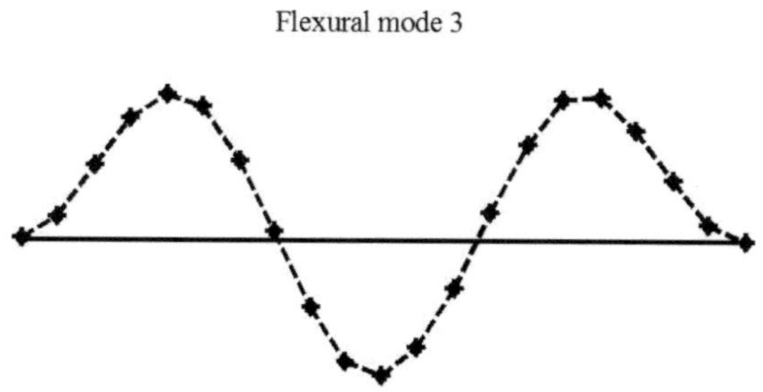

Figure 15. Modal shapes below 200 Hz: third bending mode.

3.2. Modal shapes and analysis

To understand the dynamic behavior of the adhesively bonded beams, the modal shapes are experimentally obtained from the transmissibility functions [17]. **Figure 12** represents the rigid body mode and **Figures 13–15** show the first, second, and third bending modes.

The analyzed system can be modeled as a pinned-pinned beam with torsion spring of stiffness C applied at each pinned joint (see **Figure 16**). These torsion springs represent the resistance to turn introduced by the joint to the beam.

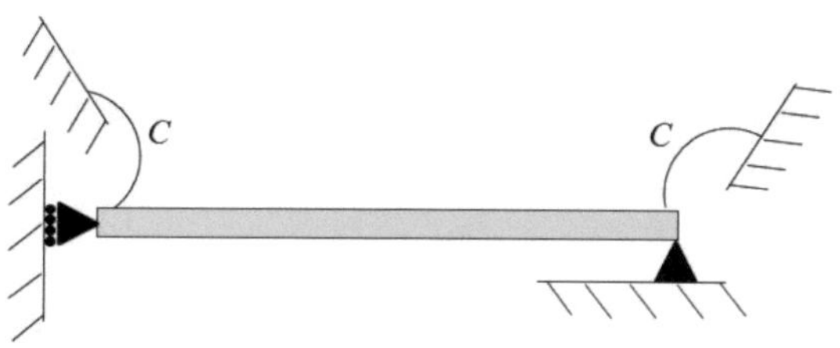

Figure 16. Model with pinned-pinned beam with torsion spring at the pinned joints.

For the analysis, only the bending modes are taken into account.

3.3. Results and discussion

Next, the influence of overlapping length and joint thickness h are presented and discussed. It should be reminded that the results shown represent the rms value of a set of 21 transmissibility functions where three different specimens are used for each configuration. The particular and average result is reported together with the standard deviation.

Table 3 collects the resonance frequencies, peak amplitudes, and modal loss factor of the each specimen tested. The loss factor η_i is computed by the Nyquist-circle method.

For the subsequently study, a distinction must be made between system stiffness and joint stiffness. System stiffness refers to the stiffness property that globally affects the mechanical behavior of the system, which is normally traduced in natural frequencies. The joint stiffness (corresponding to the previously mentioned torsion stiffness C represented in **Figure 16**) refers to a local stiffness property of the joint. Actually, it has not a significant influence on the global response (slightly affects natural frequencies), but affects to the vibration transmission mechanism.

Specimen group	Mode 1			Mode 2			Mode 3		
	f_1 (Hz)	A_1 (m/m)	$\eta_1 (\times 10^{-3})$	f_2 (Hz)	A_2 (m/m)	$\eta_2 (\times 10^{-3})$	f_3 (Hz)	A_3 (m/m)	$\eta_3 (\times 10^{-3})$
BS1	61.5/61.0/61.5 (61.4 ± 0.3)	19.8/20.2/19.5 (19.8 ± 0.3)	33.5/32.9/35.1 (33.6 ± 1.2)	162/161/162 (161.7 ± 0.6)	18.0/18.4/17.8 (18.1 ± 0.3)	12.4/9.65/10.7 (10.9 ± 1.4)	N/A	N/A	N/A
BS2	45.5/45.0/45.5 (45.3 ± 0.3)	14.1/13.8/14.5 (14.2 ± 0.3)	31.3/35.2/33.9 (33.8 ± 1.8)	127.0/126.5/126.5 (126.6 ± 0.3)	12.5/12.4/12.0 (12.3 ± 0.3)	23.8/21.4/22.0 (22.4 ± 1.2)	N/A	N/A	N/A
BS3	34.0/34.0/35.0 (34.4 ± 0.6)	9.72/8.92/9.12 (9.25 ± 0.4)	37.1/34.1/38.9 (36.7 ± 2.4)	96.0/96.5/97.0 (96.5 ± 0.5)	9.15/9.24/8.86 (9.08 ± 0.2)	24.1/22.9/25.8 (24.3 ± 1.5)	192/192/192.5 (192.1 ± 0.3)	7.92/8.03/8.7 (8.22 ± 0.4)	18.7/20.2/17.7 (18.5 ± 1.7)
BS4	57.5/57.5/57.0 (57.3 ± 0.3)	17.6/17.7/18.2 (17.8 ± 3)	17.1/18.4/14.9 (16.7 ± 1.8)	159.0/158.0/158.0 (158.3 ± 0.6)	17.7/18.0/17.2 (17.6 ± 0.4)	15.7/13.1/15.2 (14.7 ± 1.4)	N/A	N/A	N/A
BS5	55.0/54.5/54.0 (54.5 ± 0.5)	15.8/15.6/15.1 (15.5 ± 0.4)	36.5/33.1/33.8 (34.4 ± 1.8)	149.5/149.5/149.0 (149.3 ± 0.3)	11.1/11.9/11.2 (11.4± 0.4)	18.1/17.5/20.1 (18.5 ± 1.4)	N/A	N/A	N/A

Individual and mean values: (a) BS1; (b) BS2; (c) BS3; (d) BS4; (e) BS5; (f) BS6.

Table 3. Frequency domain results for the analyzed beam specimens.

3.3.1. Overlapping length influence

Three overlapping lengths are analyzed: 55, 35, and 15 mm, the specimen groups being BS1, BS2, and BS3 (see **Table 2**), respectively, the mean thickness being around 0.5 mm.

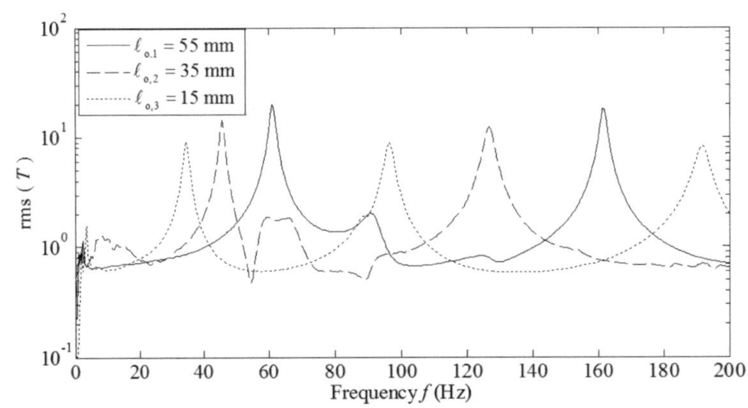

Figure 17. Transmissibility function amplitude $|T(f)|$ for three different overlapping lengths ℓ_o.

Figure 17 shows the rms($T(f)$) function for the set of the 21 transmissibility functions obtained.

From **Figure 17**, it should be pointed out that there are two resonances in the analyzed range except for, in which three can be appreciated.

- About resonance frequency. As expected, the bigger the overlapping length, the bigger the resonance frequencies. This is due to as the overlapping length increases, so does the system stiffness. This is based on the decrement of the vibrating length.
- About resonance peak amplitude. Analogously, the higher is the overlapping length, the higher are the resonance peak amplitudes. Besides, the amplitudes for the first mode are higher than those of the second mode.
- About modal loss factor. Involving modal loss factor (see **Table 3**), it should be remarked that the second modes present lower values than those of the first ones. It can be stated that the higher the overlapping length, the lower the modal loss factor. However, this difference is minimal.

3.3.2. *Thickness influence*

Three adhesive joint thickness values are analyzed. These are 0.5, 1.0, and 2.5 mm, the specimen groups being BS1, BS4, and BS5 (see **Table 2**), respectively. Small adhesive joint thickness is considered in order not to dramatically decrease the joint strength. **Figure 18** illustrates the function for the corresponding sets.

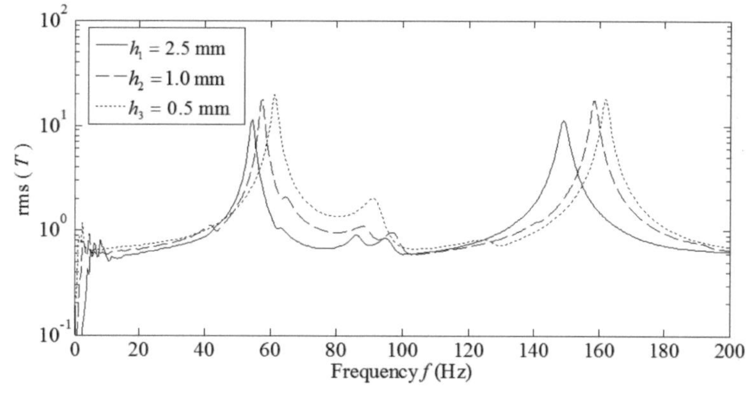

Figure 18. Transmissibility function amplitude $|T(f)|$ for three joint thicknesses h.

- About resonance frequency. It can be concluded that the thinner the thickness, the higher the joint stiffness, and therefore, the resonance frequencies increase. But, according to the previously mentioned distinction between system and joint stiffness, by decrementing five times the joint thickness, natural frequencies increase only about 10%.
- About resonance peak amplitude. The higher amplitude is found for the first resonance of each case (see **Table 3**). For each mode, the amplitude presents an inverse evolution with thickness: the minimum amplitude is given for the highest thickness h_1 = 2.5 mm. It should be noted that decreasing thickness results in increasing joint stiffness (leading to amplitude decrement) and decrementing damping material amount (leading to amplitude augmenta-

tion). Nevertheless, the effect on damping results more significant concluding that the thinner the thickness the higher the amplitude.

- About modal loss factor. Within each mode, the lower loss factor corresponds to the thinnest case (h_3=0.5 mm) due to the joint stiffness being higher, the damping capacity decreases by material deformation. However, the modal loss factor presents a minimum for the intermediate thickness.

From this analysis, the following discussion can be made:

About stiffness

- The overlapping length modifies the system stiffness, whereas the joint thickness modifies the joint stiffness.
- To decrease the system stiffness implies a decrement in the resonance frequencies and peak amplitudes A_i.
- To decrease the joint stiffness decreases the amplitudes A_i, because the vibration is worse transmitted. However, the resonance frequencies remain practically unaffected.

About damping

- The overlapping length and joint thickness h modify damping capacity because they are related to the amount of viscoelastic material used.
- To decrease overlapping length, on the one hand, decreases joint damping capacity because of the amount of material is decreased. However, on the other hand, as the joint stiffness is decreased, the viscoelastic material deformation is augmented and so is the damping capacity.
- To decrease joint thickness decreases joint damping capacity because of the joint stiffness is increased, and therefore the deformations are dismissed. Besides, damping material amount is decreased.

As a general conclusion, it can be drawn that increasing overlapping length or joint thickness can lead to opposite effects. Hence, to know and optimize the behavior of these kinds of adhesive joints, exhaustive experiments would be needed, or failing that, numerical simulation should be required in order to separate and investigate the influence of each parameter.

Author details

Jon García-Barruetabeña* and Fernando Cortés Martínez

*Address all correspondence to: jgarcia.barruetabena@deusto.es

Faculty of Engineering, University of Deusto, Bilbao, Spain

References

[1] Sondipon Adhikari. Dynamic response characteristics of a non-viscously damped oscillator. Transactions of ASME, Journal of Applied Mechanics. 2008;75(1):1–12.

[2] Milton Abramowitz. Handbook of Mathematical Functions, with Formulas, Graphs and Mathematical Tables. New York: Courier Dover Publications; 1965.

[3] Patrick Muller. Are the eigensolutions of a 1-d.o.f. system with viscoelastic damping oscillatory or not? Journal of Sound and Vibration. 2005;285:501–509.

[4] Sondipon Adhikari. Qualitative dynamic characteristics of a non-viscously damped oscillator. Proceedings of the Royal Society. 2005;461:2269–2288.

[5] Leonard Meirovitch. Dynamics and Control of Structures. New York: John Wiley and Sons; 1989.

[6] Ludwig Boltzmann. On the theory of the elastic aftereffect. Poggendorff-s Annals of Physics and Chemistry. 1876;7:624–645.

[7] Sondipon Adhikari. Damping models for non-viscously damped systems [thesis]. Cambridge University; 2000.

[8] Sondipon Adhikari. Eigenrelations for non-viscously damped systems. AIAA Journal. 2001;39(5):43–61.

[9] Sondipon Adhikari, Nils Wagner. Direct time-domain integration method for exponentially damped linear systems. Computers and Structures. 2004;82:2453–2461.

[10] Hughes T. The finite element method: linear static and dynamic finite element analysis. New Jersey: Prentice-Hall Englewood Cliffs; 1987.

[11] Fernando Cortés, Modesto Mateos, María Jesús Elejabarrieta. A direct integration formulation for exponentially damped structural systems. Computers and Structures. 2008;87:391–394.

[12] Cedrat Technologies. Technical Sheet for piezo-actators and electronic devices [Internet]. 2010 [Updated: 15 February 2010]. Available from: www.cedrat.com [Accessed: 15 February 2010]

[13] dSPACE. Available from: www.dspaceinc.com [Accessed: 23 February 2010]

[14] Brüel, Kjaer. Sound and Vibration Measurement [Internet]. Available from: www.bksc.com [Accessed: 15 January 2010]

[15] PCB electronics. Available from: www.pcb.com [Accessed: 11 January 2010]

[16] POLYTEC. Available from: www.polytec.com [Accessed: 16 January 2010]

[17] Amirhossein Aghdasi. Application of transmissibility measurement in estimation of modal parameters for a structure [thesis]. Blekinge Institute of Technology; 2012.

[18] Jon García-Barruetabeña, Fernando Cortés, José Manuel Abete, Pelayo Fernández, María José Lamela, Antonio Fernández-Cantelli. Experimental characterization and modelization of the relaxation and complex moduli of a flexible adhesive. Materials and Design. 2011;32:2783–2796.

[19] Jon García-Barruetabeña, Fernando Cortés, José Manuel Abete. A low modulus adhesive characterization by means of DMTA test. International Journal of Adhesion. 2012;88(4–6):48–498.

[20] Lucas Da Silva, Robert Adams, Stephen Gibbs. Manufacture of adhesive joints and bulk specimens with high-temperature adhesives. International Journal of Adhesion and Adhesives. 2004;24:69–83.

Chapter 12

General Stability in Viscoelasticity

Salim A. Messaoudi

Additional information is available at the end of the chapter

http://dx.doi.org/10.5772/64217

Abstract

In this chapter, we consider a problem which describes the motion of a viscoelastic body and investigate the effect of the dissipation induced by the viscoelastic (integral) term on the solution. Precisely, we show that, under reasonable conditions on the relaxation function, the system stabilizes to a stationary state. We also obtain a general decay estimate from which the usual exponential and polynomial decay rates are only special cases.

Keywords: general decay, memory, relaxation function, stability, viscoelasticity

1. Introduction

Elastic materials, when subjected to a suddenly applied loading state held constant thereafter, respond instantaneously with a state of deformation which remains constant. On the other hand, Newtonian viscous fluids respond to a suddenly applied state of uniform shear stress by a steady flow process. However, there exist materials for which any suddenly applied and maintained state of uniform stress produces an instantaneous deformation followed by a flow process which might or might not be limited in magnitude as time grows. Such materials exhibit both instantaneous elasticity effects and creep characteristics. Obviously, such a behavior cannot be described by either an elasticity theory or a viscosity theory only but it combines features of each. The most interesting examples of such materials are polymers, which can display all the intermediate range of properties (glassy, brittle solid or an elastic rubber or a viscous liquid) depending on temperature and the experimentally chosen time scale. Such materials are said to possess memories.

Many scientists, such as Maxwell, Kelvin, Voigt, and Boltzmann, have contributed in modeling these phenomena. Boltzmann, in 1874, supplied the first formulation of a three-dimensional theory of isotropic viscoelasticity. He elaborated the model of a "linear" viscoelastic solid on a basic assumption which states that at any (fixed) point x of the body, the stress at any time t depends on the strain at all the proceeding times. In addition, if the strain at all preceding times is in the same direction, then the effect is to decrease the corresponding stress. The influence of a previous strain on the stress depends on the time elapsed since that strain occurred and is weaker than those strains that occurred long ago. Such properties make the model of solid, elaborated by Boltzmann, a material with (fading) memory. These memory effects are expressed by the dependence on the deformation gradient. Therefore, for these "viscoelastic" materials, the stress at each point and at each instant does not depend only on the present value of the deformation gradient but on the entire temporal prehistory of the motion. In addition, Boltzmann made the assumption that a superposition of the influence of previous strains holds, which means that the stress-strain relation is linear. Mathematically, this is interpreted by the time convolution of a "relaxation" function with the Laplacian of the solution. As a consequence, a subtle damping effect is produced. The types of equations we intend to discuss in this chapter are of the form:

$$u_{tt}(x,t) - \Delta u(x,t) + \int_0^t g(t-s)\Delta u(x,s)\,ds = f(x,t), \quad x \in \Omega, t > 0,$$

where Ω is a bounded domain with regular boundary, g is a nonincreasing positive function, referred to as the relaxation function which describes the viscoelastic material in consideration, f is an external force, and $u(x, t)$ is the position of a point x "in the reference configuration" at a time t.

In early 1970s, Dafermos [1, 2] discussed a one-dimensional viscoelastic model, where he proved, for smooth monotonically decreasing relaxation functions, various existence and asymptotic stability results. However, no rate of decay has been given. After that, viscoelastic problems have attracted the attention of many researchers and many results of existence and long-time behavior have been established. To the best of our knowledge, the first work that studied the uniform decay of solutions was presented by Dassios and Zafirapoulos [3]. In their work, Dassios and Zafirapoulos presented a viscoelastic problem in $\mathbb{R}^3$ and proved a polynomial decay for exponentially decaying kernels. In 1994, Muñoz Rivera [4] considered, in $\mathbb{R}^n$ and in bounded domains, equations for linear isotropic homogeneous viscoelastic solids, with exponentially decaying memory kernels and showed that, in the absence of body forces, solutions decay exponentially for the bounded-domain case, whereas, for the whole space case, the decay is of a polynomial rate. After that, Cabanillas and Muñoz Rivera [5] studied problems, where the kernels are of algebraic (but not exponential) decay rates and showed that the decay of solutions is algebraic at a rate which can be determined by the rate of the decay of the relaxation function and the regularity of solutions. This result was later improved by Barreto et al. [6], where equations related to linear viscoelastic plates were treated. For viscoelastic systems with localized frictional dampings, Cavalcanti et al. [7] considered the following problem:

$$\begin{cases} u_{tt} - \Delta u + \int_0^t g(t-\tau)\Delta u(\tau)d\tau = b|u|^{m-2}u, & in\,\Omega \times (0,\infty), \\ u(x,t) = 0, \quad x \in \partial\Omega, \quad t \geq 0 \\ u(x,0) = u_0(x), \quad u_t(x,0) = u_1(x), \quad x \in \Omega \end{cases}$$

where Ω is a bounded domain of $\mathbb{R}^n$ ($n \geq 1$) with a smooth boundary $\partial\Omega$, g is a positive nonincreasing function satisfying, for two positive constants, the conditions:

$$-\xi_1 g(t) \leq g'(t) \leq -\xi_2 g(t), \quad t \geq 0,$$

and $a(x) \geq a_0 > 0$ in a subdomain $\omega \subset \Omega$, with $meas(\omega) > 0$ and satisfying some geometry restrictions. They established an exponential rate of decay. Berrimi and Messaoudi [8] improved Cavalcanti's result by weakening the conditions on both a and g. In particular, the function a can vanish on the whole domain Ω and consequently the geometry condition is no longer needed. This result has been later extended to a situation, where a source is competing with the viscoelastic dissipation, by Berrimi and Messaoudi [9]. Also, Cavalcanti et al. [10] have studied a quasilinear equation, in a bounded domain, of the form:

$$|u_t|^\rho u_{tt} - \Delta u - \Delta u_{tt} + \int_0^t g(t-\tau)\Delta u(\tau)d\tau - \gamma\Delta u_t = 0,$$

with $\rho > 0$, and a global existence result for $\gamma \geq 0$, as well as an exponential decay for $\gamma > 0$, have been established. Messaoudi and Tatar [11,12] discussed the situation when $\gamma = 0$ and established polynomial and exponential decay results in the presence, as well as in the absence, of a nonlinear source term. Fabrizio and Polidoro [13] studied a homogeneous viscoelastic equation in the presence of a linear frictional damping (au_t, $a > 0$) and showed that the exponential decay of the relaxation function g is a necessary condition for the exponential decay of the solution energy of the solution. In other words, the presence of the memory term, with a non-exponentially decaying relaxation function, may prevent the exponential decay even if the frictional damping is linear. He also obtained a similar result for the polynomial decay case.

For more general decaying kernels, Messaoudi [14,15] considered

$$\begin{cases} u_{tt} - \Delta u + \int_0^t g(t-\tau)\Delta u(\tau)d\tau = b|u|^{m-2}u, & in\,\Omega \times (0,\infty), \\ u(x,t) = 0, \quad x \in \partial\Omega, \quad t \geq 0 \\ u(x,0) = u_0(x), \quad u_t(x,0) = u_1(x), \quad x \in \Omega \end{cases} \tag{1.1}$$

with $b = 0$ and $b = 1$ and for relaxation functions satisfying

$$g'(t) \leq -\xi(t) g(t),\ \forall t \geq 0, \tag{1.2}$$

where $\xi: \mathbb{R}^+ \to \mathbb{R}^+$ is a nonincreasing differentiable function. He showed that the rate of the decay of the energy is exactly the rate of decay of g, which is not necessarily of exponential or polynomial decay type. After that, a series of papers using Eq. (1.2) have appeared. See for instance, Han and Wang [16], Liu [17,18], Park and Park [19], and Xiaosen and Mingxing [20].

In this work, we intend to study the following problem:

$$\begin{cases} u_{tt} - \Delta u + \int_0^t g(t-\tau)\Delta u(\tau) d\tau + a|u_t|^{m-2}u_t = 0, & \text{in } \Omega \times (0,\infty) \\ u(x,t) = 0, \quad x \in \partial\Omega, & t \geq 0 \\ u(x,0) = u_0(x), \quad u_t(x,0) = u_1(x), & x \in \Omega \end{cases} \tag{1.3}$$

where Ω is a bounded and regular domain of $\mathbb{R}^n$, $a > 0$ is a constant, and g is a positive nonincreasing function satisfying Eq. (1.2). We will establish some general decay results depending on the behavior of g and m.

2. Preliminary

In this section, we present some material needed in the proof of our result and state a global existence result which can be proved using the well-known Galerkin method. See, for example, [2,3]. In order to prove our main result, we make the following assumptions:

(A_1) $g: \mathbb{R}_+ \to \mathbb{R}_+$ is a bounded differentiable function such that

$$g(0) > 0,\ g'(t) \leq -\gamma(t) g(t),\ 1 - \int_0^\infty g(s) ds = l > 0,$$

where $\gamma(t)$ is a differentiable function satisfying

$$\gamma(t) > 0,\ \gamma'(t) \leq 0, \text{ and } \int_0^{+\infty} \gamma(t) dt = +\infty.$$

(A_2) Concerning the nonlinearity in the damping, we assume that

$$1 < m \le \frac{2n}{n-2}, \text{ if } n > 2 \text{ and } m > 1, \text{ if } n = 1,2$$

Remark 2.1. Examples of functions satisfying (A_1) are

$$g_1(t) = \frac{a}{(1+t)^v}, \ v > 1, \ g_2(t) = ae^{-b(t+1)p}, \ 0 < p \le 1$$

$$g_3(t) = \frac{a}{(1+t)\left[\ln(1+t)\right]^v}, \ v > 1,$$

for a and b constants to be chosen properly.

Proposition 2.1. Let $(u_0, u_1) \in H_0^1(\Omega) \times L^2(\Omega)$ be given. Assume that (A_1), (A_2) hold. Then problem (1.3) has a unique global solution:

$$u \in C\left(\left[0,\infty\right); \ H_0^1(\Omega)\right)$$

$$u_t \in C\left(\left[0,\infty\right); \ L^2(\Omega)\right) \cap L^m\left(\Omega \times (0, \infty)\right). \tag{2.1}$$

Proposition 2.2. [21] Let E: $\mathbb{R}_+ \to \mathbb{R}_+$ be a non-increasing function and φ: $\mathbb{R}_+ \to \mathbb{R}_+$ be an increasing C^2-function such that

$$\varphi(0) = 0 \text{ and } \varphi(t) \to +\infty \text{ as } t \to +\infty.$$

Assume that there exist $q \ge 0$ and $A > 0$ such that

$$\int_S^{+\infty} E^{q+1}(t)\varphi'(t)\,\mathrm{d}t \le AE(S), \ 0 \le S < +\infty,$$

then we have

$$E(t) \le cE(0)\left(1 + \varphi(t)\right)^{\frac{-1}{q}}, \ \forall t \ge 0, \ \text{if } q > 0,$$

$$E(t) \leq cE(0)e^{-\omega\varphi(t)},\ \forall t \geq 0, \text{ if } q = 0,$$

where c and ω are positive constants independent of the initial energy $E(0)$.

Next, we introduce the "modified energy":

$$\mathcal{E}(t) := \frac{1}{2}\left(1 - \int_0^t g(s)\,ds\right)\|\nabla u(t)\|_2^2 + \frac{1}{2}\|u_t\|_2^2 + \frac{1}{2}(go\nabla u)(t), \tag{2.2}$$

where

$$(gov)(t) = \int_0^t g(t-\tau)\|v(t) - v(\tau)\|_2^2\,d\tau.$$

Remark 2.2. By multiplying Eq. (1.3) by u_t and integrating over Ω, using integration by parts and hypotheses (A_1), (A_2), we get, after some manipulations, as in [3,20],

$$\begin{aligned}\mathcal{E}'(t) &\leq -\left(a\int_\Omega |u_t|^m\,dx - \frac{1}{2}(g'o\nabla u)(t) + \frac{1}{2}g(t)\|\nabla u(t)\|_2^2\right)\\ &\leq -a\int_\Omega |u_t|^m\,dx + \frac{1}{2}(g'o\nabla u)(t) \leq 0.\end{aligned} \tag{2.3}$$

3. Decay of solutions

In order to state and prove our main result, we set

$$F(t) := \varepsilon(t) + \varepsilon_1\Psi(t) + \varepsilon_2\mathcal{X}(t) \tag{3.1}$$

where ε_1 and ε_2 are positive constants to be specified later and

$$\Psi(t) := \int_\Omega uu_t\,dx, \quad \chi(t) := -\int_\Omega u_t\int_0^t g(t-\tau)(u(t)-u(\tau))\,d\tau\,dx.$$

Lemma 3.1. For ε_1 and ε_2 so small, we have

$$\alpha_1 F(t) \le E(t) \le \alpha_2 F(t) \tag{3.2}$$

holds for two positive constants α_1 and α_2.

Proof. It is straightforward to see that

$$\begin{aligned}
F(t) &\le E(t) + \frac{\varepsilon_1}{2}\int_\Omega |u_t|^2\,\mathrm{d}x + \frac{\varepsilon_1}{2}\int_\Omega |u|^2\,\mathrm{d}x \\
&+ \frac{\varepsilon_2}{2}\int_\Omega |u_t|^2\,\mathrm{d}x + \frac{\varepsilon_2}{2}\int_\Omega\left(\int_0^t g(t-\tau)(u(t)-u(\tau))\,\mathrm{d}\tau\right)^2\mathrm{d}x \\
&\le E(t) + \frac{\varepsilon_1}{2}\int_\Omega |u_t|^2\,\mathrm{d}x + \frac{\varepsilon_1}{2}C_p\int_\Omega |\nabla u|^2\,\mathrm{d}x \\
&+ \frac{\varepsilon_2}{2}\int_\Omega |u_2|^2\,\mathrm{d}x + \frac{\varepsilon_2}{2}C_p(1-l)(go\nabla u)(t) \le \alpha_2\mathcal{E}(t),
\end{aligned}$$

where C_p is the Poincaré constant. In the other hand,

$$\begin{aligned}
F(t) &\ge E(t) - \frac{\varepsilon_1}{2}\int_\Omega |u_t|^2\,\mathrm{d}x - \frac{\varepsilon_1}{2}\int_\Omega |u|^2\,\mathrm{d}x - \frac{\varepsilon_2}{2}\int_\Omega |u_t|^2\,\mathrm{d}x - \frac{\varepsilon_2}{2}C_p \\
&(1-l)(go\nabla u)(t) \ge \frac{1}{2}l\|\nabla u(t)\|_2^2 + \frac{1}{2}\|u_t\|_2^2 + \frac{1}{2}(go\nabla u)(t) + \frac{1}{\gamma+2}\|u\|_{\gamma+2}^{\gamma+2} \\
&-\left[\frac{\varepsilon_1+\varepsilon_2}{2}\right]\int_\Omega |u_t|^2\,\mathrm{d}x - \frac{\varepsilon_1}{2}C_p\int_\Omega |\nabla u|^2\,\mathrm{d}x - \frac{\varepsilon_2}{2}C_p(1-l)(go\nabla u)(t) \\
&\ge \alpha_1\mathcal{E}(t),
\end{aligned}$$

for ε_1 and ε_2 small enough. Thus, Eq. (3.2) is established.

Lemma 3.2. Assume that $m \ge 2$ and assumptions (A_1), (A_2) hold. Then, the functional $\Psi(t)$ satisfies, along the solution of Eq. (1.3), the estimate:

$$\Psi'(t) \le \int_\Omega u_t^2\,\mathrm{d}x - \frac{1}{4}\int_\Omega |\nabla u|^2\,\mathrm{d}x + \frac{1-l}{2l}(go\nabla u)(t) + C\int_\Omega |u_t|^m\,\mathrm{d}x, \tag{3.3}$$

where C is a "generic" positive constant independent of t.

Proof. By using Eq. (1.3), we easily see that

$$\Psi'(t) = \int_\Omega u_t^2 \mathrm{d}x - \int_\Omega |\nabla u|^2 \,\mathrm{d}x - a\int_\Omega |u_t|^{m-2} u_t\, u \,\mathrm{d}x + \int_\Omega \nabla u(t) \cdot \int_0^t g(t-\tau)\nabla u(\tau)\,\mathrm{d}\tau\,\mathrm{d}x. \tag{3.4}$$

We now estimate the third term of the right-hand side of Eq. (3.4), using Young's inequality and (A_2). Thus, we get

$$\begin{aligned}\int_\Omega |u_t|^{m-2} u_t\, u \,\mathrm{d}x &\le \delta\int_\Omega |u|^m \,\mathrm{d}x + c_\delta \int_\Omega |u_t|^m \,\mathrm{d}x \le \delta C\|\nabla u\|_2^{m-2}\|\nabla u\|_2^2 + \\ c_\delta \int_\Omega |u_t|^m \,\mathrm{d}x &\le \delta C \mathcal{E}^{\frac{m-2}{2}}(0)\|\nabla u\|_2^2 + c_\delta \int_\Omega |u_t|^m \,\mathrm{d}x\end{aligned} \tag{3.5}$$

where c_δ is a constant depending on δ. For the fourth term of the right-hand side of Eq. (3.4), we get

$$\begin{aligned}\int_\Omega \nabla u(t) \cdot \int_0^t g(t-\tau)\nabla u(\tau)\,\mathrm{d}\tau\mathrm{d}x &\le \frac{1}{2}\int_\Omega |\nabla u(t)|^2 \,\mathrm{d}x + \\ \frac{1}{2}\int_\Omega \left(\int_0^t g(t-\tau)|\nabla u(\tau)|\,\mathrm{d}\tau\right)^2 \mathrm{d}x &\le \frac{1}{2}\int_\Omega |\nabla u(t)|^2 \,\mathrm{d}x \\ + \frac{1}{2}\int_\Omega \left(\int_0^t g(t-\tau)\left(|\nabla u(\tau) - \nabla u(t)| + |\nabla u(t)|\right)\mathrm{d}\tau\right)^2 \mathrm{d}x.\end{aligned} \tag{3.6}$$

We then use Cauchy-Schwarz inequality, Young's inequality, and the fact that

$$\int_0^t g(\tau)\,\mathrm{d}\tau \le \int_0^{+\infty} g(\tau)\,\mathrm{d}\tau = 1 - l,$$

to obtain, for any $\eta > 0$,

$$
\begin{aligned}
&\int_\Omega\left(\int_0^t g(t-\tau)\left(|\nabla u(\tau)-\nabla u(t)|+|\nabla u(t)|\right)\mathrm{d}\tau\right)^2\mathrm{d}x \\
&\le\int_\Omega\left(\int_0^t g(t-\tau)|\nabla u(\tau)-\nabla u(t)|\mathrm{d}\tau\right)^2\mathrm{d}x+\int_\Omega\left(\int_0^t g(t-\tau)|\nabla u(t)|\mathrm{d}\tau\right)^2\mathrm{d}x \\
&+2\int_\Omega\left(\int_0^t g(t-\tau)|\nabla u(\tau)-\nabla u(t)|\mathrm{d}\tau\right)\mathrm{d}x\left(\int_0^t g(t-\tau)|\nabla u(t)|\mathrm{d}\tau\right)\mathrm{d}x \\
&\le(1+\eta)\int_\Omega\left(\int_0^t g(t-\tau)|\nabla u(t)|\mathrm{d}\tau\right)^2 dx+ \\
&\left(1+\frac{1}{\eta}\right)\int_\Omega\left(\int_0^t g(t-\tau)|\nabla u(\tau)-\nabla u(t)|\mathrm{d}\tau\right)^2\mathrm{d}x \\
&\le\left(1+\frac{1}{\eta}\right)\int_\Omega\int_0^t g(t-\tau)\mathrm{d}\tau\int_0^t g(t-\tau)|\nabla u(\tau)-\nabla u(t)|^2\mathrm{d}\tau\mathrm{d}x+ \\
&(1+\eta)\int_\Omega|\nabla u(t)|^2\left(\int_0^t g(t-\tau)\mathrm{d}\tau\right)^2\mathrm{d}x\le(1+\eta)(1-l)^2\int_\Omega|\nabla u(t)|^2\mathrm{d}x \\
&+\left(1+\frac{1}{\eta}\right)(1-l)\int_\Omega\int_0^t g(t-\tau)|\nabla u(\tau)-\nabla u(t)|^2\mathrm{d}\tau\mathrm{d}x
\end{aligned}
\tag{3.7}
$$

By combining Eqs. (3.4)–(3.7), we arrive at

$$
\begin{aligned}
&\Psi'(t)\le\int_\Omega u_t^2\mathrm{d}x-\frac{1}{2}\int_\Omega|\nabla u(t)|^2\mathrm{d}x+\delta C\mathcal{E}^{\frac{m-2}{2}}(0)\|\nabla u\|_2^2 \\
&+\frac{1}{2}(1+\eta)(1-l)^2\int_\Omega|\nabla u(t)|^2\mathrm{d}x+c_\delta\int_\Omega|u_t|^m\mathrm{d}x+\frac{1}{2}(1+\eta)(1-l)\int_\Omega\int_0^t g(t-\tau)|\nabla u(\tau)-\nabla u(t)|^2\mathrm{d}\tau\mathrm{d}x \\
&\le\int_\Omega u_t^2\mathrm{d}x-\frac{1}{2}\left[1-(1+\eta)(1-l)^2-2\delta C\mathcal{E}^{\frac{m-2}{2}}(0)\right]\int_\Omega|\nabla u(t)|^2\mathrm{d}x \\
&+\frac{1}{2}(1+\eta)(1-l)(go\nabla u)(t)+c_\delta\int_\Omega|u_t|^m\mathrm{d}x.
\end{aligned}
$$

By choosing $\eta = l/(1-l)$ and $\delta = l/4C\mathcal{E}^{\frac{(m-2)}{2}}(0)$, estimate (3.3) is established.

Lemma 3.3. Assume that $m \ge 2$ and assumptions (A_1), (A_3) hold. Then, the functional $\mathcal{X}(t)$ satisfies, along the solution of Eq. (1.3) and for any $\delta > 0$, the estimate

$$\mathcal{X}'(t) \leq \delta\left[1+2(1-l)^2\right]\|\nabla u\|_2^2 + c_\delta(go\nabla u)(t)$$
$$+c_\delta\int_\Omega |u_t|^m\,\mathrm{d}x + \left(\delta - \int_0^t g(s)\mathrm{d}s\right)\int_\Omega u_t^2\mathrm{d}x + \frac{g(0)}{4\delta}C_p\left(-(g'o\nabla u)(t)\right). \tag{3.8}$$

Proof: By using Eq. (1.3), we easily see that

$$\mathcal{X}'(t) = \int_\Omega \nabla u(t)\cdot\left(\int_0^t g(t-\tau)(\nabla u(t)-\nabla u(\tau))\mathrm{d}\tau\right)\mathrm{dx}$$
$$-\int_\Omega\left(\int_0^t g(t-\tau)\nabla u(\tau)\mathrm{d}\tau\right)\cdot\left(\int_0^t g(t-\tau)(\nabla u(t)-\nabla u(\tau))\mathrm{d}\tau\right)\mathrm{dx} \tag{3.9}$$
$$+a\int_\Omega |u_t|^{m-2}u_t(t)\int_0^t g(t-\tau)(u(t)-u(\tau))\mathrm{dzdx}$$
$$-\int_\Omega u_t\int_0^t g'(t-\tau)(u(t)-u(\tau))\mathrm{d}\tau\mathrm{dx} - \left(\int_0^t g(s)\mathrm{d}s\right)\int_\Omega u_t^2\mathrm{dx}$$

Similarly to Eq. (3.4), we estimate the right-hand side terms of Eq. (3.9). So for any $\delta > 0$, we have

$$-\int_\Omega \nabla u(t)\cdot\left(\int_0^t g(t-\tau)(\nabla u(t)-\nabla u(\tau))\mathrm{d}\tau\right)\mathrm{d}x$$
$$\leq \delta\int_\Omega |\nabla u|^2\,\mathrm{d}x + \frac{1-l}{4\delta}(g\circ\nabla u)(t), \tag{3.10}$$

$$
\begin{aligned}
&\int_\Omega \left(\int_0^t g(t-s)\nabla u(s)\,\mathrm{d}s \right) \cdot \left(\int_0^t g(t-s)\left(\nabla u(t)-\nabla u(s)\right)\mathrm{d}s \right) \mathrm{d}x \\
&\le \delta \int_\Omega \left(\int_0^t g(t-s)\left|\nabla u(s)\right| \mathrm{d}s \right)^2 \mathrm{d}x \\
&+ \frac{1}{4\delta} \int_\Omega \left(\int_0^t g(t-s)\left|\nabla u(t)-\nabla u(s)\right| \mathrm{d}s \right)^2 \mathrm{d}x \\
&\le \delta \int_\Omega \left(\int_0^t g(t-s)(\left|\nabla u(t)-\nabla u(s)\right| + \left|\nabla u(t)\right|)\mathrm{d}s \right)^2 \mathrm{d}x \\
&+ \frac{1}{4\delta} \left(\int_0^t g(t-s)\,\mathrm{d}s \right) \int_\Omega \int_0^t g(t-s)\left|\nabla u(t)-\nabla u(s)\right|^2 \mathrm{d}s\,\mathrm{d}x \\
&\le 2\delta \int_\Omega \left(\int_0^t g(t-s)\left|\nabla u(t)-\nabla u(s)\right| \mathrm{d}s \right)^2 \mathrm{d}x \\
&+ 2\delta(1-l)^2 \int_\Omega |\nabla u|^2\,\mathrm{d}x + \frac{1}{4\delta}(1-l)(g \circ \nabla u)(t) \\
&\le \left(2\delta + \frac{1}{4\delta} \right)(1-l)(g \circ \nabla u)(t) + 2\delta(1-l)^2 \int_\Omega |\nabla u|^2\,\mathrm{d}x,
\end{aligned}
\tag{3.11}
$$

$$
\begin{aligned}
&\int_\Omega |u_t|^{m-2} u_t(t) \int_0^t g(t-\tau)\left(u(t)-u(\tau)\right)\mathrm{d}\tau\,\mathrm{d}x \\
&\le \delta \int_\Omega \left| \int_0^t g(t-\tau)\left(u(t)-u(\tau)\right)\mathrm{d}\tau \right|^m \mathrm{d}x + c_\delta \int_\Omega |u_t|^m\,\mathrm{d}x \\
&\le \delta \left(\int_0^t g(\tau)\,\mathrm{d}\tau \right)^{m-1} \int_\Omega \int_0^t g(t-\tau)\left|u(t)-u(\tau)\right|^m \mathrm{d}\tau\,\mathrm{d}x + \\
&c_\delta \int_\Omega |u_t|^m\,\mathrm{d}x \le \delta(1-l)^{m-1} C \int_0^t g(t-\tau)\left\|\nabla u(t)-\nabla u(\tau)\right\|_2^m \mathrm{d}\tau \\
&+ c_\delta \int_\Omega |u_t|^m\,\mathrm{d}x \le \delta(1-l)^{m-1} C_p \left(\frac{2\mathcal{E}(0)}{l} \right)^{\frac{m-2}{2}} \int_0^t g(t-\tau)\left\|\nabla u(t)-\nabla u(\tau)\right\|_2^2 \mathrm{d}\tau \\
&+ c_\delta \int_\Omega |u_t|^m\,\mathrm{d}x \le \delta(1-l)^{m-1} C_p \left(\frac{2\mathcal{E}(0)}{l} \right)^{\frac{m-2}{2}} (g \circ \nabla u)(t) + c_\delta \int_\Omega |u_t|^m\,\mathrm{d}x,
\end{aligned}
\tag{3.12}
$$

$$
\begin{aligned}
-\int_\Omega u_t \int_0^t g'(t-\tau)\big(u(t)-u(\tau)\big)\mathrm{d}\tau\,\mathrm{d}x &\le \delta\int_\Omega |u_t|^2\,\mathrm{d}x \\
+\frac{g(0)}{4\delta}C_p\int_\Omega\int_0^t -g'(t-s)\left|\nabla u(t)-\nabla u(s)\right|^2\,\mathrm{d}s\mathrm{d}x.
\end{aligned} \tag{3.13}
$$

A combination of Eqs. (3.9)–(3.13), then, yields Eq. (3.8).

Theorem 3.4. Let $(u_0, u_1)\in H_0^1(\Omega)\times L^2(\Omega)$ be given. Assume that (A_1), (A_2) hold. Then, for any $t_0 > 0$, there exist positive constants K and λ such that the solution of Eq. (1.3) satisfies

$$
\mathcal{E}(t)\le Ke^{-\lambda\int_{t_0}^{t}\gamma(s)\mathrm{d}s},\quad \forall t\ge t_0,\ \text{ if } m\ge 2 \tag{3.14}
$$

$$
\mathcal{E}(t)\le K\big(1+\gamma(t)\big)^{\frac{-(2m-2)}{2-m}},\quad \forall t\ge t_0,\ \text{ if } 1<m<2. \tag{3.15}
$$

Proof: We start with the case $m \ge 2$. Since $g(0) > 0$, then there exists $t_0 > 0$ such that

$$
\int_0^t g(s)\mathrm{d}s \ge \int_0^{t_0} g(s)\mathrm{d}s = g_0 > 0,\quad \forall t\ge t_0.
$$

By using Eqs. (2.3), (3.1), (3.3), and (3.8), we obtain

$$
\begin{aligned}
F'(t) &\le -\big(1-(\varepsilon_1+\varepsilon_2)c_\delta\big)a\int_\Omega |u_t|^m\,\mathrm{d}x \\
&-\big[\varepsilon_2(g_0-\delta)-\varepsilon_1\big]\int_\Omega u_t^2\mathrm{d}x-\left[\frac{\varepsilon_1 l}{4}-\varepsilon_2\delta\left(1+2(1-l)^2\right)\right]\|\nabla u\|_2^2 \\
&+\left[\frac{1}{2}-\varepsilon_2\frac{g(0)}{4\delta}C_p\right](g'o\nabla u)(t)+ \\
&\left[\varepsilon_1\frac{1-l}{2l}+\varepsilon_2 c_\delta\right](go\nabla u)(t).
\end{aligned}
$$

At this point, we choose δ so small that

$$
g_0-\delta>\frac{1}{2}g_0,\quad \frac{4}{l}\delta\left(1+2(1-l)^2\right)<\frac{1}{4}g_0.
$$

Whence δ is fixed, the choice of any two positive constants ε_1 and ε_2 satisfying

$$\frac{1}{4}g_0\varepsilon_2 < \varepsilon_1 < \frac{1}{2}g_0\varepsilon_2 \tag{3.16}$$

makes

$$k_1 = \varepsilon_2\left(g_0 - \delta\right) - \varepsilon_1 > 0, \;\; k_2 = \frac{\varepsilon_1 l}{4} - \varepsilon_2\delta\left(1 + 2\left(1 - l\right)^2\right) > 0.$$

We then pick ε_1 and ε_2 so small that Eqs. (3.2) and (3.16) remain valid and, further,

$$1 - \left(\varepsilon_1 + \varepsilon_2\right)c_\delta > 0, \; \frac{1}{2} - \varepsilon_2\frac{g(0)}{4\delta}C_p > 0.$$

Therefore, we arrive at

$$F'(t) \leq -\beta\varepsilon(t) + c\left(go\nabla u\right)(t), \quad \forall t \geq t\ 0, \tag{3.17}$$

for two constants $c, \beta > 0$. We multiply (3.17) by $\gamma(t)$ and use Eqs. (1.2) and (2.3), to get

$$\begin{aligned}\gamma(t)F'(t) &\leq -\beta\gamma(t)\mathcal{E}(t) - c\left(g'o\nabla u\right)(t)\\ &\leq -\beta\gamma(t)\mathcal{E}(t) - c\mathcal{E}'(t), \;\; \forall t \geq t_0.\end{aligned}$$

This implies that

$$\gamma(t)F'(t) + c\mathcal{E}'(t) \leq -\beta\gamma(t)\mathcal{E}(t), \;\; \forall t \geq t_0.$$

Hence,

$$\left(\gamma(t)F(t) + c\mathcal{E}(t)\right)' - \gamma'(t)F(t) \leq -\beta\gamma(t)\mathcal{E}(t), \;\; \forall t \geq t_0.$$

Again, by using the fact that $\gamma'(t) \leq 0$, letting

$$\mathcal{L}(t) = \gamma(t)F(t) + c\mathcal{E}(t),$$

and noting that $\mathcal{L}\sim\mathcal{E}$, we arrive at

$$\mathcal{L}'(t) \le -\beta\gamma(t)\mathcal{E}(t) \le -\lambda\gamma(t)\mathcal{L}(t), \ \forall t \ge t_0. \tag{3.18}$$

A simple integration of Eq. (3.18) over (t_0, t) leads to

$$\mathcal{L}(t) \le L(t_0)e^{-\lambda\int_{t_0}^{t}\gamma(s)\mathrm{d}s}, \quad \forall t \ge t_0.$$

We obtain, then, Eq. (3.14) by virtue of equivalence of $\mathcal{E}$ and $\mathcal{L}$.

To establish Eq. (3.15), we re-estimate Eqs. (3.5) and (3.12), for $m < 2$, as follows

$$\begin{aligned}&\int_\Omega |u_t|^{m-2} u_t u \mathrm{d}x \le \delta\int_\Omega |u|^2 \,\mathrm{d}x + c_\delta \int_\Omega |u_t|^{2m-2} \,\mathrm{d}x\\ &\le \delta C_p \|\nabla u\|_2^2 + c_\delta C\left(\int_\Omega |u_t|^m \,\mathrm{d}x\right)^{\frac{2m-2}{m}}.\end{aligned} \tag{3.19}$$

Similarly, we have

$$\begin{aligned}&\int_\Omega |u_t|^{m-2} u_t(t)\int_0^t g(t-\tau)(u(t)-u(\tau))\mathrm{d}\tau\mathrm{d}x\\ &\le \delta C_p (go\nabla u)(t) + c_\delta\left(\int_\Omega |u_t|^m \,\mathrm{d}x\right)^{\frac{2m-2}{m}}.\end{aligned} \tag{3.20}$$

By repeating all above steps and using Eqs. (3.19), (3.20) instead of Eqs. (3.5), (3.12), we arrive at

$$F'(t) \le -\beta\mathcal{E}(t) + c(go\nabla u)(t) + c_1\left(\int_\Omega |u_t|^m \,\mathrm{d}x\right)^{\frac{2m-2}{m}}, \quad \forall t \ge t_0;$$

which gives

$$\mathcal{E}(t) \le -\beta_1 F'(t) + c(go\nabla u)(t) + c_1 \left(\int_\Omega |u_t|^m \, dx \right)^{\frac{2m-2}{m}}, \quad \forall t \ge t_0 \tag{3.21}$$

By multiplying Eq. (3.21) by $\gamma(t)\mathcal{E}^q(t)$, for $q > 0$ to be specified later, and using (A_1), Eq. (2.3), and Young's inequality, we get

$$\begin{aligned} &\gamma(t)\mathcal{E}^{q+1}(t) \le -\beta_2 \gamma(t) F^q F'(t) + c\mathcal{E}^q(t)(-\mathcal{E}'(t)) + \\ &c\gamma(t)\mathcal{E}^q(t)(-\mathcal{E}'(t))^{\frac{2m-2}{m}} \le \frac{-\beta_2 \gamma(t)}{q+1} \frac{dF^{q+1}(t)}{dt} - \frac{c}{2} \frac{d\mathcal{E}^{q+1}(t)}{dt} \\ &+ \mu \mathcal{E}^{\frac{qm}{2-m}}(t) + c_\mu(-\mathcal{E}'(t)). \end{aligned} \tag{3.22}$$

By choosing $q = (2 - m)/(2m - 2)$ (hence, $qm/(2 - m) = q + 1$] and taking μ small enough, Eq. (3.22) yields

$$\begin{aligned} &\gamma(t)\mathcal{E}^{q+1}(t) \le \frac{-\beta_2}{q+1} \gamma(t) \frac{dF^{q+1}(t)}{dt} \\ &- \frac{c}{2} \frac{d\mathcal{E}^{q+1}(t)}{dt} + c(-\mathcal{E}(t)) \le \frac{-\beta_2}{q+1} \frac{d}{dt}\left(\gamma(t) F^{q+1}(t)\right) \\ &+ \frac{\beta_2 \gamma'(t)}{q+1} - \frac{c}{q+1} \frac{d\mathcal{E}^{q+1}(t)}{dt} + c(-\mathcal{E}'(t)). \end{aligned} \tag{3.23}$$

By recalling that $\gamma'(t) \le 0$ and integrating (3.23) over (S, T), $S \ge t_0$, we get

$$\int_S^T \gamma(t)\mathcal{E}^{q+1}(t) dt \le \frac{\beta_2}{q+1} \gamma(S) F^{q+1}(S) + \frac{c}{q+1} \mathcal{E}^{q+1}(S) + 2\mathcal{E}(S) \le A\mathcal{E}(S), \tag{3.24}$$

for some positive constant A. Therefore, Proposition 2.2 gives (3.15). This completes the proof.

Remark 3.1. Estimates (3.14) and (3.15) also hold for $t \in [0, t_0]$ by virtue of continuity and boundedness of $\mathcal{E}$ and γ.

Remark 3.2. This result generalizes and improves many results in the literature. In particular, it allows some relaxation functions which satisfy

$$g'(t) \le -ag^\rho(t), \quad 1 \le \rho < 2,$$

instead of the usual assumption $1 \le \rho < 3/2$.

Remark 3.3. Note that the exponential and the polynomial decay estimates, given in early works, are only particular cases of Eq. (3.14). More precisely, we obtain exponential decay for $\gamma(t)\equiv a$ and polynomial decay for $\gamma(t) = a(1 + t)^{-1}$, where $a > 0$ is a constant.

Acknowledgements

The author would like to express his sincere thanks to King Fahd University of Petroleum and Minerals for its support. This work has been funded by KFUPM under project # IN151002.

Author details

Salim A. Messaoudi

Address all correspondence to: messaoud@kfupm.edu.s

Department of Mathematics and Statistics, KFUPM, Dhahran, Saudi Arabia

References

[1] Dafermos C.M., Asymptotic stability in viscoelasticity, Arch. Rational Mech. Anal. 37 (1970), 297–308.

[2] Dafermos C.M., On abstract Volterra equations with applications to linear viscoelasticity, J. Differ. Equations 7 (1970), 554–569.

[3] Dassios G. and Zafiropoulos F., Equipartition of energy in linearized 3-*d* viscoelasticity, Quart. Appl. Math. 48(4) (1990), 715–730.

[4] Munoz Rivera J.E, Asymptotic behavior in linear viscoelasticity, Quart. Appl. Math. 52(4) (1994), 628–648.

[5] Cabanillas E.L. and Munoz Rivera J.E., Decay rates of solutions of an anisotropic inhomogeneous n-dimensional viscoelastic equation with polynomial decaying kernels, Comm. Math. Phys. 177 (1996), 583–602.

[6] Barreto R., Lapa E.C., and Munoz Rivera J.E., Decay rates for viscoelastic plates with memory, J. Elasticity 44(1) (1996), 61–87.

[7] Cavalcanti M.M., Domingos Cavalcanti V. N., and Soriano J.A., Exponential decay for the solution of semilinear viscoelastic wave equations with localized damping, Electron. J. Differential Equations 2002(44) (2002), 1–14.

[8] Berrimi S. and Messaoudi S.A., Exponential decay of solutions to a viscoelastic equation with nonlinear localized damping, Elect J. Diff. Eqns. 2004 (88) (2004), 1–10.

[9] Berrimi S. and Messaoudi S.A., Existence and decay of solutions of a viscoelastic equation with a nonlinear source, Nonlinear Anal. TMA 64 (2006), 2314–2331.

[10] Cavalcanti M.M., Domingos Cavalcanti V. N., and Ferreira J, Existence and uniform decay for nonlinear viscoelastic equation with strong damping, Math. Meth. Appl. Sci. 24 (2001), 1043–1053.

[11] Messaoudi S.A. and Tatar N.E., Exponential and polynomial decay for a quasilinear viscoelastic equation, Nonlinear Anal. TMA 68 (2007), 785–793.

[12] Messaoudi S.A. and Tatar N.E., Global existence and uniform stability of solutions for a quasilinear viscoelastic problem, Math. Meth. Appl. Sci. 30 (2007), 665–680.

[13] Fabrizio M. and Polidoro S., Asymptotic decay for some differential systems with fading memory, Appl. Anal. 81(6) (2002), 1245–1264.

[14] Messaoudi S.A. General decay of solutions of a viscoelastic equation, J. Math. Anal. Appl. 341 (2008), 1457–1467.

[15] Messaoudi S.A., General decay of the solution energy in a viscoelastic equation with a nonlinear source, Nonlinear Anal. TMA 69 (2008), 2589–2598.

[16] Han Xi. and Wang M., General decay of energy for a viscoelastic equation with nonlinear damping, Math. Meth. Appl. Sci. 32(3) (2009), 346–358.

[17] Liu W.J., General decay of solutions to a viscoelastic wave equation with nonlinear localized damping, Ann. Acad. Sci. Fenn. Math. 34(1) (2009), 291–302.

[18] Liu W.J., General decay rate estimate for a viscoelastic equation with weakly nonlinear time-dependent dissipation and source terms, J. Math. Phys. 50(11), art. no. 113506 (2009).

[19] Park J. and Park S., General decay for quasilinear viscoelastic equations with nonlinear weak damping, J. Math. Phys. 50, art. 083505 (2009).

[20] Xiaosen H. and Mingxing W., General decay of energy for a viscoelastic equation with nonlinear damping, Math. Methods Appl. Sci. 32(3) (2009), 346–335.

[21] Komornik V., Exact Controllability and Stabilization. The Multiplier Method, Masson-John Wiley, Paris (1994).

Chapter 13

Non-Classical Memory Kernels in Linear Viscoelasticity

Sandra Carillo and Claudio Giorgi

Additional information is available at the end of the chapter

http://dx.doi.org/10.5772/64251

Abstract

In linear viscoelasticity, a large variety of regular kernels have been classically employed, depending on the mechanical properties of the materials to be modeled. Nevertheless, new viscoelastic materials, such as viscoelastic gels, have been recently discovered and their mechanical behavior requires convolution integral with singular kernels to be described. On the other hand, when the natural/artificial aging of the viscoelastic material has to be taken into account, time-dependent kernels are needed. The aim of this chapter is to present a collection of nonstandard viscoelastic kernels, with special emphasis on singular and time-dependent kernels, and discuss their ability to reproduce experimental behavior when applied to real materials. As an application, we study some magneto-rheological elastomers, where viscoelastic and magnetic effects are coupled.

Keywords: Materials with memory, Viscoelasticity, Unbounded memory kernels, Existence of solutions, Asymptotic solutions', Behavior

1. Introduction

The stress-strain relation in linear viscoelasticity involves a convolution integral with a memory kernel. The fading memory principle requires that the memory kernel decays quickly as the elapsed time goes to infinity, but no limitation is imposed to its behavior near zero. So, a wide range of kernels may be used depending on the nature of the materials to be modeled. Starting from the rheological model of a standard viscoelastic solid, whose kernel involves a single exponential, a large variety of regular kernels have been classically employed: discrete and continuous Prony series, completely monotonic functions, etc. Recently, new viscoelastic materials, such as viscoelastic gels, have been described by virtue of convolution integral with singular kernels: for instance, fractional and hypergeometric kernels [1]. On the other hand, when the natural/artificial aging of the viscoelastic material has to be taken into account, time-dependent kernels are needed. Furthermore, the behavior of some new materials, for instance, ferrogel and magneto-rheological elastomers, can be determined by coupling viscoelastic and magnetic effects.

The material of this chapter is organized as follows. First we present the model of a viscoelastic body which represents the basis for our study. It is assumed to be homogeneous and isotropic, and its crucial feature is that the stress response at time t linearly depends on the whole *past history* of the strain up to t. Then, we look for the modeling of *aging isothermal viscoelasticity*, assuming that the viscoelastic structural parameters are time dependent while the material is subject to chemical or physical agents at constant temperature. Finally, singular kernel problems are addressed to, at first, in the case of a viscoelastic body and, later, when the viscoelastic behavior is coupled with magnetization. In particular, the case of magneto-viscoelastic bodies is considered. Indeed, the idea of coupling the viscoelastic behavior with magnetic effects is suggested by new materials which are obtained by inserting magnetic defects into a solid body to have the opportunity to influence the mechanical properties of the body when a magnetic field is applied.

2. Preliminary notions and notations

This section is devoted to provide the key notions concerning the model of isothermal viscoelastic body with memory. For sake of simplicity, the body is supposed homogeneous. In order to briefly introduce the subject, at the beginning, we restrict our attention to one-dimensional processes. Let ε denote the uniaxial strain and σ the corresponding tensile stress at every point x of the reference configuration of the sample. According to Boltzmann's formulation of hereditary elasticity [2], a linear viscoelastic solid may be described by a stress-strain relation in the Riemann-Stieltjes integral form

$$\sigma(x,t) = \int_{-\infty}^{t} G(t-s)\mathrm{d}_s\varepsilon(x,s) \tag{1}$$

where G is named *Boltzmann function* (or *memory kernel*) and $\varepsilon(\cdot)$ is a fading strain history, namely

$$\lim_{s\to-\infty}\varepsilon(x,s) = 0, \qquad \varepsilon(x,s) = \int_{-\infty}^{s} \mathrm{d}_s\varepsilon(x,\zeta). \tag{2}$$

In particular, when the strain history vanishes from $-\infty$ to 0, then Eq. (1) reduces to

$$\sigma(x,t) = \int_0^t G(t-s)\mathrm{d}_s\varepsilon(x,s). \tag{3}$$

A peculiar behavior of viscoelastic solid materials is named *relaxation property*: if the solid is held at a constant strain starting from a given time $t_0 \geq 0$, the stress tends (as $t \to \infty$) to a constant value which is "proportional" to the applied constant strain. Indeed, if $\varepsilon(x, \cdot)$ is continuous on $(-\infty, t_0]$ and

$$\varepsilon(x,t) = \varepsilon(x,t_0) = \varepsilon_0(x), \quad \forall t \geq t_0,$$

it follows that

$$\lim_{t\to\infty}\sigma(x,t) = \lim_{t\to\infty} G_\infty\varepsilon(x,t) + \lim_{t\to\infty}\int_{-\infty}^{t_0}[G(t-s)-G_\infty]\mathrm{d}_s\varepsilon(x,s) = G_\infty\varepsilon_0(x),$$

where the *relaxation modulus*

$$G_\infty = \lim_{\tau\to\infty} G(\tau)$$

is assumed to be positive. Then, using Eq. (2) and letting

$$\hat{G}(\tau) = G(\tau) - G_\infty \tag{4}$$

the stress-strain relation (1) may be rewritten as

$$\sigma(x,t) = G_\infty\varepsilon(x,t) + \int_{-\infty}^{t}\hat{G}(t-s)\mathrm{d}_s\varepsilon(x,s). \tag{5}$$

Of course, the choice of G is required to satisfy some basic principles, like the *fading memory principle* and the *dissipation principle*, a thermostatic version of the second law of thermodynamics (see [3], for instance). In general, these conditions allow the memory kernel to be unbounded at the origin.

In the terminology of Dautray and Lions [4], hereditary effects with *long memory range* are represented by a convolution integral, where

$$G \in L^1(0,T) \cap C^2(0,T), \qquad \forall T > 0 \tag{6}$$

whereas a *short memory range* is related to singular kernels of the Dirac delta type. In the latter case, letting $\hat{G} = \Gamma\delta_0$, where δ_0 denotes the Dirac mass at 0^+, from Eq. (1) it follows:

$$\sigma(x,t) = G_\infty \varepsilon(x,t) + \Gamma \partial_t \varepsilon(x,t) \tag{7}$$

where ∂_t denotes partial derviation with respect to 't'.

which is named the *Kelvin-Voigt model*. On the other hand, assumption (6) may be strengthened by letting G be bounded along with its derivatives

$$G \in L^\infty(0,T) \cap C_b^2(0,T), \qquad \forall T > 0. \tag{8}$$

If this is the case,

$$G_0 = \lim_{\tau \to 0^+} G(\tau)$$

and an integration by parts changes Eq. (1) into the alternate forms

$$\sigma(x,t) = G_0 \varepsilon(t) + \int_{-\infty}^{t} G'(t-s)\varepsilon(x,s)ds = G_0 \varepsilon(x,t) + \int_0^\infty G'(\tau)\varepsilon(x,t-\tau)d\tau, \tag{9}$$

where the *relaxation function* $G'(\tau)$ is the derivative with respect to τ of the Boltzmann function G. This constitutive stress-strain relation is based on the Lebesgue representation of linear functionals in the history space theory devised by Volterra [5]. Provided that Eq. (2) holds true, the Boltzmann and the Volterra constitutive relations are equivalent. The latter approach, however, can be applied to a wilder class of strain histories (uniformly bounded, for instance), in which Eq. (2) is no longer needed.

In the three-dimensional case, all fields depend on the space-time pair $(\boldsymbol{x}, t) \in \Omega \times \mathbb{R}$, where $\Omega \subset \mathbb{R}^3$ is the reference configuration. The displacement vector $\boldsymbol{u}(\boldsymbol{x}, t)$ is given by

$$\boldsymbol{u}(\boldsymbol{x},t) = \boldsymbol{\mu}(\boldsymbol{x},t) - \boldsymbol{x},$$

where $\boldsymbol{\mu}(\boldsymbol{x}, \cdot)$ is the motion of $\boldsymbol{x}$, and

$$\boldsymbol{E} = \frac{1}{2}\left[\nabla \boldsymbol{u} + \nabla \boldsymbol{u}^T\right],$$

is the infinitesimal strain tensor. Borrowing from Eq. (9), the viscoelastic Cauchy stress tensor $\boldsymbol{T}$ is given by

$$T(x,t) = \mathbb{G}_0 E(x,t) - \int_{-\infty}^{t} \mathbb{G}'(t-s)E(x,s)\mathrm{d}s = \mathbb{G}_0 E(x,t) + \int_0^{\infty} \mathbb{G}'(\tau)E(x,t-\tau)\mathrm{d}\tau. \tag{10}$$

where $\mathbb{G}:\mathbb{R}^+ \to \mathrm{Lin(Sym)}$ stands for the relaxation function and

$$\mathbb{G}_0 = \mathbb{G}(0), \qquad \mathbb{G}_\infty = \lim_{\tau\to\infty} \mathbb{G}(\tau), \qquad \mathbb{G}'(\tau) = \partial_\tau \mathbb{G}(\tau).$$

A simple manipulation of Eq. (10) yields an alternate stress-strain relation:

$$T(x,t) = \mathbb{G}_\infty E(x,t) - \int_0^{\infty} \mathbb{G}'(\tau)[E(x,t) - E(x,t-\tau)]\mathrm{d}\tau. \tag{11}$$

For fading strain histories obeying Eq. (2), an integration by parts allows Eq. (10) to be rewritten as

$$T(x,t) = \mathbb{G}_\infty E(x,t) + \int_{-\infty}^{t} \hat{\mathbb{G}}(t-s)\mathrm{d}_s E(x,s) = \mathbb{G}_\infty E(x,t) - \int_0^{\infty} \hat{\mathbb{G}}(\tau)\mathrm{d}_\tau E(x,t-\tau), \tag{12}$$

where $\hat{\mathbb{G}}$ is defined as $\hat{G}$ in Eq. (4). The material is said to enjoy the *fading memory principle* when, for every $\varepsilon > 0$ there exists a positive time shift $S_0(\varepsilon)$, possibly dependent on the strain history, such that

$$\left| \int_0^{\infty} \mathbb{G}'(\tau+s)E(x,t-\tau)\,\mathrm{d}\tau \right| = \left| \int_0^{\infty} \hat{\mathbb{G}}(\tau+s)\mathrm{d}_\tau E(x,t-\tau) \right| < \varepsilon, \qquad \forall s > s_0. \tag{13}$$

Note that Eq. (13) does work even if $G(t)$ is allowed to be singular and non-integrable at the origin. Indeed, the fading memory property requires that the memory kernel decays quickly as the *elapsed time* τ go to infinity, but no limitation is imposed to its behavior near zero.

3. Aging models in linear viscoelasticity

Aging is a gradual process in which the properties of a material change, over time or with use, due to chemical or physical agents. Corrosion, obsolescence, and weathering are examples of aging. In metallurgical processes, aging may be induced by a heat treatment (age hardening). Consequences of aging are of various types. For instance, the damages caused by melting or time-deteriorating processes are examples for decreasing stiffness in elastic springs. Instead, solidification of concrete is an irreversible transition process where the system increases its stiffness and releases a large amount of energy per volume. As pointed out in the sequel, the former type of aging is compatible with thermodynamics under isothermal conditions, while the latter involves a latent heat and then requires a non-isothermal framework. For definiteness, in this section, we investigate viscoelastic solids and assume that the viscoelastic model holds while the material is subject to chemical or physical agents at constant temperature. It is then understood that we look for the modeling of *aging isothermal viscoelasticity*.

In modeling aging effects, we might think that in Eq. (1), the dependence of the Boltzmann function G on t and s is not merely through the difference $t - s$ but involves t and s separately. It is a central problem to understand how to model G and we would like to argue as far as possible on physical grounds. The recourse to physical arguments to model aging properties is not new in the literature (see, e.g., [6, 7]). Quite naturally one may refer to the classical rheological models with regular kernels (8) and hence to express the aging properties in terms of time-dependent elasticity and viscosity coefficients.

To this purpose, we first address attention to rheological models and, in particular, we consider the standard solid and the Wiechert-Maxwell model [8, 9]. Hence, we establish the functional providing the stress in terms of the strain. This procedure has the advantage of showing how the dependence on the present value and that on the history of ε are influenced by the rheological parameters. Next we generalize the model and look for the corresponding three-dimensional version. For a generic time-dependent relaxation function, a free energy is found to hold for the stress functional as a suitable Graffi-Volterra functional [3, 5, 10]. As a consequence, the stress functional is found to be compatible with thermodynamics subject to weak restrictions on the relaxation function.

3.1. Insights from a rheological model

To get some insights about the modeling of aging viscoelastic solids, we start from the classical standard linear solid where a Maxwell unit, consisting of a spring and a dashpot connected in series, is set in parallel with a lone spring. While we have in mind the behavior of the model in terms of elongation and forces, we extend the formulae to the continuum framework by the standard analogies stress-force and strain-elongation. It is understood that the model is framed

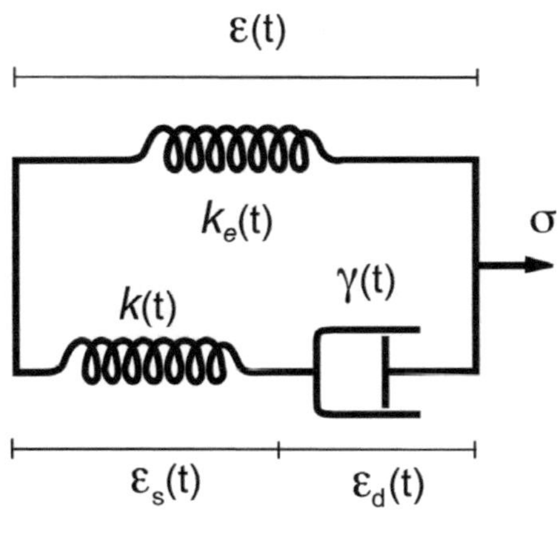

Figure 1. Mechanical scheme of an aging standard viscoelastic solid.

within a one-dimensional picture, so that both strains and stresses are scalar fields depending on *(x,t)*. Since the elastic and Maxwell elements are in parallel, the strain is the same for every element and the applied stress is the sum of the stress in each element (see **Figure 1**).

Hereafter, the dependence on x of all the fields involved is understood and not written. For the Maxwell element, let ε_s and ε_d be the strain of the spring and that of the dashpot. Hence, denoting by ε the common strain we have

$$\varepsilon = \varepsilon_s + \varepsilon_d, \tag{14}$$

Let σ_e be the stress on the isolated spring while σ_m the stress on the Maxwell element. Then, the total applied stress is given by

$$\sigma = \sigma_e + \sigma_m.$$

Moreover let k and k_e be the elastic modulus (or rigidity) of the spring of the Maxwell element and of the spring in parallel, respectively, and γ the viscosity of the dashpot (**Figure 1**). It is the essential feature of the aging effect that k, k_e and γ are positive functions of the time t. In the Maxwell unit, the spring and the dashpot are in series and hence they are subject to the same stress so that, according to the Hook's law,

$$\sigma_e = k_e\varepsilon, \qquad \sigma_m = k\varepsilon_s = \gamma\partial_t\varepsilon_d, \tag{15}$$

where ∂_t denotes partial differentiation with respect to time t. Using the last equality, from Eq. (14) we have

$$\frac{1}{\alpha}\partial_t\varepsilon_d + \varepsilon_d = \varepsilon, \qquad \alpha = \frac{k}{\gamma} \tag{16}$$

Incidentally, if k_e and k are time independent then time differentiation of Eq. (14) and use of Eq. (15) give

$$\partial_t\varepsilon = \frac{1}{k}(\partial_t\sigma - k_e\partial_t\varepsilon) + \frac{1}{\gamma}(\sigma - k_e\varepsilon),$$

which holds for any viscoelastic standard element. Letting $g_0 = k_e + k$ and $g_\infty = k_e$, this differential equation is equivalent to

$$\partial_t\sigma = g_0\partial_t\varepsilon - \alpha(\sigma - g_\infty\varepsilon),$$

which is commonly used in the literature. So far it is only assumed that

(A1) $k_e, k, \gamma \in \mathbb{C}^1(\mathbb{R})$ and $k_e(t), k(t) \geq 0$, $\gamma(t) \geq \gamma_0 > 0$ for every $t \in \mathbb{R}$.

(A2) $\int_{-\infty}^{t} \alpha(\xi) d\xi = \infty$

The last condition is fulfilled when $\alpha = k/\gamma$ is a constant function, for instance.

We may regard Eq. (16) as a differential equation in the unknown $\varepsilon_d(t)$. Then, integration over $[t_0, t]$ yields

$$\varepsilon_d(t) = \varepsilon_d(t_0) \exp\left(-\int_{t_0}^{t} \alpha(y) dy \right) + \int_{t_0}^{t} \exp\left(-\int_{s}^{t} \alpha(y) dy \right) \alpha(s) \varepsilon(s) ds.$$

It is convenient to let $t_0 \to -\infty$. By assuming that ε_d is uniformly bounded on $(-\infty, t]$, assumption (A2) allows us to take

$$\lim_{t_0 \to -\infty} \varepsilon_d(t_0) \exp\left(-\int_{t_0}^{t} \alpha(s) ds \right) = 0.$$

Hence, we have

$$\varepsilon_d(t) = \int_{-\infty}^{t} \exp\left(-\int_{s}^{t} \alpha(y) dy \right) \alpha(s) \varepsilon(s) ds,$$

and from the representation

$$\sigma = k_e \varepsilon + k \varepsilon_s = [k_e + k] \varepsilon - k \varepsilon_d,$$

we obtain the stress-strain relation

$$\sigma(t) = [k_e(t) + k(t)] \varepsilon(t) - \int_{-\infty}^{t} k(t) \exp\left(-\int_{s}^{t} \alpha(y) dy \right) \alpha(s) \varepsilon(s) ds. \quad (17)$$

which involves both the present value $\varepsilon(t)$ and the past history $\varepsilon(s), s \in -\infty, t$. Since

$$\exp\left(-\int_s^t \alpha(y)\mathrm{d}y\right)\alpha(s) = \partial_s \exp\left(-\int_s^t \alpha(y)\mathrm{d}y\right),$$

an integration by parts allows (17) to be rewritten as

$$\sigma(t) = k_e(t)\varepsilon(t) + \int_{-\infty}^{t} k(t)\exp\left(-\int_s^t \alpha(y)\mathrm{d}y\right)\partial_s\varepsilon(s)ds. \tag{18}$$

provided that ε is uniformly bounded on $[-\infty,t)$. A change of variables $\tau = t-s$ within Eq. (17) leads to the alternate form

$$\sigma(t) = [k_e(t) + k(t)]\varepsilon(t) - \int_0^{\infty} k(t)\exp\left(-\int_0^{\tau} \alpha(t-\xi)\mathrm{d}\xi\right)\alpha(t-\tau)\varepsilon(t-\tau)d\tau. \tag{19}$$

Finally, after introducing the so-called *relative history*,

$$\eta^t(\tau) = \varepsilon(t) - \varepsilon(t-\tau),$$

the stress-strain relation may be rewritten as

$$\sigma(t) = k_e(t)\varepsilon(t) + \int_0^{\infty} k(t)\exp\left(-\int_0^{\tau} \alpha(t-\xi)\mathrm{d}\xi\right)\alpha(t-\tau)\eta^t(\tau)d\tau. \tag{20}$$

3.2. Some remarks on the aging effect

To give some evidence to the aging effects, we fix a time $t_0 < t$ and we let

$$k_e = k_e(t_0), \qquad k = k(t_0), \qquad \gamma = \gamma(t_0).$$

This statement holds even if $t_0 = -\infty$ provided that we identify the constant values with the limits as $t \to -\infty$. If no aging affects the material, then

$$k_e(t) = k_e, \qquad k(t) = k, \qquad \gamma(t) = \gamma \qquad \forall t \in \mathbb{R}.$$

Otherwise, remembering that $\alpha = k/\gamma$, we introduce the functions

$$\kappa(t) = k_e(t)/k_e, \qquad \varkappa(t) = k(t)/k, \qquad w(y) = \alpha(y)/\alpha.$$

In particular, κ, $\varkappa$ and w equal unity for non-aging materials. This approach leads to identify κ and w with the aging factors of the elastic and the Maxwell elements, respectively. Moreover, Eq. (18) becomes

$$\sigma(t) = k_e\kappa(t)\varepsilon(t) - \int_{-\infty}^{t} k\exp[-\alpha(t-s)]H(t,s)\partial_s\varepsilon(s)ds \tag{21}$$

where

$$H(t,s) = \varkappa(t)\exp\left[\alpha\int_s^t[1-w(y)]dy\right].$$

This suggests that aging effects can be modeled by means of two functions: κ and H. In our notation, the present value $\varepsilon(t)$ is affected by the factor $\kappa(t)$, whereas the history of ε is affected by the function $H(t,s)$. Letting

$$J(\tau) = k_e + k\exp[-\alpha\tau], \qquad J_\infty = \lim_{\tau\to\infty} J(\tau) = k_e, \qquad \hat{J}(\tau) = J(\tau) - J_\infty$$

the stress-strain relation (21) may be rewritten as

$$\sigma(t) = J_\infty\kappa(t)\varepsilon(t) + \int_{-\infty}^{t} \hat{J}(t-s)H(t,s)\partial_s\varepsilon(s)ds \tag{22}$$

For non-aging materials, $\kappa(t) = H(t,s) \equiv 1$, and this relation reduces to Eq. (5).

We end by observing that in [11] fatigue effects are modeled by using the convolution form (3) modified by the occurrence of a reduced time t_r in place of time t, that is

$$\sigma(t_r) = \int_0^{t_r} G(t_r - s)\partial_s\varepsilon(s)ds = G_\infty\varepsilon(t_r) + \int_0^{t_r}\hat{G}(t_r - s)\partial_s\varepsilon(s)ds, \tag{23}$$

where

$$t_r = \int_0^t \frac{1}{a_T(\tau)} d\tau,$$

a_T being named the *time-temperature shift factor*. A similar approach may equally well model other aging effects. If we denote by *f(t)* the function associated with the aging process applied to the body, then we may introduce a *rescaled time* t_r which is given by

$$t_r = \int_0^t f(\tau) d\tau.$$

To our mind the use of a rescaled time t_r is an operative way of accounting for aging effects. Hereafter, we show that Eq. (18) may be represented as a linear convolution integral after introducing a suitable rescaled time. We start by letting

$$\exp\left(-\int_s^t \alpha(y) dy\right) = \exp(-[A(t) - A(s)])$$

where, for every fixed t,

$$A(s) = \int_0^s \alpha(y) dy, \qquad s \leq t,$$

is positive and nondecreasing because of (A1). Moreover, from (A2) $\lim_{s \to -\infty} A(s) = -\infty$, so that $t_r = A(t)$ plays the role of a rescaled time. Letting $\hat{\varepsilon}(A(s)) = \varepsilon(s)$, we have

$$\partial_{A(s)} \hat{\varepsilon}(A(s)) dA(s) = \partial_s \hat{\varepsilon}(A(s)) ds = \partial_s \varepsilon(s) ds$$

and the stress-strain relation (18) may be rewritten as

$$\sigma(t) = k_e(t)\varepsilon(t) + k(t) \int_{-\infty}^{A(t)} \exp(-[A(t) - A(s)]) \partial_{A(s)} \hat{\varepsilon}(A(s)) dA(s).$$

This expression suggests that aging effects may be partly represented by a suitable change of the time scale within the memory integral. Indeed,

$$\hat{\sigma}(t_r) = k_e(t)\hat{\varepsilon}(t_r) + k(t) \int_{-\infty}^{t_r} \exp[-(t_r - s_r)] \partial_{s_r} \hat{\varepsilon}(s_r) ds_r, \tag{24}$$

where t_r=$A(t)$, s_r=$A(s)$, and $\hat{\sigma}(t_r)$=$\sigma(t)$. This expression completely matches with Eq. (23) only if k_e and k are constants. For non-aging materials, the scaling turns out to be linear, t_r=$A(t)$=αt, and Eq. (24) becomes

$$\sigma(t) = k_e\varepsilon(t) + \int_{-\infty}^{t} k\exp[-\alpha(t-s)]\partial_s\varepsilon(s)\mathrm{d}s.$$

3.3. From long to short memory: a possible aging effect

Instead of (A1) and (A2), we assume here

(B1) k_e, $\gamma > 0$ are constants.

(B2) $k \in \mathbb{C}^1(\mathbb{R})$ is positive, nondecreasing and such that

$$\lim_{t\to-\infty} k(t) = \beta > 0, \qquad \lim_{t\to\infty} k(t) = \infty.$$

From (B1), the viscosity of the damper and the rigidity of the lone spring are constants, whereas (B2) translates the fact that the spring in the Maxwell element becomes completely rigid in the longtime. Under the additional very mild assumption[1]

(B3) $\lim_{t\to\infty} \frac{k'(t)}{[k(t)]^2}=0,$

we can prove that the Kelvin-Voigt viscoelastic model (7) is recovered when $t \to \infty$. Namely, letting

$$k_t(s) = k(t)\exp\left[-\frac{1}{\gamma}\int_s^t k(y)\mathrm{d}y\right],$$

within (B1)-(B3) the distributional convergence

$$k_t \to \gamma\delta_0 \tag{25}$$

occurs as $t \to \infty$, so that Eq. (18) collapses into the Kelvin-Voigt stress-strain relation

$$\sigma_{\mathrm{KV}} = k_e\varepsilon + \gamma\partial_t\varepsilon.$$

[1] It is easily seen that (B3) always holds, for instance, when k is eventually concave down as $t \to \infty$.

The rigorous proof can be found in [12]. Since the function $k_t(\cdot)$ is nonnegative for every t, Eq. (25) follows by showing that, for every fixed $v \geq 0$,

$$\lim_{t\to\infty}\int_v^{\infty} k_t(s)\mathrm{d}s = \begin{cases} \gamma & \text{if } v = 0, \\ 0 & \text{if } v > 0. \end{cases}$$

Assumptions (B1)-(B3) comply with the dissipation principle, as proved by Example 2 in Section 3.7.

3.4. The Wiechert-Maxwell model with aging

The Wiechert-Maxwell model (or Generalized Maxwell model) is composed by a bunch of (say N) Maxwell elements, assembled in parallel, and a further spring in parallel with the whole array. Since all elements are in parallel the strain is the same for every element and the applied stress is the sum of the stress in each element.

ε denotes the common strain and σ_e denotes the stress on the isolated spring, while $\sigma_1,\ldots,\sigma_N$ are the stresses on the Maxwell pairs. Moreover, let k_e, $k_1,\ldots$, k_N be the elastic modulus (or rigidity) of the $N+1$ springs and $\gamma_1,\ldots,\gamma_N$ the viscosity coefficients of the dashpots. It is the essential feature of the aging effect that k_e, $k_1,\ldots$, k_N and $\gamma_1,\ldots,\gamma_N$ are functions of the time t (see **Figure 2**).

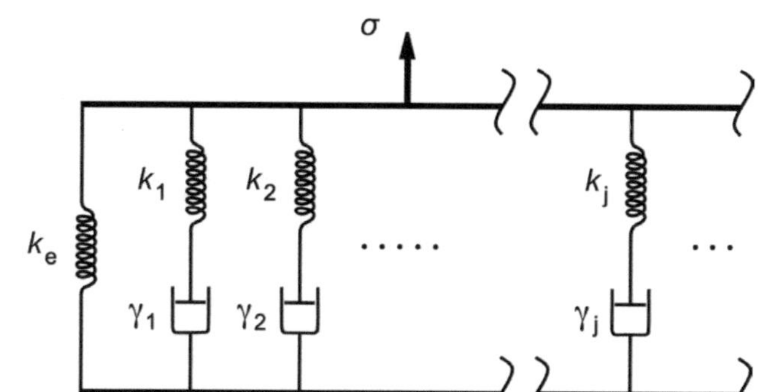

Figure 2. Mechanical scheme of a Wiechert-Maxwell model.

The dependence of k and γ on time requires that we review the elementary arguments to determine the relations among σ_e, $\sigma_1,\ldots,\sigma_N$ and ε. For each jth Maxwell element, let ε_{sj} and ε_{dj} be the strain of the spring and that of the dashpot. Hence, we have

$$\varepsilon = \varepsilon_{sj} + \varepsilon_{dj}, \qquad \sigma_e = k_e\varepsilon, \qquad \sigma_j = k_j\varepsilon_{sj} = \gamma_j\partial_t\varepsilon_{dj}.$$

As a consequence, $\gamma_j\partial_t\varepsilon_{dj} + k_j\varepsilon_{dj} = k_j\varepsilon$, and then

$$\partial_t \varepsilon_{dj} + \alpha_j \varepsilon_{dj} = \alpha_j \varepsilon, \qquad \alpha_j = k_j / \gamma_j. \tag{26}$$

Previous assumptions are generalized for any $j = 1,\dots,N$, as follows:

(C1) $k_e, k_j, \gamma_j \in \mathbb{C}^1(\mathbb{R})$ and $k_e(t) \geq 0$, $k_j(t) \geq 0$, $\gamma_j(t) \geq \gamma 0 > 0$, for every $t \in \mathbb{R}$,

(C2) $\int_{-\infty}^{t} \alpha_j(\xi) d\xi = \infty$, for every $t \in \mathbb{R}$.

By regarding Eq. (26) as a differential equation in the unknown $\varepsilon_{dj}(t)$, for the jth Maxwell element we have

$$\sigma_j(t) = k_j(t)\varepsilon(t) - k_j(t)\int_{-\infty}^{t} \exp\left(-\int_s^t \alpha_j(y)dy\right)\alpha_j(s)\varepsilon(s)ds.$$

The whole stress on the Wiechert-Maxwell model is then given by

$$\sigma(t) = \left[k_e(t) + \sum_{j=1}^{N} k_j(t)\right]\varepsilon(t) - \sum_{j=1}^{N} k_j(t)\int_{-\infty}^{t} \exp\left(-\int_s^t \alpha_j(y)dy\right)\alpha_j(s)\varepsilon(s)ds. \tag{27}$$

To give some evidence to the aging effects as in Section 3.2, we assume that all springs and dashpots within the Maxwell elements have common aging factors.

(C3) There exist a time $t_0 \in \mathbb{R}$ (possibly, $t_0 = -\infty$) and two functions $\varkappa$, w: $\mathbb{R} \to \mathbb{R}$ such that $\varkappa(t_0) = w(t_0) = 1$ and for every $j = 1,\dots,N$

$$k_j(t) = k_j \varkappa(t), \quad \alpha_j(y) = \alpha_j - \bar{\alpha}[1 - w(y)], \text{ where } \bar{\alpha} = \frac{1}{N}\sum_{j=1}^{N} \alpha_j$$

In particular, $\sum_{j=1}^{N} \alpha_j(y) = \sum_{j=1}^{N} \alpha_j w(y)$. Defining the aging factors as follows:

$$\kappa(t) = k_e(t) / k_e(t_0), \qquad H(t,s) = \varkappa(t)\exp\left[\bar{\alpha}\int_s^t [1 - w(y)]dy\right],$$

the stress-strain relation (27) may be rewritten in the form (22) by letting

$$J_\infty = k_e(t_0), \qquad \hat{J}(\tau) = \sum_{j=1}^{N} k_j \exp[-\alpha_j \tau].$$

As in the standard solid model, the present value $\varepsilon(t)$ is affected by the factor $\kappa(t)$ only, whereas the history of ε is affected by the function $H(t, s)$.

3.5. Time-dependent linear viscoelasticity

Borrowing from the Wiechert-Maxwell solid developed above, we now state the uniaxial stress-strain constitutive equation that allows for time-dependent properties. If we introduce the function

$$G(t,s) = k_e(t) + \sum_{j=1}^{N} k_j(t) \exp\left(-\int_s^t \alpha_j(y) \mathrm{d}y \right)$$

which is defined on the half plane $\mathcal{D} = \{(t, s) \in \mathbb{R}^2 : s \le t\}$, the Wiechert-Maxwell constitutive law (27) may be rewritten into the general form

$$\sigma(x,t) = G_0(t)\varepsilon(x,t) - \int_{-\infty}^{t} \partial_s G(t,s)\varepsilon(x,s)\mathrm{d}s, \tag{28}$$

where ∂_s denotes partial differentiation with respect to the variable s and

$$G_0(t) := G(t,t) = k_e(t) + \sum_{j=1}^{N} k_j(t),$$

for all $t \in \mathbb{R}$. In addition, from (C1)–(C2) we have

$$G_\infty(t) := \lim_{s \to -\infty} G(t,s) = k_e(t) > 0, \qquad G_0(t) - G_\infty(t) = \sum_{j=1}^{N} k_j(t) \ge 0.$$

For further convenience, we define $\breve{G}: \mathbb{R} \times \mathbb{R}^+ \to \mathbb{R}$ as $\breve{G}(t, \tau) = G(t, t - \tau)$ so that

$$G_0(t) = \breve{G}(t,0) > 0, \quad G_\infty(t) = \lim_{\tau \to \infty} \breve{G}(t,\tau) > 0, \quad \partial_\tau G(t, t-\tau) = \partial_\tau \breve{G}(t,\tau).$$

Finally, remembering that $G_0(t) = G(t, t)$, an integration by part of (28) yields

$$\sigma(x,t)=\int_{-\infty}^{t} G(t,s)\partial_s\varepsilon(x,s)ds = G_\infty(t)\varepsilon(t)+\int_{-\infty}^{t}\hat{G}(t,s)\partial_s\varepsilon(x,s)ds \tag{29}$$

provided that Eq. (2) holds and

$$\hat{G}(t,s)=G(t,s)-G_\infty(t).$$

The classical expressions (5) and (9) are recovered from Eqs. (29) and (28), respectively, by simply assuming that $G(t, s) = G(t\text{-}s)$, $s \le t$. If this is the case, G_0 and G_∞ turn out to be constants.

We now look for a general, though linear, time-dependent three-dimensional model. According to Eq. (28), the Cauchy stress tensor $\boldsymbol{T}$ is given by

$$\boldsymbol{T}(\boldsymbol{x},t)=\mathbb{G}_0(t)\boldsymbol{E}(\boldsymbol{x},t)-\int_{-\infty}^{t}\partial_s\mathbb{G}(t,s)\boldsymbol{E}(\boldsymbol{x},s)\mathrm{d}s, \tag{30}$$

where $\mathbb{G}$ stands for the t-dependent relaxation function and

$$\mathbb{G}:\mathcal{D}\to \mathrm{Lin(Sym)}, \qquad \mathbb{G}_0(t):=\mathbb{G}(t,t).$$

Letting $\breve{\mathbb{G}}$: $\mathbb{R}\times\mathbb{R}^+\to$ Lin(Sym) such that

$$\breve{\mathbb{G}}(t,\tau)=\mathbb{G}(t,t-\tau),$$

a change of the integration variable into Eq. (30) yields an alternate stress-strain relation

$$\boldsymbol{T}(\boldsymbol{x},t)=\mathbb{G}_0(t)\boldsymbol{E}(\boldsymbol{x},t)+\int_0^{\infty}\partial_\tau\breve{\mathbb{G}}(t,\tau)\boldsymbol{E}(\boldsymbol{x},t-\tau)\mathrm{d}\tau, \tag{31}$$

or equivalently

$$\boldsymbol{T}(\boldsymbol{x},t)=\mathbb{G}_\infty(t)\boldsymbol{E}(\boldsymbol{x},t)-\int_0^{\infty}\partial_\tau\breve{\mathbb{G}}(t,\tau)[\boldsymbol{E}(\boldsymbol{x},t)-\boldsymbol{E}(\boldsymbol{x},t-\tau)]\mathrm{d}\tau. \tag{32}$$

Moreover,

$$\mathbb{G}_0(t)=\breve{\mathbb{G}}(t,0),\qquad \mathbb{G}_\infty(t)=\lim_{\tau\to\infty}\breve{\mathbb{G}}(t,\tau),\qquad \partial_\tau\breve{\mathbb{G}}(t,\tau)=\partial_\tau\mathbb{G}(t,t-\tau).$$

For non-aging materials, $\mathbb{G}(t, s)$ and $\breve{\mathbb{G}}(t, \tau)$ reduce to $\mathbb{G}(t-s)$ and $\mathbb{G}(\tau)$, respectively.

Hereafter, for ease in writing, we introduce the function

$$\mathcal{G}(t,\tau)=-\partial_\tau\breve{\mathbb{G}}(t,\tau),\qquad (t,\tau)\in\mathbb{R}\times\mathbb{R}^+, \tag{33}$$

which is assumed to satisfy the following properties.

(M1) $(t,\tau)\mapsto\mathcal{G}(t,\tau)\in L^\infty(\mathcal{C})$ for every compact set $\mathcal{C}\subset\mathbb{R}\times\mathbb{R}^+$.

(M2) For every fixed $t\in\mathbb{R}$, the map $\tau\mapsto\mathcal{G}(t,\tau)$ is positive semi-definite, absolutely continuous and summable on $\mathbb{R}^+$. Then, for every $t\in\mathbb{R}$

$$\int_0^\infty\mathcal{G}(t,\tau)d\tau=\mathbb{G}_0(t)-\mathbb{G}_\infty(t)\geq 0.$$

Besides, it is differentiable for all $\tau\in\mathbb{R}^+$ and

$$(t,\tau)\mapsto\partial_\tau\mathcal{G}(t,\tau)\in L^\infty(\mathcal{C})$$

for every compact set $\mathcal{C}\subset\mathbb{R}\times\mathbb{R}^+$.

(M3) For every fixed $\tau>0$, the map $t\mapsto\mathcal{G}(t,\tau)$ is differentiable for all $t\in\mathbb{R}$. Besides,

$$(t,\tau)\mapsto\partial_\tau\mathcal{G}(t,\tau)\in L^\infty(\mathcal{C})$$

for every compact set $\mathcal{C}\subset\mathbb{R}\times\mathbb{R}^+$.

(M4) There exists a nonnegative scalar function M: $\mathbb{R}\to\mathbb{R}^+$, bounded on bounded intervals, such that

$$\partial_t\mathcal{G}(t,\tau)+\partial_\tau\mathcal{G}(t,\tau)\leq -M(t)\mathcal{G}(t,\tau)$$

for every $(t,\tau)\in\mathbb{R}\times\mathbb{R}^+$.

According to (M2), the t-dependent relaxation function $\breve{\mathbb{G}}$ may be represented as

$$\breve{\mathbb{G}}(t,\tau) = \mathbb{G}_0(t) - \int_0^\tau \mathcal{G}(t,\sigma)d\sigma.$$

Borrowing from the scalar case, $\mathbb{G}_\infty(t)$ is assumed to be positive definite for every $t \in \mathbb{R}$, namely

$$\mathbb{G}_\infty(t)\boldsymbol{E} \cdot \boldsymbol{E} > 0 \quad \forall \boldsymbol{E} \in \mathrm{Sym}.$$

Finally, an integration by part of Eq. (30) yields

$$\boldsymbol{T}(\boldsymbol{x},t) = \int_{-\infty}^{t} \mathbb{G}(t,s)\partial_s \boldsymbol{E}(\boldsymbol{x},s)\mathrm{d}s = \mathbb{G}_\infty(t)\boldsymbol{E}(\boldsymbol{x},t) + \int_{-\infty}^{t} \hat{\mathbb{G}}(t,s)\partial_s \boldsymbol{E}(\boldsymbol{x},s)\mathrm{d}s \tag{34}$$

provided that Eq. (2) holds for $\boldsymbol{E}$ and

$$\hat{\mathbb{G}}(t,s) = \mathbb{G}(t,s) - \mathbb{G}_\infty(t).$$

As an advantage, within Eq. (34), $\mathbb{G}$ may be unbounded at the origin.

In order to stress the aging effects, we might assume the following *factorization of the memory kernel G*.

(M5) There exist three functions, $\kappa : \mathbb{R} \to \mathbb{R}^+$, $\mathbb{H} : \mathcal{D} \to \mathrm{Lin}^+(\mathrm{Sym})$ and $\mathbb{J} : \mathbb{R}^+ \to \mathrm{Lin}(\mathrm{Sym})$, such that $\mathbb{H}$ is uniformly bounded, $\lim_{\tau\to\infty} \mathbb{J}(\tau) = \mathbb{J}_\infty$ and, for every $t \in \mathbb{R}$ and $s < t$,

$$\hat{\mathbb{G}}(t,s) = \mathbb{H}(t,s)[\mathbb{J}(t-s) - \mathbb{J}_\infty], \qquad \mathbb{G}_\infty(t) = \mathbb{J}_\infty \kappa(t).$$

Accordingly, the stress-strain relation (34) may be rewritten into the form (22). The aging factors κ and $\mathbb{H}$ reduces to unit when non-aging materials are considered.

So far, we restrict our attention to scrutinize stress-strain relations in the form (30). In particular, for isotropic materials $\breve{\mathbb{G}}$ takes the special form

$$\breve{\mathbb{G}}(t,\tau) = \lambda(t,\tau)\mathbf{1}\otimes\mathbf{1} + 2\mu(t,\tau)\mathbb{I},$$

where $\mathbf{1}$ is the unit second-order tensor, $\mathbb{I}$ is the symmetric fourth-order identity tensor, and λ, μ: $\mathbb{R} \times \mathbb{R}^+$ are named *Lamé relaxation functions*. Accordingly,

$$\mathcal{G}(t,\tau) = -\partial_\tau \lambda(t,\tau)\mathbf{1}\otimes\mathbf{1} - 2\partial_\tau \mu(t,\tau)\mathbb{I},$$

3.6. A Wiechert-type three-dimensional model

In the sequel, we scrutinize the special isotropic vector-valued kernel $\mathcal{G} = \mathcal{G}_1 \mathbf{1}\otimes\mathbf{1} + \mathcal{G}_2\mathbb{I}$, where $\mathcal{G}_1$ and $\mathcal{G}_2$ are given by

$$\mathcal{G}_i(t,\tau) = \sum_{j=1}^{N} k_{ji}(t)\alpha_{ji}(t-\tau)\exp\left(-\int_0^\tau \alpha_{ji}(t-y)\mathrm{d}y\right) \qquad i = 1,2, \tag{35}$$

as in the rheological Wiechert-Maxwell model devised in Section 3.1. We first prove that properties (M1)-(M4) hold provided that some additional restrictions are imposed on the material functions k_{ji} and α_{ji}. Finally, we give some examples of these functions that fulfill these conditions.

- (M1) Starting from (C1), it is quite trivial to prove this property.
- (M2) By virtue of Eq. (35) and (C1), G_i, i=1,2, are positive and continuously differentiable with respect to t and τ. Moreover,

$$\int_0^\infty \mathcal{G}(t,\tau)\mathrm{d}\tau = \int_0^\infty \mathcal{G}_1(t,\tau)\mathrm{d}\tau\, \mathbf{1}\otimes\mathbf{1} + \int_0^\infty \mathcal{G}_2(t,\tau)\mathrm{d}\tau\, \mathbb{I} = \sum_{j=1}^{N} k_{j1}(t)\, \mathbf{1}\otimes\mathbf{1} + \sum_{j=1}^{N} k_{j2}(t)\, \mathbb{I}. \tag{36}$$

Hence, $\mathcal{G}(t)$, is summable and vanishing at infinity for every $t \in \mathbb{R}$. In addition, we have

$$\partial_\tau \mathcal{G}_i(t,\tau) = \sum_{j=1}^{N} [\alpha'_{ji}(t-\tau) + \alpha_{ji}^2(t-\tau)]\exp\left(-\int_0^\tau \alpha_{ji}(t-y)\mathrm{d}y\right), \quad i = 1,2,$$

Hence (M2) is fulfilled.

- (M3) It is obviously true as $\alpha_{ji} \in \mathbb{C}^1(\mathbb{R})$ by virtue of (C1)-(C2). In particular,

$$\partial_t \mathcal{G}_i(t,\tau) = \sum_{j=1}^{N}[k'_{ji}(t)\alpha_{ji}(t-s) + k_{ji}(t)\alpha'_{ji}(t-\tau)$$

$$-k_{ji}(t)\alpha_{ji}(t-\tau)[\alpha_{ji}(t) - \alpha_{ji}(t-\tau)]]\exp\left(-\int_0^{\tau}\alpha_{ji}(t-y)\mathrm{d}y\right)$$

$$= \sum_{j=1}^{N}[k'_{ji}(t) - k_{ji}(t)\alpha_{ji}(t)]\alpha_{ji}(t-\tau)\exp\left(-\int_0^{\tau}\alpha_{ji}(t-y)\mathrm{d}y\right) - \partial_\tau \mathcal{G}_i(t,\tau).$$

- (M4) In order to prove this property we need more restrictive conditions. Since

$$\partial_t \boldsymbol{\mathcal{G}} + \partial_\tau \boldsymbol{\mathcal{G}} = (\partial_t \mathcal{G}_1 + \partial_\tau \mathcal{G}_1)\mathbf{1}\otimes\mathbf{1} + (\partial_t \mathcal{G}_2 + \partial_\tau \mathcal{G}_2)\mathbb{I}$$

a sufficient condition to ensure (M4) is given by

$$\partial_t \mathcal{G}_i + \partial_\tau \mathcal{G}_i \le k_{ji}^2(t)\mathcal{G}_i, \qquad i = 1,2. \tag{37}$$

In order to prove these inequalities, we now assume

$$k'_{ji}(t)\gamma_{ji}(t) \le k_{ji}^2(t), \quad \forall t \in \mathbb{R}. \tag{38}$$

and for every $t \in \mathbb{R}$ we let

$$M(t) = \min_{i=1,2}\ \min_{j=1,..,N}\left[\alpha_{ji}(t) - \frac{k'_{ji}(t)}{k_{ji}(t)}\right].$$

It is apparent that $M(t) \ge 0$ and then from (35) it follows:

$$\partial_t \mathcal{G}_i(t,\tau) + \partial_\tau \mathcal{G}_i(t,\tau) = -\sum_{j=1}^{N}\left[\alpha_{ji}(t) - \frac{k'_{ji}(t)}{k_{ji}(t)}\right]k_{ji}(t)\alpha_{ji}(t-\tau)\exp\left(-\int_0^{\tau}\alpha_{ji}(t-y)\mathrm{d}y\right)$$

$$\le -M(t)\sum_{j=1}^{N}k_{ji}(t)\alpha_{ji}(t-\tau)\exp\left(-\int_0^{\tau}\alpha_{ji}(t-y)\mathrm{d}y\right) = -M(t)\mathcal{G}_i(t,\tau)$$

When non-aging material parameters are involved, Eq. (37) reduces to $\partial_\tau \mathcal{G}_i + M\mathcal{G}_i \le 0$, $i = 1, 2$, which implies the exponential decay of the kernels.

3.7. Some examples

We present here some special expressions of material functions k_{ji} and α_{ji}, $j = 1, 2,\ldots, N$, $i = 1, 2$, which fulfill properties (M1)-(M4).

- *Example 1.*

  For simplicity, we restrict our attention to a single Maxwell element. Letting $j = 1$. and $i = 1, 2$, we choose

$$k_{1i}(t) = \kappa_i, \quad \gamma_{1i}(t) = \frac{\eta_i}{e^{\beta_i t} + 1}, \qquad \beta_i, \kappa_i, \eta_i > 0$$

  and then

$$\alpha_{1i}(t) = \frac{\kappa_i}{\eta_i}(e^{\beta_i t} + 1),$$

  so that (A1) and (A2) hold true. Condition (38) is fulfilled for all $t \in \mathbb{R}$ and for every choice of the parameters, so that

$$M(t) = \min_{i=1,2}\left[\frac{\kappa_i}{\eta_i}(e^{\beta_i t} + 1)\right] > 0 \qquad \forall t \in \mathbb{R}.$$

- *Example 2.*

  Otherwise, for $j = 1$. and $i = 1,2$, we can choose

$$k_{1i}(t) = \kappa_i(e^{\omega_i t} + 1), \quad \gamma_{1i}(t) = \eta_i, \qquad \forall t \in \mathbb{R}, \quad i = 1,2,$$

  where ω_i, k_i, $\eta_i > 0$. Accordingly,

$$\alpha_{1i}(t) = \frac{\kappa_i}{\eta_i}(e^{\omega_i t} + 1), \qquad \forall t \in \mathbb{R}, \quad i = 1,2,$$

  so that (A1) and (A2) hold true. On the other hand, condition (38) reduces to

$$\omega_i \eta_i e^{\omega_i t} \leq \kappa_i (e^{\omega_i t} + 1)^2, \quad i = 1,2.$$

  which is equivalent to

$$\frac{\omega_i\eta_i - 2\kappa_i}{\kappa_i} \le e^{\omega_i t} + e^{-\omega_i t} = \cosh\omega_i t, \quad i=1,2,$$

and is fulfilled for all $t \in \mathbb{R}$ provided that $\omega_i \le 3\, K_i / \eta_i$. If this is the case,

$$M(t) = \min_{i=1,2}\left[\frac{\kappa_i(e^{\omega_i t}+1)^2 - \omega_i\eta_i e^{\omega_i t}}{\eta_i(e^{\omega_i t}+1)}\right] > 0 \qquad \forall t \in \mathbb{R}.$$

3.8. Motion, free energies, and thermodynamics

We now derive the motion equation related to the time-dependent viscoelastic stress-strain relation (32) and we examine its compatibility with thermodynamics. The displacement field $\boldsymbol{u}: \Omega\times\mathbb{R} \to \mathbb{R}^3$, relative to the reference configuration $\Omega \subset \mathbb{R}^3$, is subject to the *equation of motion*

$$\rho\partial_{tt}\boldsymbol{u} = \nabla\cdot\boldsymbol{T} + \boldsymbol{f},$$

where $\boldsymbol{f}$ is the body force, per unit volume. Hence, from Eqs. (32)–(33), we obtain

$$\rho\partial_{tt}\boldsymbol{u}(\boldsymbol{x},t) - \nabla\cdot\mathbb{G}_\infty(t)\nabla\boldsymbol{u}(\boldsymbol{x},t) - \nabla\cdot\int_0^\infty \mathcal{G}(t,s)\nabla[\boldsymbol{u}(\boldsymbol{x},t) - \boldsymbol{u}(\boldsymbol{x},t-s)]\mathrm{d}s = \boldsymbol{f}(\boldsymbol{x},t). \tag{39}$$

In order to introduce the initial boundary value problem for this equation, we have to take in mind that it is not invariant under time shift.

Consistent with linear viscoelasticity, we restrict attention to isothermal processes, namely those where the temperature is constant and uniform. Hence, the local form of the second law inequality reduces to the *dissipation inequality*

$$-\rho\frac{\mathrm{d}}{\mathrm{d}t}\psi + \boldsymbol{T}\cdot\boldsymbol{D} \ge 0,$$

where ρ is the mass density, Ψ is the Helmholtz free energy density per unit volume, and $\boldsymbol{D}$ is the stretching tensor. Again for consistency with the linearity of the model, we let the mass density ρ be constant and take the approximation

$$\boldsymbol{D} \simeq \partial_t\boldsymbol{E} = \frac{1}{2}\left[\nabla\partial_t\boldsymbol{u} + \nabla\partial_t\boldsymbol{u}^T\right].$$

Accordingly, we take the dissipation inequality in the form

$$\rho \frac{\mathrm{d}}{\mathrm{d}t}\psi \leq \boldsymbol{T}\cdot\nabla\partial_t \boldsymbol{u}. \tag{40}$$

In materials with memory, the motion equations are required to rule both the displacement instantaneous value $\boldsymbol{u}$(t) and its history up to t. Letting $t_0 \in \mathbb{R}$ be arbitrarily fixed, we define the *relative displacement history* $\zeta^t(\boldsymbol{x},s)$, with $(t,s) \in [t_0,T] \times \mathbb{R}^+$, by

$$\zeta^t(\boldsymbol{x},s) = \begin{cases} \boldsymbol{u}(\boldsymbol{x},t) - \boldsymbol{u}(\boldsymbol{x},t-s), & s \leq t - t_0, \\ \zeta_{t_0}(\boldsymbol{x},s-t+t_0) + \boldsymbol{u}(\boldsymbol{x},t) - \boldsymbol{u}(\boldsymbol{x},t_0), & s > t - t_0, \end{cases} \tag{41}$$

where ζ_{t_0} is the prescribed initial (relative) past history of $\boldsymbol{u}$ up to t_0,

$$\zeta_{t_0}(\boldsymbol{x},s) = \boldsymbol{u}(\boldsymbol{x},t_0) - \boldsymbol{u}(\boldsymbol{x},t_0 - s) \qquad s \in [0,+\infty).$$

Accordingly, $\zeta^t(\boldsymbol{x},0) = 0$ and the motion equation (39) becomes a system

$$\begin{cases} \rho\partial_{tt}\boldsymbol{u}(\boldsymbol{x},t) - \nabla\cdot \mathsf{G}_\infty(t)\nabla\boldsymbol{u}(\boldsymbol{x},t) - \nabla\cdot\int_0^\infty \mathcal{G}(t,s)\nabla\zeta^t(\boldsymbol{x},s)\mathrm{d}s = \boldsymbol{f}(\boldsymbol{x},t), \\ \partial_t\zeta^t(\boldsymbol{x},s) = \partial_t\boldsymbol{u}(\boldsymbol{x},t) - \partial_s\,\zeta^t(\boldsymbol{x},s) \end{cases} \tag{42}$$

where $\boldsymbol{u} : \Omega\times[t_0, +\infty)$ and $\zeta^t : \Omega\times\mathbb{R}^+ \to \mathbb{R}^3$, $T \in [t_0, +\infty)$ are the unknown variables. Their initial conditions are prescribed at $t_0 \in \mathbb{R}$ as follows

$$\begin{cases} \boldsymbol{u}(\boldsymbol{x},t_0) = \boldsymbol{u}_{t_0}(\boldsymbol{x}), \\ \partial_t\boldsymbol{u}(\boldsymbol{x},t_0) = \boldsymbol{v}_{t_0}(\boldsymbol{x}), \\ \boldsymbol{\zeta}^{t_0}(\boldsymbol{x},s) = \boldsymbol{\zeta}_{t_0}(\boldsymbol{x},s), \quad s \in [0,+\infty). \end{cases} \tag{43}$$

Let $H^0 = [\mathrm{L}^2(\Omega)]^3$ and $H^1 = [\mathrm{H}^1_0(\Omega)]^3$, and let $\langle\cdot,\cdot\rangle_j$ denote the usual inner product in H^j, $j = 0,1$. For every $t \geq t_0$, we introduce the family of memory spaces

$$\mathcal{M}_t = L^2_{\mathcal{G}}(\mathbb{R}^+;H^1), \qquad \langle \zeta,\xi\rangle_{\mathcal{M}_t} = \int_0^\infty \langle \mathcal{G}(t,s)\zeta(s),\xi(s)\rangle_1 \mathrm{d}s,$$

where $\langle\cdot,\cdot\rangle_{\mathcal{M}_t}$ denotes the t-dependent weighted L^2 inner product equipping each $\mathcal{M}_t$. In this functional framework, the motion equation admits a unique regular solution. The proof of this result can be found in [12, Th. 4.5].

Theorem 1. *Let* $\mathcal{H}_t = H^1 \times H^0 \times \mathcal{M}_t$ and $\boldsymbol{f} \in H^0$. Under assumptions (M1)-(M4), for every $T > t_0$ and every initial datum $z_{t_0} = (\boldsymbol{u}_{t_0}, \boldsymbol{v}_{t_0}, \zeta_{t_0}) \in \mathcal{H}_{t_0}$, problem *(42)-(43)* admits a unique solution $z(t) = (\boldsymbol{u}(t), \partial_t\boldsymbol{u}(t), \zeta^t)$ on the interval $[t_0,T]$ *such that*

$$\boldsymbol{u} \in C([t_0,T],H^1) \cap C^1([t_0,T],H^0), \qquad \zeta^t \in \mathcal{M}_t, \quad \forall t \in [t_0,T],$$

and

$$\sup_{t\in[t_0,T]} \|z(t)\|_{\mathcal{H}_t} < C,$$

for some $C > 0$ *depending only on* T, t_0 *and the size of the initial datum* $\|z\|_{\mathcal{H}_{t_0}}$.

Now, we introduce a time-dependent free energy density borrowing its expression from the Graffi's single-integral quadratic form (see [13] and references therein). Let

$$\psi(\nabla\boldsymbol{u}(t),\nabla\zeta^t,t) = \frac{1}{2}\mathbb{G}_\infty(t)\nabla\boldsymbol{u}(t)\cdot\nabla\boldsymbol{u}(t) + \frac{1}{2}\int_0^\infty \mathcal{G}(t,s)\nabla\zeta^t(s)\cdot\nabla\zeta^t(s)\mathrm{d}s.$$

For ease in writing, hereafter the dependence on x is understood and not written. In addition, we assume $\rho = 1$. After integrating over Ω, we end up with the total free energy functional

$$\Psi_t(\boldsymbol{u}(t),\zeta^t) = \int_\Omega \psi(\nabla\boldsymbol{u}(t),\nabla\zeta^t,t)\mathrm{d}v = \frac{1}{2}\langle\mathbb{G}_\infty(t)\boldsymbol{u}(t),\boldsymbol{u}(t)\rangle_1 + \frac{1}{2}\|\zeta^t\|^2_{\mathcal{M}_t}.$$

Theorem 2. For an aging viscoelastic material, the dissipation inequality (40) is fulfilled provided that (M4) holds and

$$\mathbb{G}'_\infty(t) \le 0, \quad \forall t \in \mathbb{R}. \tag{44}$$

Proof. First we observe that

$$\frac{\mathrm{d}}{\mathrm{d}t}\langle \mathbb{G}_\infty(t)\boldsymbol{u}(t),\boldsymbol{u}(t)\rangle_1 = 2\langle \mathbb{G}_\infty(t)\boldsymbol{u}(t),\partial_t\boldsymbol{u}(t)\rangle_1 + \langle \mathbb{G}'_\infty(t)\boldsymbol{u}(t),\boldsymbol{u}(t)\rangle_1$$

Then, by virtue of (42), and some integration by parts, we obtain

$$\frac{\mathrm{d}}{\mathrm{d}t}\|\boldsymbol{\zeta}^t\|^2_{\mathcal{M}_t} = \int_0^\infty \langle [\partial_t \boldsymbol{\mathcal{G}}(t,s) + \partial_s \boldsymbol{\mathcal{G}}(t,s)]\,\boldsymbol{\zeta}^t(s),\,\boldsymbol{\zeta}^t(s)\rangle_1 \mathrm{d}s + 2\langle\,\boldsymbol{\zeta}^t,\partial_t\boldsymbol{u}(t)\rangle_{\mathcal{M}_t},$$

and, taking into account (41),

$$\begin{aligned}\langle \boldsymbol{T}(t),\nabla\partial_t\boldsymbol{u}(t)\rangle_0 &= \langle \mathbb{G}_\infty(t)\boldsymbol{u}(t) + \int_0^\infty \boldsymbol{\mathcal{G}}(t,s)\boldsymbol{\zeta}^t(s)\mathrm{d}s,\partial_t\boldsymbol{u}(t)\rangle_1 \\ &= \mathbb{G}_\infty(t)\langle \boldsymbol{u}(t),\partial_t\boldsymbol{u}(t)\rangle_1 + \langle \boldsymbol{\zeta}^t,\partial_t\boldsymbol{u}(t)\rangle_{\mathcal{M}_t}\end{aligned}$$

In summary, we end up with

$$\begin{aligned}\frac{\mathrm{d}}{\mathrm{d}t}\Psi_t(\boldsymbol{u}(t),\,\boldsymbol{\zeta}^t) &= \langle \boldsymbol{T}(t),\nabla\partial_t\boldsymbol{u}(t)\rangle_0 + \frac{1}{2}\langle \mathbb{G}'_\infty(t)\boldsymbol{u}(t),\boldsymbol{u}(t)\rangle_1 \\ &\quad + \frac{1}{2}\int_0^\infty \langle [\partial_t \boldsymbol{\mathcal{G}}(t,s) + \partial_s \boldsymbol{\mathcal{G}}(t,s)]\,\boldsymbol{\zeta}^t(s),\,\boldsymbol{\zeta}^t(s)\rangle_1 \mathrm{d}s\end{aligned}$$

Owing to (M4), this yields

$$\frac{\mathrm{d}}{\mathrm{d}t}\Psi_t(\boldsymbol{u}(t),\,\boldsymbol{\zeta}^t) \le \langle \boldsymbol{T}(t),\nabla\partial_t\boldsymbol{u}(t)\rangle_0 - \frac{1}{2}M(t)\,\|\nabla\,\boldsymbol{\zeta}^t\|^2_{\mathcal{M}_t} + \frac{1}{2}\langle \mathbb{G}'_\infty(t)\nabla\boldsymbol{u}(t),\nabla\boldsymbol{u}(t)\rangle$$

where $M(t) \ge 0$, and (44) finally implies the dissipation inequality

$$\frac{\mathrm{d}}{\mathrm{d}t}\Psi_t(\boldsymbol{u}(t),\,\zeta^t) \le \langle \boldsymbol{T}(t),\nabla\partial_t\boldsymbol{u}(t)\rangle_0.$$

4. Singular kernel models in linear viscoelasticity

The study of singular kernel problems is motivated by the modeling of new materials and, in particular, of the mechanical behavior of some new viscoelastic polymers and bio-inspired materials. As noticed in [14], the appropriate way to handle the response of certain time-dependent systems exhibiting long tail memories is to account for power laws, both for creep and relaxation, leading to the occurrence of fractional hereditariness. Another example encountered in natural materials is mineralized tissues as bones, ligaments, and tendons. They exhibit a marked power-law time-dependent behavior under applied loads (see e.g. [15]), since the high stiffness of the crystals in such tissues is combined with the exceptional hereditariness of the collagen protein-based matrix. In all these cases, we are forced to abandon the regularity assumptions (8) and assume the memory kernels obey Eq. (6) and are unbounded at the origin.

The idea of singular kernels to model particular cases of viscoelastic behaviors was introduced by Boltzmann [2] in the nineteenth century. The fast growth of polymer science motivated further developments of viscoelasticity in the middle of the twentieth century [16, 17], but a Volterra-type integro-differential equation with a regular kernel (typically, a finite sum of exponentials) was preferred to the Boltzmann approach in the modeling of the mechanical response [5, 18]. Later, however, many authors addressed their interest to singular kernel problems, both under the analytical as well as the model point of view [19–24], and their thermodynamical admissibility was analyzed in [25]. In modern viscoelasticity, it is a central problem to understand how to model the memory kernels, and it should be argued as far as possible on physical grounds. So, the first question to answer to is why do we consider singular kernel models. More recently, new viscoelastic materials, such as viscoelastic gels, have been discovered and their mechanical properties are well described by virtue of convolution integral with singular kernels: for instance, fractional and hypergeometric kernels [1]. This applicative interest gave rise to a wide research activity concerning singular kernel problems, both in rigid thermodynamics with memory as well as in viscoelasticity (see, for instance, [26–31], and especially concerning applications of fractional calculus to the theory of viscoelasticity and the study of new bio-inspired materials [15, 32–35]. A recent book [36] provides an overview on this subject. In this framework, Fabrizio [37] analyzes the connection between Volterra and fractional derivatives models and shows how experimental results motivate us to adopt, as in this present article, less restrictive functional requirements on the kernel representing the relaxation modulus.

4.1. Singular isothermal viscoelastic body with memory

To start with, the one-dimensional classical viscoelasticity problem is recalled. It reads

$$u_{tt} = G(0)u_{xx} + \int_0^t G'(t-\tau)u_{xx}(\tau)d\tau + f \tag{45}$$

$$u(\cdot,0) = u_0,\ u_t(\cdot,0) = u_1 \text{ in } \Omega;\ u = 0 \text{ on } \Sigma = \partial\Omega \times (0,T) \tag{46}$$

where $\Omega = (0,1)$. When, to model the physical behavior of new materials or polymers, the regularity assumptions on the relaxation modulus are relaxed, G is assumed to satisfy the following functional requirements

$$G \in L^1(0,T) \cap C^2(0,T), \quad G' \not\in L^1(0,T),\ \forall T \in \mathbb{R} \tag{47}$$

http://dx.doi.org/10.5772/64251

that is, now, the relaxation function $G(t)$ is not required to be finite at $t = 0$ and then Eq. (45) loses its meaning and, hence, needs to be replaced by a different one. The method to overcome this difficulty, devised in [28], consists in the introduction of a suitable sequence of regular problems, depending on a *small parameter* $0 < \varepsilon \ll 1$ which, in the limit $\varepsilon \to 0$ reduce to the singular problem under investigation. The key steps of the approximation strategy can be sketched as follows.

- Let K, termed *integrated relaxation function*, denote

$$K(\xi) := \int_0^{\xi} G(\tau)d\tau \ , \ K(0) = 0; \tag{48}$$

  it is well defined, since $G \in L^1(0,T), \forall T \in \mathbb{R}^+$.

- Then, introduce the regular problems:

$$P^{\varepsilon}: u_{tt}^{\varepsilon} = G^{\varepsilon}(0)u_{xx}^{\varepsilon} + \int_0^t G'^{\varepsilon}(t-\tau)u_{xx}^{\varepsilon}(\tau)d\tau + f \quad \text{where} \quad G^{\varepsilon}(\cdot) := G(\varepsilon + \cdot) \tag{49}$$

together with the initial and boundary conditions

$$u^{\varepsilon}|_{t=0} = u_0(x), \quad u_t^{\varepsilon}|_{t=0} = u_1(x), \quad u^{\varepsilon}|_{\partial\Omega\times(0,T)} = 0 \ , \ t < T. \tag{50}$$

- For each ε, the problem P^{ε} is a *regular* approximated problem since $G^{\varepsilon}(0)$ is finite and, therefore, the initial boundary value problem (49)-(50) admits a unique solution:
- then, find approximated solutions u^{ε}, $0 < \varepsilon \ll 1$,
- show the existence of the limit solution $u := \lim_{\varepsilon \to 0} u^{\varepsilon}$
- prove the uniqueness of the limit solution u which represents a weak solution admitted by the *singular* problem.

Note that, corresponding to each value of ε, the problem P^{ε} is equivalent to the *integral* equation:

$$P^{\varepsilon}: u^{\varepsilon}(t) = \int_0^t K^{\varepsilon}(t-\tau)u_{xx}^{\varepsilon}(\tau)\mathrm{d}\tau + u_1 t + u_0 + \int_0^t \mathrm{d}\tau \int_0^{\tau} f(\xi)\mathrm{d}\xi, \tag{51}$$

Partial derivation w.r. to t, twice, of Eq. (51) delivers Eq. (49) together with initial and boundary conditions (50). Furthermore, when $\varepsilon = 0$, we obtain the well-defined problem

$$P: u(t) = \int_0^t K(t-\tau)u_{xx}(\tau)\mathrm{d}\tau + u_1 t + u_0 + \int_0^t \mathrm{d}\tau \int_0^{\tau} f(\xi)\mathrm{d}\xi \tag{52}$$

where the superscripts, in the case $\varepsilon = 0$, are omitted for notational simplicity. Hence, the following theorems can be proved. Here only the outlines of the proofs are given; the details are comprised in [28] when homogeneous Dirichlet b.c.s (50) are imposed and in [27] when homogeneous Neumann b.c.s are considered.

Theorem 1*Given u^{ε} solution to the integral problem P^{ε} (51), then*

$$\exists u(t) = \lim_{\varepsilon \to 0} u^{\varepsilon}(t) \;\; in \;\; L^2(Q), Q = \Omega \times (0,T). \tag{53}$$

Proof's outline:

- weak formulation, on introduction of test functions $\varphi \in H^1(\Omega \times (0,T)$ s.t. $\varphi_x = 0$, on $\partial\Omega$,
- consider separately the terms without ε,
- the terms with u^{ε} and K^{ε},
- prove convergence via Lebesgue's theorem.

Furthermore, the weak solution, as stated in the following theorem, is unique.

Theorem 2*The integral problem (52) admits a unique weak solution.*

Proof's outline: The result is proved by contradiction, see [28] for details, assuming there are two different solution and, then, showing that such an assumption leads to a contradiction.

As a final remark, we wish to emphasize that, since the isothermal rigid viscoelasticity model exhibits remarkable analogies, under the analytical point of view [38], with rigid thermodynamics with memory, then, analogous results can be obtained also in the study of singular kernel problems in such a framework [29].

4.2. Magneto-viscoelasticity problems

This section is concerned about a problem in magneto-viscoelasticity, again under the assumption of a memory kernel singular at the origin. The interest in magneto-viscoelastic material finds its motivation in the growing interest in new materials such as magneto-rheological elastomers or, in general, magneto-sensitive polymeric composites (see [39–41] and references therein). The model adopted here to describe the magneto-elastic interaction is introduced in [42]. Evolution problems in magneto-elasticity are studied in [43] and, later magneto-viscoelasticity problems are considered in [44, 45]. Notably, under the analytical viewpoint, when the coupling with magnetization is considered, the problem to study is modeled via a nonlinear integro-differential system while the purely viscoelastic problem is linear.

To understand the model equations, a brief introduction on the model magnetization here adopted, based on [46], who revisited the Gilbert magnetization model. Accordingly, when $\Omega \subset \mathbb{R}^3$ denotes the body configuration, the related magnetization changes according to the

http://dx.doi.org/10.5772/64251

Landau Lifshitz equation, which, in Gilbert form, where m represents the magnetization vector reads

$$\gamma^{-1}\boldsymbol{m}_t - \boldsymbol{m}\times\left(a\Delta\boldsymbol{m} - \boldsymbol{m}_t\right) = 0 \ , |\boldsymbol{m}| = 1, \gamma, a \in \mathbb{R}^+. \tag{54}$$

The quantities of interest, in the general three-dimensional case, are the following ones:

$\boldsymbol{u} := \boldsymbol{u}(\boldsymbol{x}, t)$	*displacement vector*
$\boldsymbol{m} := \boldsymbol{m}(\boldsymbol{x}, t)$	*magnetization vector*
$\mathbb{G}(s) = \{G_{klmn}(s)\},\ s \in [0, T]$	*visco-elasticity tensor*
$\mathbb{L} = \{\lambda_{klmn}\}$	*magneto-elasticity tensor*
$\mathbb{E} = \{\epsilon_{lm}\}$	*strain tensor*
$\boldsymbol{\epsilon}(\boldsymbol{u}) := \{\epsilon_{lm}(\boldsymbol{u})\} = \frac{1}{2}\{\boldsymbol{u}_{l,m} + \boldsymbol{u}_{m,l}\}$	*deformation tensor*

$\mathbb{G}\,\nabla\boldsymbol{u}\cdot\nabla\boldsymbol{v} = G_{klmn}\epsilon_{kl}(\boldsymbol{u})\epsilon_{mn}(\boldsymbol{v})$
$\mathbb{L}\boldsymbol{m}\otimes\boldsymbol{m} = \{\lambda_{klmn}m_k m_l\}$
$\mathbb{L}\boldsymbol{m}\otimes\nabla\boldsymbol{u} = \{\lambda_{klmn}m_k\epsilon_{lm}(\boldsymbol{u})\}$
$\mathbb{L}\boldsymbol{m}\otimes\boldsymbol{m}\cdot\nabla\boldsymbol{u} = \lambda_{klmn}m_k m_l\epsilon_{mn}(\boldsymbol{u})$

where the coefficients λ_{klmn} are subject to the condition

$$\lambda_{ijkl} = \lambda_1\delta_{ijkl} + \lambda_2\delta_{ij}\delta_{kl} + \lambda_3(\delta_{ik}\delta_{jl} + \delta_{il}\delta_{jk}) \tag{55}$$

Then, the following constitutive assumptions are assumed. Thus, the exchange magnetization energy is given by

$$E_{ex}(\mathbf{m}) = \frac{1}{2}\int_\Omega a_{ij}m_{k,i}m_{k,j}d\Omega \tag{56}$$

where

- $a_{ij} = a_{ji}$ symmetric positive definite matrix
- $a_{ij} = a\,\delta_{ji},\ a \in \mathbb{R}^+$ diagonal matrix (most materials).

Then, the magneto-elastic energy is given by

$$E_{em}(\mathbf{m}, H) = \frac{1}{2}\int_\Omega \boldsymbol{\lambda}_{ijkl}\, m_i\, m_j\, \varepsilon_{kl}\,(H)\,d\Omega \tag{57}$$

- $\lambda_{ijkl} = \lambda_1 \delta_{ijkl} + \lambda_2 \delta_{ij} \delta_{kl} + \lambda_3 (\delta_{ik} \delta_{jl} + \delta_{il} \delta_{jk})$,
- $\delta_{ijkl} = 1$ if $i = j = k = l$ and $\delta_{ijkl} = 0$ otherwise,
- $\lambda_1, \lambda_2, \lambda_3$ $\mathbb{R}_{\text{constants}}$.

The viscoelastic energy is given by

$$E_{ve}(\boldsymbol{u}) = \frac{1}{2}\int_\Omega G_{klmn}(0)\varepsilon_{kl}\varepsilon_{mn} \mathrm{d}\Omega + \frac{1}{2}\int_0^t d\tau \left(\int_\Omega G'_{klmn}(t-\tau)\varepsilon_{kl}(\tau)\varepsilon_{mn}(\tau) d\Omega \right) \tag{58}$$

where the tensor's entries of $\mathbb{G}$ satisfy

- $G_{klmn} = G_{mnkl} = G_{lkmn}$
- $G_{klmn} e_{kl} e_{mn} \geq \beta e_{kl} e_{kl}, \qquad \beta > 0, e_{kl} = e_{lk}$
- $G'_{klmn} e_{kl} e_{mn} \leq 0$
- $G''_{klmn} e_{kl} e_{mn} \geq 0$

Then, the total energy of the system is given by

$$E(\mathbf{m}, u) = E_{ex}(\mathbf{m}) + E_{em}(\mathbf{m}, u) + E_{ve}(u), \tag{59}$$

taking into account, further to the single magnetic and viscoelastic contribution, of the exchange energy.

4.2.1. A regular magneto-viscoelasticity problem

The problem we are concerned about is the behavior of a viscoelastic body subject also to the presence of a magnetic field; in the one-dimensional case, it is modeled by the nonlinear system

$$\begin{cases} u_{tt} - G(0)\, u_{xx} - \int_0^t G'(t-\tau)\, u_{xx}(\tau)\, \mathrm{d}\tau - \frac{\lambda}{2}(\Lambda(\mathbf{m})\cdot\mathbf{m})_x = f, \\ \qquad\qquad\qquad\qquad\qquad\qquad\qquad\qquad\qquad\qquad \text{in } \mathcal{Q} \\ \mathbf{m}_t + \mathbf{m}\frac{|\mathbf{m}|^2 - 1}{\varepsilon} + \lambda\Lambda(\mathbf{m})u_x - \mathbf{m}_{xx} = 0, \end{cases} \tag{60}$$

where $\Omega = (0,1)$, $\mathcal{Q} := \Omega \times (0, T)$ and $\mathcal{M} \equiv (0, \boldsymbol{m}$, where $\boldsymbol{m} = (m_1, m_2)$, denotes the magnetization vector, orthogonal to the conductor, since $\boldsymbol{u} \equiv (u, 0, 0)$, when both quantities are written in $\mathbb{R}^3$

in addition, v is the outer unit normal at the boundary $\partial\Omega$, Λ is a linear operator defined by $\Lambda(\boldsymbol{m}) = (m_1, m_2)$ the scalar function u is the displacement in the direction of the conductor itself, here identified with the x axis and λ is a positive parameter. In addition, the term f represents an external force which also includes the deformation history.

In [44], the existence and uniqueness of the solution to the problem given by (60), together with the following initial and boundary conditions, is proved

$$u(\cdot,0) = u_0 = 0, \; \mathbf{m}(\cdot,0) = \boldsymbol{m}_0, \quad |\boldsymbol{m}_0| = 1 \quad \text{in } \Omega, \tag{61}$$

$$u = 0, \qquad \frac{\partial \boldsymbol{m}}{\partial \nu} = 0 \qquad \text{on} \quad \Sigma = \partial\Omega \times (0,T), \tag{62}$$

under the assumptions

$$\begin{cases} u_0 \in H_0^1(\Omega), \quad u_1 \in L^2(\Omega), \quad \mathrm{m}_0 \in \boldsymbol{H}^1(\Omega), \\ f \in L^2(\Omega \times (0,T)), \quad G(t) \in C^2(0,T), \end{cases} \tag{63}$$

Then, the following existence and uniqueness result [44] holds.

Theorem 3 Given the problem (60)-(63), it admits a unique solution for any given $T > 0$ and ε small enough (i.e., $\varepsilon\lambda^{-2}GT$), s.t.

- $u \in C^0([0,T];\, H_0^1(\Omega)) \cap C^1([0,T];\, L^2(\Omega))$,
- $\boldsymbol{m} \in \mathbf{C}^0([0,T]; H^1(\Omega)) \cap \mathbf{L}^2(0,T; H^2(\Omega))$,
- $\boldsymbol{m}_t \in \mathbf{L}^2(0,T; L^2(\Omega))$.

The proof, is based on the a priori estimate on the viscoelastic term:

$$\frac{1}{2}\int_\Omega |\varphi_x|^2 \, dx + \frac{1}{2}\int_\Omega |\varphi_t|^2 \, dx \le \alpha e^T C(f,\varphi_0,\varphi_1), \quad \alpha, C \in \mathbb{R}^+ \tag{64}$$

A result of existence, in a three-dimensional regular magneto-viscoelasticity problem, is given in [45].

4.2.2. A singular magneto-viscoelasticity problem

Now, as in the purely viscoelastic case, when the requirement $G' \in L^1(0,T)$ is removed, the magneto-viscoelasticity problem cannot be written under the form (60); however, since $G \in L^1(0,T)$ via integration with respect to time of the integro-differential equation, it can be formulated in the following equivalent form

$$\begin{cases} u_t(t) - \int_0^t G(t-\tau)u_{xx}(\tau)d\tau - u_1 - \int_0^t \frac{\lambda}{2}(\Lambda(\mathbf{m})\cdot\mathbf{m})_x d\tau = \int_0^t f(\tau)d\tau \\ \mathbf{m}_t + \mathbf{m}\frac{|\mathbf{m}|^2 - 1}{\delta} + \lambda\Lambda(\mathbf{m})u_x - \mathbf{m}_{xx} = 0, \end{cases} \quad \text{in} Q \tag{65}$$

The strategy to prove the existence result [47], relies on the fact that the classical problem (60) as soon as the initial time is $t_0 = \varepsilon$, for any arbitrary $\varepsilon > 0$, the relaxation modulus satisfies the classical regularity requirements, namely, as in subSection 4.0.1, $G^{\varepsilon}(\cdot) := G(\varepsilon + \cdot)$ implies that $G^{\varepsilon} \in C^2[0,\ T]$ Hence, each *time-translated* approximated problems

$$P^{\varepsilon}: \begin{cases} u_{tt}^{\varepsilon} - G^{\varepsilon}(0)u_{xx}^{\varepsilon} - \int_0^t G'^{\varepsilon}(t-\tau)u_{xx}^{\varepsilon}(\tau)d\tau - \frac{\lambda}{2}(\Lambda(\mathbf{m}^{\varepsilon})\cdot\mathbf{m}^{\varepsilon})_x = f \\ \mathbf{m}_t^{\varepsilon} + \mathbf{m}^{\varepsilon}\frac{|\mathbf{m}^{\varepsilon}|^2 - 1}{\delta} + \lambda\Lambda(\mathbf{m}^{\varepsilon})u_x^{\varepsilon} - \mathbf{m}_{xx}^{\varepsilon} = 0, \end{cases} \quad \text{in} Q \tag{66}$$

with the assigned initial and boundary conditions

$$u^{\varepsilon}(\cdot,0) = u_0 = 0,\ u_t^{\varepsilon}(\cdot,0) = u_1,\quad \mathbf{m}^{\varepsilon}(\cdot,0) = \mathbf{m}_0,\quad \text{in } \Omega, \tag{67}$$

$$u^{\varepsilon} = 0, \qquad \frac{\partial \mathbf{m}^{\varepsilon}}{\partial \nu} = 0 \qquad \text{on} \quad \Sigma = \partial\Omega \times (0,T), \tag{68}$$

is regular. Then, according to [44], the problem P^{ε} admits a unique strong solution. According to [47], where all the needed proofs are given, the following existence result can be stated.

Theorem 3.1*For all$T > 0$, there exists a weak solution $(u, \boldsymbol{m})$ to the problem* (65)-(61)-(62), *that is a vector function* $(u, \boldsymbol{m})$ *s.t.*

- $u \in L^{\infty}(0,T;H_0^1(\Omega))$;
- $u_t \in L^{\infty}(0,T;L^2(\Omega))$;
- $\mathrm{m} \in L^{\infty}(0,T;H^1(\Omega))$;
- $\mathrm{m}_t \in L^2(\mathcal{Q})$.

which satisfies

$$
\begin{aligned}
&-\int_Q \phi_t\, u^{\varepsilon}(t)\,\mathrm{d}x\,\mathrm{d}t + \int_Q \int_0^t G^{\varepsilon}(t-\tau)\,u_x^{\varepsilon}(\tau)\,\phi_x\,\mathrm{d}\tau\,\mathrm{d}x\,\mathrm{d}t + \int_Q \int_0^t \frac{\lambda}{2}\Lambda(\mathbf{m}^{\varepsilon})\cdot\mathbf{m}^{\varepsilon}\,\phi_t\,\mathrm{d}\tau\,\mathrm{d}x\mathrm{d}t \\
&-\int_Q \left[u_1 + \int_0^t f(\tau)\,\mathrm{d}\tau\right]\phi\,\mathrm{d}x\,\mathrm{d}t + \int_Q \psi_t\cdot\mathbf{m}^{\varepsilon}\,\mathrm{d}x\,\mathrm{d}t + \\
&\int_Q \mathbf{m}_0\cdot\psi(\cdot,0)\,\mathrm{d}x\,\mathrm{d}t + \int_Q \left(\frac{|\,\mathbf{m}^{\varepsilon}\,|^2 - 1}{\delta}\right)\psi\cdot\mathbf{m}^{\varepsilon}\,\mathrm{d}x\,\mathrm{d}t \\
&-\int_Q \lambda\, u_x^{\varepsilon}\,\Lambda(\mathbf{m}^{\varepsilon})\cdot\psi\,\mathrm{d}x\,\mathrm{d}t - \int_Q \mathbf{m}_x^{\varepsilon}\cdot\psi_x\,\mathrm{d}x\,\mathrm{d}t = 0.
\end{aligned}
\tag{69}
$$

$\forall\phi$ smooth s.t. $\phi(0,t)=\phi(1,t)=0,\ \phi(\cdot,T)=0$, and $\forall\psi\equiv(\psi_1,\psi_2)$ s.t. $\psi(x,T)=0$, where ϕ and ψ arbitrarily chosen test functions in $\mathcal{Q}$.

The proof, not included here, is provided in [47].

Proof's Outline:

- consider the viscoelastic energy associated to the problem to obtain a suitable *a priori estimate*
- consider the energy connected to interaction between magnetic and viscoelastic effects to obtain further suitable estimates
- consider the total energy together with smooth enough initial data to estimate the energy at the generic time t
- introduce an appropiate weak formulation and suitable test functions
- consider separately the limit process when $\varepsilon \to 0$

As a closing remark, we can note that, under the applicative point of view as well as under the analytical one, the free energy associated to the model plays a crucial role. Indeed, the proof relies on estimates which are based on the free energies connected to the model here adopted. Specifically, the viscoelastic energy allows [47], also in the magneto-viscoelastic case, to prove an a priori estimate on which the subsequent results are based. This is not surprising since the connection relating free energies and evolution problems is well known; see for instance [48] and references therein.

Acknowledgements

S. Carillo wishes to acknowledge the partial financial support of GNFM-INDAM, INFN, and SAPIENZA Università di Roma.

Author details

Sandra Carillo[1,2*] and Claudio Giorgi[3]

*Address all correspondence to: sandra.carillo@sbai.uniroma1.it

1 Department, SBAI, Basic and Applied Sciences in Engineering, Sapienza University of Rome, Rome, Italy

2 I.N.F.N., Sezione Roma1, Gr. IV, M.M.N.L.P, Rome, Italy

3 DICATAM, Sezione di Matematica, University of Brescia, Brescia, Italy

References

[1] J. Ciambella, A. Paolone and S. Vidoli, Memory decay rates of viscoelastic solids: not too slow, but not too fast either, Rheol. Acta 50, 661–674 (2011).

[2] L. Boltzmann, Zur Theorie der elastichen Nachrwirkung, Sitzber. Kaiserl. Akad. Wiss. Wien, Math.-Naturw. Kl. 70, 275–300 (1874).

[3] M. Fabrizio and A. Morro, Mathematical Problems in Linear Viscoelasticity, SIAM Studies Appl. Math. 12, Philadelphia, PA, 1992.

[4] R. Dautray and J.-L. Lions, Mathematical analysis and numerical methods for science and technology. Vol. 5. Evolution problems I, Springer-Verlag, Berlin, 1992.

[5] V. Volterra, Sur la théorie mathématique des phénomènes héréditaires, J. Math. Pures Appl. 7, 249–298 (1928).

[6] I. Carol and Z.P. Bažant, Viscoelasticity with aging caused by solidification of nonaging constituent, J. Eng. Mech. 119, 2252–2269 (1993).

[7] A.D. Drozdov and V.B. Kolmanovskii, Stability in viscoelasticity, North-Holland, Amsterdam, 1994.

[8] J.C. Maxwell, Constitution of bodies, Sci. Papers, College Arts Sci. Univ. Tokyo 2, 616–624 (1877).

[9] E. Wiechert, Gesetze der Elastichen Nachwirkung für constante Temperatur, Ann. Phys. 286, 335–348, 546–570 (1893).

[10] D. Graffi, Sull'espressione dell'energia libera nei materiali visco-elastici, Ann. Mat. Pura Appl. 98, 273–279 (1974).

[11] B. S. Underwood, A continuum damage for asphalt cement and asphalt mastic fatigue, Int. J. Fatigue 82, 387–401 (2016).

[12] M. Conti, V. Danese, C. Giorgi and V. Pata, A model of viscoelasticity with time-dependent memory kernels, Am. J. Math., to appear.

[13] M. Fabrizio, C. Giorgi and V. Pata, A new approach to equations with memory, Arch. Ration. Mech. Anal. 198, 189–232 (2010).

[14] A. Jaishankar and G.H. McKinley, Power-law rheology in the bulk and at the interface: Quasi-properties and fractional constitutive equations, Proc. R. Soc. A, 469 (2013) DOI: 10.1098/rspa.2012.0284

[15] L. Deseri, M. Zingales and P. Pollaci, The state of fractional hereditary materials (FHM), Discrete Contin. Dyn. Syst. Ser. B 19, 2065–2089 (2014).

[16] P.E. Rouse, A theory of the linear viscoelastic properties of dilute solutions of coiling polymers, J. Chem. Phys. 21, 1272–1280 (1953).

[17] B.H. Zimm, Dynamics of polymer molecules in dilute solutions: viscoelasticity, flow birefringence and dielectric loss, J. Chem. Phys. 24, 269–278 (1956).

[18] C.M. Dafermos, An abstract Volterra equation with applications to linear viscoelasticity, J. Diff. Equ. 7, 554–569 (1970).

[19] W. Desch and R. Grimmer, Singular relaxation moduli and smoothing in three-dimensional viscoelasticity, Trans. Am. Math. Soc. 314, 381–404 (1989).

[20] G. Gentili, Regularity and stability for a viscoelastic material with a singular memory kernel, J. Elasticity 37, 139–156 (1995).

[21] M. Grasselli and A. Lorenzi, Abstract nonlinear Volterra integro-differential equations with nonsmooth kernels. Atti Accad. Naz. Lincei Cl. Sci. Fis. Mat. Natur. Rend. Lincei 9, 1 Mat. Appl. 2 (1991).

[22] J. Janno and L. von Wolfersdorf, Identification of weakly singular memory kernels in viscoelasticity, Z. Angew. Math. Mech. 78, 391–403 (1998).

[23] R.K. Miller and A. Feldstein, Smoothness of solutions of Volterra integral equations with weakly singular kernels, SIAM J. Math. Anal. 2, 242–258 (1971).

[24] M. Renardy, W.J. Hrusa and J.A. Nohel, Mathematical problems in viscoelasticity, Longman Scientific & Technical, Harlow John Wiley & Sons Inc., New York, 1987.

[25] C. Giorgi and A. Morro, Viscoelastic solids with unbounded relaxation function, Continuum Mech. Thermodyn., 4, 151–165 (1992).

[26] V. Berti, Existence and uniqueness for an integral-differential equation with singular kernel, Boll. Un. Mat. Ital. 9-B, 299–309 (2006).

[27] S. Carillo, Singular kernel problems in materials with memory, Meccanica, 50, 603–615 (2015). doi: 10.1007/s11012-014-0083-y

[28] S. Carillo, V. Valente and G. Vergara Caffarelli, A linear viscoelasticity problem with a singular memory kernel: an existence and uniqueness result, Differ. Integral Equ. 26, 1115–1125 (2013). http://projecteuclid.org/euclid.die/1372858565

[29] S. Carillo, V. Valente and G. Vergara Caffarelli, Heat conduction with memory: a singular kernel problem. Evolution Equations and Control Theory 3, 399–410 (2014). doi:10.3934/eect.2014.399

[30] A. Hanyga, Wave propagation in media with singular memory, Math. Comput. Model., 34, 1329–1421 (2001).

[31] A. Hanyga and M. Seredynska, Asymptotic and exact fundamental solutions in hereditary media with singular memory kernels, Quart. Appl. Math., 60, 213–244 (2002).

[32] R.C. Koeller, Applications of fractional calculus to the theory of viscoelasticity, ASME J. Appl. Mech. 51, 299–307 (1984).

[33] K. Adolfsson, M. Enelund, and P. Olsson, On the fractional order model of viscoelasticity, Mech. Time Depend. Mater., 9, 15–34 C (2005).

[34] M. Enelund, L. Mähler, K. Runesson and B.L. Josefson, Formulation and integration of the standard linear viscoelastic solid with fractional order rate laws, Int. J. Sol. Struct. 36, 2417–2442 (1999).

[35] M. Enelund and P. Olsson, Damping described by fading memory, Analysis and application to fractional derivative models, Int. J. Sol. Struct. 36, 939–970 (1999).

[36] F. Mainardi, Fractional Calculus and Waves in Linear Viscoelasticity. An Introduction to Mathematical Models, Imperial College, Press, London, 2010.

[37] M. Fabrizio, Fractional rheological models for thermomechanical systems. Dissipation and free energies. Fract. Calc. Appl. Anal. 17, 206–223 (2014).

[38] S. Carillo, Some Remarks on Materials with Memory: Heat Conduction and Viscoelasticity, Supplement 1, 12, 163-178, (2005). doi :10.2991/jnmp.2005.12.s1.14; (see also S. Carillo, Erratum to: Some Remarks on Materials with Memory: Heat Conduction and Viscoelasticity, 3, 22, i-iii, (2015); doi : 10.1080/14029251.2014.971573)

[39] M. Hossain, P. Saxena and P. Steinmann, Modelling the curing process in magneto-sensitive polymers: rate-dependence and shrinkage, Int. J. Non-Linear Mech., 74, 108–121 (2015). ISSN 0020-7462, http://dx.doi.org/10.1016/j.ijnonlinmec.2015.04.008.

[40] M. Hossain, P. Saxena and P. Steinmann, Modelling the mechanical aspects of the curing process of magneto-sensitive elastomeric materials, Int. J. Sol. Struct., 58, (2015), 257–269. ISSN 0020-7683, http://dx.doi.org/10.1016/j.ijsolstr.2015.01.010.

[41] Matsumura, K., Yamaguchi, H. Elastic modulus of viscoelastic magnetic silicone gel body. Polym. Eng. Sci., 50 (5), 857–862 (2010).

[42] M. Chipot, I. Shafrir, V. Valente and G. Vergara Caffarelli, *A nonlocal problem arising in the study of magneto-elastic interactions*. Boll. UMI Serie IX, I, 197–222 (2008).

[43] M. Chipot, I. Shafrir, V. Valente and G. Vergara Caffarelli, On a hyperbolic-parabolic system arising in magnetoelasticity. J. Math. Anal. Appl., 352, 120–131 (2009).

[44] S. Carillo, V. Valente, and G. Vergara Caffarelli, A result of existence and uniqueness for an integro-differential system in magneto-viscoelasticity, Appl. Anal., 90, 1791–1802 (2010). doi: 10.1080/00036811003735832

[45] S. Carillo, V. Valente and G. Vergara Caffarelli, An existence theorem for the magneto-viscoelastic problem. Discrete Contin. Dyn. Syst. Ser. S., 5, 3, (2012), 435–447. doi: 10.3934/dcdss.2012.5.435

[46] M. Bertsch, P. Podio Guidugli and V. Valente, On the dynamics of deformable ferromagnets. I. Global weak solutions for soft ferromagnets at rest, Ann. Mat. Pura Appl. 179(4), 331–360 (2001).

[47] S. Carillo, M. Chipot, V. Valente and G. Vergara Caffarelli, A magneto-viscoelasticity problem with a singular memory kernel, arXiv: 1601.06276, (2016), 13 pages.

[48] G. Amendola, S. Carillo, J.M. Golden and A. Manes, Viscoelastic fluids: free energies, differential problems and asymptotic behaviour. Discrete Contin. Dyn. Syst. Ser. B, 19(7), 1815–1835 (2014). doi :10.3934/dcdsb.2014.19.1815

Finite Element Modeling and Experiments of Systems with Viscoelastic Materials for Vibration Attenuation

Antonio Marcos G. de Lima,
Luiz Fernando F. Rodovalho and Romes A. Borges

Additional information is available at the end of the chapter

http://dx.doi.org/10.5772/64532

Abstract

This chapter addresses the finite element modeling methodologies intended for performance evaluation, analysis, and design of viscoelastic systems. The mathematical models widely used to represent the frequency and temperature dependent behavior of viscoelastic materials are also considered, namely the complex modulus approach, the fractional derivative model, the Golla-Hughes-McTavish (GHM) model, and the anelastic displacement fields (ADFs) model. The straightforward strategies to integrate the viscoelastic effects into finite element matrices of structural systems such as three-layer sandwich plates, intended for the modeling of medium and large-scale engineering structures, are presented. In the same context, emphasis is placed on the condensation methods for the reduction of the order of the finite element matrices to perform frequency-response functions, complex eigenvalue problem, and time domain analyses. Based on the fact that for viscoelastic materials subjected to dynamic loadings superimposed on static preloads, the classical modeling assuming isothermal conditions can lead to poor designs, since the energy dissipated within the volume of the material leads to temperature rises, an experimental investigation of the self-heating phenomenon is also addressed.

Keywords: viscoelastic damping, finite element, reduction, self-heating phenomenon

1. Introduction

It is well known that viscoelastic materials can be used with advantage to mitigate undesirable vibrations [1, 2] and, consequently, to increase the fatigue life of engineering structures to avoid catastrophes [3, 4]. As a result, they have been applied in a number of engineering applications such as robots, automobiles, airplanes, communication satellites, tall buildings, and space structures [5, 6].

However, one of the major difficulties regarding the use of viscoelastic materials for vibration mitigation is the fact that their mechanical properties vary strongly with environment and operational factors for which, the excitation frequency and temperature are usually considered to be the most relevant parameters [1]. This is a reason for which in the last decades, a number of viscoelastic models exist in the open literature to be used in conjunction with the finite element (FE) discretization procedure. Among those viscoelastic models, the fractional derivative model (FDM) is a more complex relationship between the stress and strain than the standard linear viscoelastic model [1], which is based on the use of fractional derivatives [7–9] in order to reduce the number of terms required by the generalized standard viscoelastic model. The so-called Golla-Hughes-McTavish (GHM) model initially proposed by Golla and Hughes [10] is based on the use of internal variables to represent the dissipation mechanism of viscoelastic materials. In the Laplace domain, the resulting GHM complex modulus function is a series of mini-oscillators terms similar to that of a damped single degree of freedom system. McTavish and Hughes [11] have demonstrated later the FE modeling strategy of a truss structure incorporating a structural viscoelastic damper by using the GHM model. Another viscoelastic model normally used is the so-called anelastic displacement field (ADF) model suggested by Lesieutre and his co-authors [12–14]. The strategy is the use of anelastic fields to represent the dissipative effects of viscoelastic materials similar to the GHM model. However, as opposed to the GHM, the ADF model can be formulated directly in the time domain.

For large FE models of real-world engineering structures incorporating viscoelastic materials typically composed by many thousands of degrees of freedom (DOFs), the inclusion of internal variables by using ADF or GHM models leads to global systems of equations of motion whose numbers of DOFs largely exceeds the order of the associated undamped system. As a result, if such evaluations are made based on response computations performed on the full finite element matrices, the computational time required to obtain the solutions is high [15–18]. It must be also reminded that if the interest is to perform active control techniques by using the resulting viscoelastic models, the dimension is a typical problem and must be not disregarded [19–23]. Thus, it is interesting to perform model condensation techniques especially adapted to viscoelastic systems with the aim of reducing the computational burden [24, 25].

The simplest model reduction method very useful to reduce static systems is the well-known Guyan static method [19], where the condensation is performed by separating the physical coordinates of the static system in master and slaves coordinates. However, as discussed in [20], this strategy is not adapted to viscoelastic systems [23], even when the reduced coordinates are a subset of the initial physical coordinates. In this case, the internal balancing method proposed by Moore [24] is more interesting to be used, but leads to reduced coordinates that are not necessarily a subset of the original physical coordinates. In order to circumvent this problem, Yae and Inman [19] have proposed a modified internal balancing method adapted to viscoelastic systems.

The other very simple reduction strategy, is the so-called modal reduction method [20], where in the construction of the reduction basis it is considered only the most relevant eigenvectors characterizing the dynamic behavior of the system in the frequency band of interest. However, it is shown here that since the viscoelastic stiffness matrix is frequency- and temperature-dependent, this strategy must be modified to generate a constant reduction basis formed by the static residues associated with the external loads and the viscoelastic damping forces. Clearly, the dimension of the reduced viscoelastic system of matrices corresponds to the number of eigenvectors and the static residues kept in the truncated series to form the reduction basis [18].

A natural extension of the modeling capability of engineering systems incorporating a viscoelastic damping device is the optimization aiming the reduction of cost and/or maximization of its performance. In the quest for optimization, the engineers are frequently faced with large FE models of real-world systems that require a great number of evaluations of the cost functions involved [19]. Consequently, if such computations are performed based on the full FE matrices, the time required to obtain the dynamic responses may be high. Here, a general strategy to construct a constant reduction basis for viscoelastic systems is suggested, composed by the eigenmodes of the associated conservative viscoelastic behavior enriched by the static residues associated with the external loadings and the viscoelastic damping forces. Also, the reduction basis can be easily extended to the case of robust condensation of viscoelastic systems subjected to parametric uncertainties if the static residues due to the small modifications are included into the basis. This robust condensation strategy in the frequency domain is a very interesting tool to be integrated into numerical optimization and/or model updating processes [25].

Another problem regarding the practical application of a viscoelastic damper to mitigate unwanted vibrations is the fact that the assumption of assuming a constant and uniform temperature distribution within the viscoelastic material can lead to a poor design and to unexpected damping performance of it due to the self-heating phenomenon [26–29]. As a result, it is expected that the temperature changes induced by the self-heating when the viscoelastic damper is subjected to cyclic excitations have a strong influence on the stiffness degradation and damping capacity of it. Moreover, in applications such as engine mounts, it must be considered the effects of the cyclic loadings superimposed on the static strains in the self-heating modeling procedure in which the prediction of the temperature evolution inside the viscoelastic material volume is an interesting thermoviscoelastic problem to be solved. Here, the influence of the cyclic loading superimposed on static preloads on the self-heating phenomenon is investigate numerically and experimentally for a three-dimensional translational viscoelastic mount used for vibration attenuation.

In the remainder, after the presentation of the theoretical foundations of the methodology, the description of some numerical studies of engineering systems incorporating passive constrained viscoelastic layers and discrete viscoelastic damping devices is addressed. The main interest is to illustrate those viscoelastic models and topics described in the methodology, intended to design and performance analyses of viscoelastically damped systems. In addition, the results of some experimental investigations with a freely suspended plate partially treated by passive constraining damping layer are carried out to validate the viscoelastic models and to confirm the effectiveness of the viscoelastic materials applied as a passive control strategy.

Finally, an experimental investigation of the self-heating phenomenon on a three-dimensional translational viscoelastic mount subjected to dynamic loadings superimposed on static preloads is also addressed.

2. Review of viscoelastic models

2.1. The concept of complex modulus approach

According to the linear theory of viscoelasticity [30], the complex modulus of viscoelastic materials in the frequency domain is expressed as follows:

$$G(\omega) = G'(\omega) + iG''(\omega) = G'(\omega)[1 + i\eta(\omega)] \tag{1}$$

where $G'(\omega)$, $G''(\omega)$ and $\eta(\omega) = G''(\omega)/G'(\omega)$ designates the storage modulus, loss modulus and loss factor, respectively. As reported by [31], these parameters can be used to characterize the dynamic behavior of linear viscoelastic materials.

This model is adopted in the study reported herein since it enables the direct use of the data of viscoelastic materials commonly provided by the manufacturers [1]. Within this context, in the open literature, various mathematical models have been developed in order to represent the material modulus function, $G(\omega)$, as summarized later in this chapter.

However, it is widely recognized that the temperature also exerts a strong influence upon the properties of viscoelastic materials. Hence, it becomes important to account its variations in the modeling procedures of engineering systems incorporating viscoelastic materials. According to Nashif et al. [1] and Christensen [30], this can be done by making use of the frequency–temperature superposition principle (FTSP), where the damping properties of linear viscoelastic materials as functions of frequency at various temperatures can be collapsed on a single

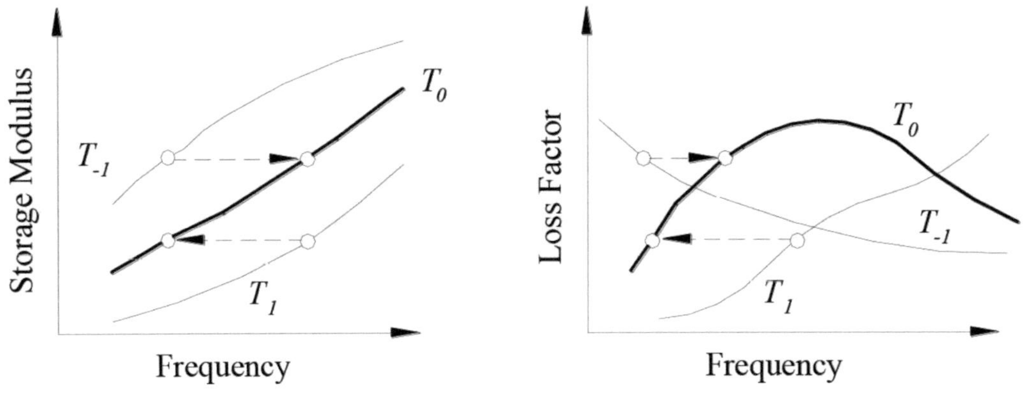

Figure 1. Illustration of the frequency–temperature superposition principle.

master curve, as illustrated in **Figure 1**, if appropriate horizontal shifts along the frequency axes are used. This establishes the well-known *shift factor* and *reduced frequency* concepts, symbolically expressed by the following relations [18]:

$$G(\omega_r, T_0) = G(\alpha_T \omega, T_0), \quad \eta(\omega_r, T_0) = \eta(\alpha_T \omega, T_0) \tag{2}$$

where T is the temperature, $\omega_r = \alpha_T(T)\omega$ is the reduced frequency, ω is the excitation frequency, $\alpha_T(T)$ is the shift factor, and T_0 is a reference value of temperature.

Functions $G(\omega_r)$ and $\alpha_T(T)$ can be obtained from experimental tests for specific viscoelastic materials [1]. Drake and Soovere [32] suggest analytical expressions for the complex modulus and shift factor for various commercially available viscoelastic materials, as functions of reduced frequency and temperature. The following equation represents the complex modulus for the 3M ISD112™ viscoelastic material [33], as provided by the authors:

$$\eta$$

$$G(\omega_r) = 0.4307 + \frac{1200}{1 + 3.24 \times \left(\dfrac{i\omega_r}{1543000}\right)^{-0.18} + \left(\dfrac{i\omega_r}{1543000}\right)^{-0.6847}} \; [MPa] \tag{3}$$

where $\alpha(T) = 10^{\left(-3758.4 \times \left(\frac{1}{T} - 0.00345\right) - 225.06 \times \log(0.00345 \times T) + 0.23273 \times (T - 290)\right)}$ and $\omega_r = \alpha(T)\omega$.

Figure 2 shows the curves representing the variations of the storage modulus and loss factor as functions of the reduced frequency, as obtained from Eq. (3).

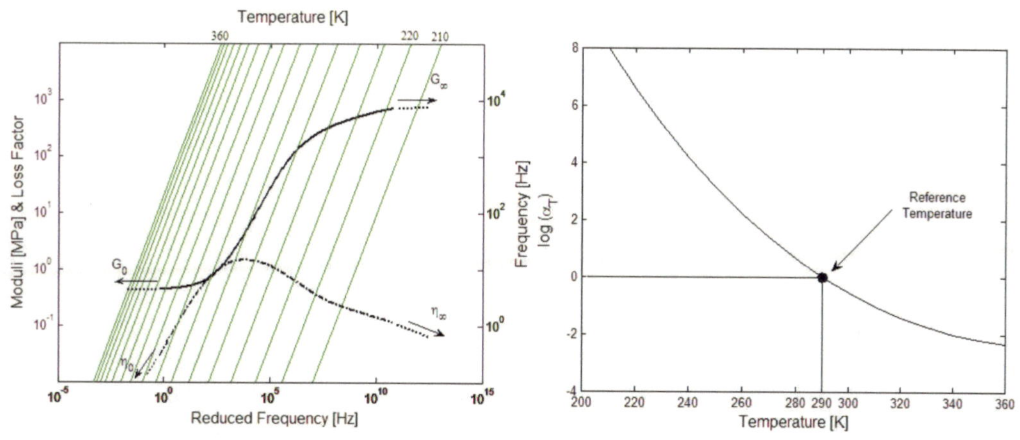

Figure 2. Master curves (—— G' ; — · · η) and shift factor for the 3M™ ISD112 material.

2.2. Golla-Hughes-McTavish model (GHM)

According to the developments made by Golla and Hughes [10], and McTavish and Hughes [11], the viscoelastic material modulus function is expressed under the form:

$$G(s) = G_0\left(1 + \sum_{i=1}^{N_G} \alpha_i \frac{s^2 + 2\zeta_i\omega_i s}{s^2 + 2\zeta_i\omega_i s + \omega_i^2}\right) \tag{4}$$

where G_0 is the static modulus, s designates the Laplace variable and $G_\infty = G_0\left(1 + \sum_{i=1}^{N_G} \alpha_i\right)$ is the higher frequency modulus.

It can be clearly seen the similarity of each N_G term of the series appearing in Eq. (4) with the transfer function of a damped single DOF system known as mini-oscillator formed by the design variables (α_i, ω_i, ζ_i), identified for a specific viscoelastic material [20].

2.3. Anelastic displacement field model (ADF)

The ADF model, developed by Lesieutre and co-workers [12–14], is similar, in some aspects, to the GHM model. The modulus function is represented in Laplace domain by:

$$G(s) = G_0\left(1 + \sum_{i=1}^{N_A} \Delta_i \frac{s}{\omega^2 + \Omega_i^2}\right) \tag{5}$$

where G_0 is the low frequency modulus, N_A is the number of anelastic displacement fields, each of which is represented by parameters Ω_i, the inverse of the characteristic relaxation time at constant deformation, Δ_i is the relaxation magnitude, and $G_\infty = G_0\left(1 + \sum_{i=1}^{N_A} \Delta_i\right)$ is the higher frequency modulus.

2.4. Fractional derivative model (FDM)

The FDM model proposed by Bagley and Torvik [7–9] can be viewed as the generalization of the standard viscoelastic model, by the introduction of non-integer time derivatives in the differential constitutive equation relating the stresses to strains as follows:

$$\sigma(t) + \sum_{m=1}^{M} b_m D^{\beta_m}\sigma(t) = G_0\varepsilon(t) + \sum_{n=1}^{P} G_n D^{\alpha_n}\varepsilon(t) \tag{6}$$

where σ_n and β_m are non-integer numbers. As stated by Bagley and Torvik [7], experimental observations have indicated that the behavior of many commercially viscoelastic materials can be adequately modeled by retaining only one term in each of the series appearing in Eq. (6). Through this simplified form, the modulus requires only five parameters, resulting in the following expression:

http://dx.doi.org/10.5772/64532

$$G(s) = \left(G_0 + G_\infty s^\alpha\right) / \left(1 + bs^\beta\right) \tag{7}$$

3. Curve fitting of material parameters

One important aspect regarding the use of GHM, ADF, and FDM models presented in previously section is the identification of the model parameters from experimental data. In most cases, manufacturers provide data sheets containing the material storage modulus $G'(\omega, T)$ and loss factor $\eta(\omega, T)$, defined in Eq. (1), as functions of excitation frequency ω and temperature T. Thus, it becomes necessary to express, for each model, the material modulus expressions as complex functions by making, $s = i\omega$:

$$\text{(GHM): } \operatorname{Re}\left[G(\omega)\right] = G_0\left[1 + \sum_{i=1}^{N_G} \alpha_i \frac{\left[\omega^2\left(\omega^2 - \omega_i^2\right) + 4\zeta_i^2\omega_i^2\omega^2\right]}{\left(\omega^2 - \omega_i^2\right)^2 + 4\zeta_i^2\omega_i^2\omega^2}\right] \tag{8.a}$$

$$\text{(GHM): } \operatorname{Im}\left[G(\omega)\right] = G_0\left[\sum_{i=1}^{N_G} \alpha_i \frac{2\zeta_i\omega_i^3\omega}{\left(\omega^2 - \omega_i^2\right)^2 + 4\zeta_i^2\omega_i^2\omega^2}\right] \tag{8.b}$$

$$\text{(ADF): } \operatorname{Re}\left[G(\omega)\right] = G_0\left[1 + \sum_{i=1}^{N_A} \Delta_i \frac{\left(\omega/\Omega_i\right)^2}{1 + \left(\omega/\Omega_i\right)^2}\right] \tag{9.a}$$

$$\text{(ADF): } \operatorname{Im}\left[G(\omega)\right] = G_0\left[\sum_{i=1}^{N_A} \Delta_i \frac{\left(\omega/\Omega_i\right)^2}{1 + \left(\omega/\Omega_i\right)^2}\right] \tag{9.b}$$

$$\text{(FDM): } \operatorname{Re}\left[G(\omega)\right] = \frac{G_0\left[1 + b\omega^\beta \cos\left(\pi\beta/2\right)\right] + G_\infty\left[\omega^\alpha \cos\left(\pi\alpha/2\right) + b\omega^{\alpha+\beta}\cos\left(\pi(\alpha-\beta)/2\right)\right]}{1 + 2b\omega^\beta\cos\left(\pi\beta/2\right) + \left(b\omega^\beta\right)^2} \tag{10.a}$$

$$\text{(FDM): } \operatorname{Im}\left[G(\omega)\right] = \frac{G_\infty\left[\omega^\alpha \sin\left(\pi\alpha/2\right) + b\omega^{\alpha+\beta}\cos\left(\pi(\alpha-\beta)/2\right)\right] - G_0 b\omega^\beta \sin\left(\pi\beta/2\right)}{1 + 2b\omega^\beta\cos\left(\pi\beta/2\right) + \left(b\omega^\beta\right)^2} \tag{10.b}$$

From the equations above, for each viscoelastic model, the determination of the material parameters can be carried out by formulating a deterministic optimization problem in which the objective function represents the difference between the experimental data points and the corresponding model predictions. Clearly, the number of design variables depends on the previous choice of a model order, which is assumed to be sufficient to represent the frequency-dependent behavior in the frequency band of interest: $N_{par}^G = 1 + 3N_G$ for the GHM; $N_{par}^A = 1 + 2N_A$ for the ADF; and $N_{par}^F = 5$ for the FDM.

4. Incorporation of the viscoelastic behavior into FE models

Based on the formulation presented in the previous sections, and including external viscous damping, the FE equations of motion in the frequency domain of a viscoelastic structure containing N DOFs can be expressed as [18]:

$$\boldsymbol{M}\ddot{\boldsymbol{q}}(t)+\boldsymbol{D}\dot{\boldsymbol{q}}(t)+\boldsymbol{K}^{*}\boldsymbol{q}(t)=\boldsymbol{f}(t),\quad \boldsymbol{f}(t)=\boldsymbol{bu}(t)\quad \text{and}\quad \boldsymbol{y}(t)=\boldsymbol{cq}(t) \tag{11}$$

where $\boldsymbol{M},\ \boldsymbol{D},\ \boldsymbol{K}^{*}\in R^{N\times N}$ are, respectively, the mass, viscous damping, and complex stiffness matrices. $\boldsymbol{q}(t)\in R^{N}$ and $\boldsymbol{f}(t)\in R^{N}$are the vectors of displacements and external loads. $\boldsymbol{y}(t)\in R^{c}$ and $\boldsymbol{u}(t)\in R^{f}$ are the vectors of responses external loads. $\boldsymbol{b}\in R^{N\times f}$ and $\boldsymbol{c}\in R^{c\times N}$ are matrices, which enable to select, among the FE DOFs, those in which the external excitations are applied and responses are computed, respectively.

It should be emphasized that the dependency of the stiffness matrix on frequency and temperature is a consequence of the dependency of the material *moduli* with respect to these parameters as expressed by Eq. (1). Thus, by assuming harmonic excitation conditions, $\mathbf{f}(t)=\mathbf{F}e^{i\omega t}$, $\boldsymbol{u}(t)=\boldsymbol{U}e^{i\omega t}$, and steady-state harmonic responses, $\boldsymbol{q}(t)=\boldsymbol{Q}e^{i\omega t}$, $\boldsymbol{y}(t)=\boldsymbol{Y}e^{i\omega t}$, the following equations in the frequency domain are obtained:

$$\left[\boldsymbol{K}(\omega,T)+i\omega\boldsymbol{D}-\omega^{2}\boldsymbol{M}\right]\boldsymbol{Q}(\omega,T)=\boldsymbol{F},\quad \boldsymbol{F}=\boldsymbol{bU}\quad \text{and}\ \boldsymbol{Y}(\omega,T)=\boldsymbol{cQ}(\omega,T) \tag{12}$$

It is assumed that the model contains both elastic and viscoelastic elements. Thus, the elastic–viscoelastic correspondence principle [1] is applied leading to:

$$\boldsymbol{K}(\omega,T)=\boldsymbol{K}_{e}+\boldsymbol{K}_{v}(\omega,T) \tag{13}$$

where $\boldsymbol{K}_e$ and $\boldsymbol{K}_v(\omega, T)$ are the stiffness matrices associated with the elastic and viscoelastic substructures, respectively. Also, taking into account the stress state for the specific viscoelastic element assumed in the analysis, one of the moduli (through the relation, $G(\omega, T) = E(\omega, T)/2(1+\nu)$) can be factored out of, $K_v(\omega, T)=G(\omega, T)\overline{K}_v$, where $\overline{K}_v$ is the frequency- and temperature-independent matrix. Then, Eqs. (12) and (13) can be combined to give the following complex *receptance matrix* for the viscoelastic system:

$$\boldsymbol{H}(\omega,T)=\boldsymbol{cZ}(\omega,T)^{-1}\boldsymbol{b} \tag{14}$$

where $Z(\omega, T)=K_e+G(\omega, T)\overline{K}_v+i\omega D-\omega^2 M$ is the complex dynamic stiffness matrix.

Clearly, the difficulty in predicting the dynamic responses for viscoelastic systems comes from the fact that the stiffness matrix depends on frequency and temperature. As a result, one has an eigenvalue problem that must be solved iteratively [34]. Some other procedures for dealing with this problem have been suggested, such as the contributions by Palmeri and Ricciardelli [35] and by Palmeri et al. [36], where the eigenvalue problem of viscoelastic systems has been derived in the time domain with the help of the novel concept of modal relaxation functions. In the papers proposed by Yuan and Agrawal [37] and Wagner and Adhikari [38], an alternative state-space approach has been proposed for the time-domain analysis of viscoelastic structures. Others alternatives have been suggested based on the adoption of particular representations for the frequency-dependent behavior of the viscoelastic materials [39]. Such an approach is used in the FDM, GHM, and ADF models, which enable to transform the equations of motion of a viscoelastic system in the time-domain into state-space forms, with frequency-independent state matrices, at the expense of a typically high increase in the order of the system matrices.

4.1. GHM model

Applying the Laplace Transformation to Eq. (11) and replacing $G(s)$ by the modulus function given by Eq. (4), the following governing equation of motion is obtained:

$$\left[s^2 \boldsymbol{M} + s\boldsymbol{D} + \boldsymbol{K}_e + G_0 \overline{\boldsymbol{K}}_v \left(1 + \sum_{i=1}^{N_A} \Delta_i \frac{s}{\omega^2 + \Omega_i^2} \right) \right] \boldsymbol{Q}(\omega, T) = \boldsymbol{F} \tag{15}$$

A series of dissipation coordinates $Q_i^G (i=1, \ldots, N_G)$ are defined as:

$$\boldsymbol{Q}(\omega, T) = \boldsymbol{Q}^e + \sum_{i=1}^{N_A} \boldsymbol{Q}_i^A \tag{16}$$

Upon introduction of Eq. (16) into (15), after some mathematical manipulations and back transformation to time domain, the following coupled system of equations is obtained:

$$\boldsymbol{M}_A \ddot{\boldsymbol{q}}_A(t) + \boldsymbol{D}_A \dot{\boldsymbol{q}}_A(t) + \boldsymbol{K}_A \boldsymbol{q}_A(t) = \boldsymbol{f}_A(t) \tag{17}$$

where

$$\boldsymbol{M}_G = \begin{bmatrix} \boldsymbol{M} & \boldsymbol{0} & \cdots & \boldsymbol{0} \\ \boldsymbol{0} & \frac{\alpha_1}{\omega_1^2} \boldsymbol{K}_v^0 & \boldsymbol{0} & \vdots \\ \vdots & \boldsymbol{0} & \ddots & \boldsymbol{0} \\ \boldsymbol{0} & \cdots & \boldsymbol{0} & \frac{\alpha_{N_G}}{\omega_{N_G}^2} \boldsymbol{K}_v^0 \end{bmatrix}, \quad \boldsymbol{D}_G = \begin{bmatrix} \boldsymbol{D} & \boldsymbol{0} & \cdots & \boldsymbol{0} \\ \boldsymbol{0} & \frac{2\zeta_1 \alpha_1}{\omega_1} \boldsymbol{K}_v^0 & \boldsymbol{0} & \vdots \\ \vdots & 0 & \ddots & \boldsymbol{0} \\ \boldsymbol{0} & \cdots & \boldsymbol{0} & \frac{2\zeta_{N_G} \alpha_{N_G}}{\omega_{N_G}} \boldsymbol{K}_v^0 \end{bmatrix},$$

$$K_G = \begin{bmatrix} K_e + K_v^{\infty} & -\alpha_1 K_v^0 & \cdots & -\alpha_{N_G} K_v^0 \\ -\alpha_{N_G} K_v^{0^T} & \alpha_1 K_v^0 & 0 & \vdots \\ \vdots & 0 & \ddots & 0 \\ -\alpha_{N_G} K_v^{0^T} & \cdots & 0 & \alpha_{N_G} K_v^0 \end{bmatrix}, \quad q_G = \left[q, q_1^G, \ldots, q_{N_G}^G\right]^T \text{ and } f_G = \left[f, 0, \ldots, 0\right]^T$$

where M_G, D_G, $K_G \in R^{T_G x T_G}$ with $T_G = N(1 + N_G)$, $K_v^0 = G_0 \overline{K}_v$ and $K_v^{\infty} = G_{\infty}^G \overline{K}_v$..

4.2. ADF model

Following the procedure outlined above for the GHM model, the equations of motion obtained by considering the ADF model is expressed by Eq. (5) [12]:

$$\left[s^2 M + s D + K_e + G_0 \overline{K}_v \left(1 + \sum_{i=1}^{N_A} \Delta_i \frac{s}{\omega^2 + \Omega_i^2} \right) \right] Q(\omega, T) = F \tag{18}$$

In the ADF model the coordinates are separated into an elastic part, Q^e, which is instantaneously proportional to the stress, and an anelastic part, Q_i^A (i=1, …, N_A), that represents the dissipation effects of the viscoelastic materials [13]:

$$Q(\omega, T) = Q^e + \sum_{i=1}^{N_A} Q_i^A \tag{19}$$

The system of equations is adopted for describing the time-evolution of the anelastic fields:

$$\frac{C_i}{\Omega_i} K_v^{\infty} \dot{Q}_i^A + C_i K_v^{\infty} Q_i^A - K_v^{\infty} Q = 0 \tag{20}$$

Introducing Eq. (19) into (18), transforming the resulting equation to time domain and combining it with Eq. (20), the following coupled system of equations is obtained:

$$M_A \ddot{q}_A(t) + D_A \dot{q}_A(t) + K_A q_A(t) = f_A(t) \tag{21}$$

where:

$$M_A = \begin{bmatrix} M & \cdots & 0 \\ \vdots & \ddots & \vdots \\ 0 & \cdots & 0 \end{bmatrix}, \quad K_A = \begin{bmatrix} K_e + K_v^{\infty} & -K_v^{\infty} & \cdots & -K_v^{\infty} \\ -K_v^{\infty^T} & C_1 K_v^{\infty} & 0 & \vdots \\ \vdots & 0 & \ddots & 0 \\ -K_v^{\infty^T} & \cdots & 0 & C_{N_A} K_v^{\infty} \end{bmatrix}$$

$$\boldsymbol{D}_A = \begin{bmatrix} \boldsymbol{D} & \boldsymbol{0} & \cdots & \boldsymbol{0} \\ \boldsymbol{0} & \frac{C_1}{\Omega_1}\boldsymbol{K}_v^{\infty} & \boldsymbol{0} & \vdots \\ \vdots & \boldsymbol{0} & \ddots & \boldsymbol{0} \\ \boldsymbol{0} & \cdots & \boldsymbol{0} & \frac{C_{N_A}}{\Omega_{N_A}}\boldsymbol{K}_v^{\infty} \end{bmatrix}, \quad \boldsymbol{q}_A = \left[\boldsymbol{q}, \boldsymbol{q}_1^A, \ldots, \boldsymbol{q}_{N_G}^A\right]^T, \quad \boldsymbol{f}_A = [\boldsymbol{f}, \boldsymbol{0}, \ldots, \boldsymbol{0}]^T$$

where, $M_A, \ D_A, \ K_A \in R^{T_A x T_A}$ with $T_A = N(1+N_A)$, $K_v^0 = G_0\overline{K}_v$ and $K_v^{\infty} = G_{\infty}^A\overline{K}_v$.

4.3. FDM model

Introducing Eq. (7) into the Laplace transform of Eq. (11) and multiplying the resulting equation by, $(1 + bs^{\beta})$, the following system of equations of motion is obtained:

$$\left(s^{2+\beta}b\boldsymbol{M} + s^{1+\beta}b\boldsymbol{D} + s^2\boldsymbol{M} + s\boldsymbol{D} + s^{\beta}b\boldsymbol{K}_e + s^{\alpha}G_{\infty}\overline{\boldsymbol{K}}_v + \boldsymbol{K}_e + G_0\overline{\boldsymbol{K}}_v\right)\boldsymbol{Q} = \left(1 + bs^{\beta}\right)\boldsymbol{F} \tag{22}$$

According to Bagley and Torvik [8], the system of Eq. (22) can be written under the compact form, $\sum_{j=0}^{J} s^{j/m} A_j Q = (1 + bs^{\beta})F$, where $J = m(2 + \beta)$, m is the smallest common multiple of the denominators of the fractional terms α and β, matrices A_j are computed by direct comparison between Eq. (22) and the compact representation:

$$\left(s^{1/m}\boldsymbol{M}_F + \boldsymbol{K}_F\right)\boldsymbol{Q}_F = \boldsymbol{F}_F \tag{23}$$

where

$$\boldsymbol{M}_F = \begin{bmatrix} \boldsymbol{0} & \boldsymbol{0} & \cdots & \boldsymbol{0} & \boldsymbol{A}_j \\ \boldsymbol{0} & \boldsymbol{0} & \cdots & \boldsymbol{A}_j & \boldsymbol{A}_{j-1} \\ \vdots & \vdots & \iddots & \vdots & \vdots \\ \boldsymbol{0} & \boldsymbol{A}_j & \cdots & \boldsymbol{A}_3 & \boldsymbol{A}_2 \\ \boldsymbol{A}_j & \boldsymbol{A}_{j-1} & \cdots & \boldsymbol{A}_2 & \boldsymbol{A}_1 \end{bmatrix}, \quad \boldsymbol{K}_F = \begin{bmatrix} \boldsymbol{0} & \boldsymbol{0} & \cdots & -\boldsymbol{A}_j & \boldsymbol{0} \\ \boldsymbol{0} & \boldsymbol{0} & \cdots & -\boldsymbol{A}_{j-1} & \boldsymbol{0} \\ \vdots & \vdots & \iddots & \vdots & \vdots \\ -\boldsymbol{A}_j & -\boldsymbol{A}_{j-1} & \cdots & -\boldsymbol{A}_2 & \boldsymbol{0} \\ \boldsymbol{0} & \boldsymbol{0} & \cdots & \boldsymbol{0} & A_0 \end{bmatrix},$$

$$\boldsymbol{Q}_F = \left[s^{(j-1)/m}\boldsymbol{Q}, s^{(j-2)/m}\boldsymbol{Q}, \ldots, s^{1/m}\boldsymbol{Q}, \boldsymbol{Q}\right]^T, \quad \boldsymbol{F}_F = \left[\boldsymbol{0}, \ldots, \left(1 + bs^{\beta}\right)\boldsymbol{F}\right]^T, {}_F, \boldsymbol{K}_F \in R^{T_F x T_F}, \ T_F = JN.$$

5. Modal reduction methods for viscoelastic systems

It can be seen that the internal non-physical coordinates used by the GHM and ADF models to represent the viscoelastic dissipation mechanism lead to large system of equations of motion. Thus, it requires a high computational cost in the computation of dynamic responses of the viscoelastic system [18]. To develop the formulation pertaining the modal reduction method, it is convenient to transform the Eqs. (17) and (21) into an equivalent first-order form (space-state model) with an output equation added as follows:

$$\dot{\boldsymbol{x}} = \boldsymbol{A}\boldsymbol{x} + \boldsymbol{B}\boldsymbol{u} \ , \ \boldsymbol{y} = \boldsymbol{C}\boldsymbol{x} \tag{24}$$

where A assumes the following forms for the GHM and ADF models, respectively:

$$\boldsymbol{A} = \begin{bmatrix} \boldsymbol{0} & \boldsymbol{0} & \cdots & \boldsymbol{0} & \boldsymbol{I} & \boldsymbol{0} & \cdots & \boldsymbol{0} \\ \boldsymbol{0} & \boldsymbol{0} & \cdots & \boldsymbol{0} & \boldsymbol{0} & \boldsymbol{I} & \cdots & \boldsymbol{0} \\ \vdots & \vdots & \cdots & \vdots & \vdots & \vdots & \cdots & \vdots \\ \boldsymbol{0} & \boldsymbol{0} & \cdots & \boldsymbol{0} & \boldsymbol{0} & \boldsymbol{0} & \cdots & \boldsymbol{I} \\ -\boldsymbol{M}^{-1}\boldsymbol{K}_T & \alpha_1 \boldsymbol{M}^{-1}\boldsymbol{R} & \cdots & \alpha_{N_G} \boldsymbol{M}^{-1}\boldsymbol{R} & \boldsymbol{0} & \boldsymbol{0} & \cdots & \boldsymbol{0} \\ \omega_1^2 \boldsymbol{\Lambda}^{-1}\boldsymbol{R}^T & \omega_1^2 \boldsymbol{I} & \cdots & \boldsymbol{0} & \boldsymbol{0} & -2\zeta_1\omega_1 \boldsymbol{I} & \cdots & \boldsymbol{0} \\ \vdots & \vdots & \ddots & \vdots & \vdots & \vdots & \ddots & \vdots \\ \omega_{N_G}^2 \boldsymbol{\Lambda}^{-1}\boldsymbol{R}^T & \boldsymbol{0} & \cdots & \omega_{N_G}^2 \boldsymbol{I} & \boldsymbol{0} & \boldsymbol{0} & \cdots & -2\zeta_{N_G}\omega_{N_G} \boldsymbol{I} \end{bmatrix}, \quad \boldsymbol{x} = \begin{Bmatrix} \boldsymbol{q}^{T} \\ \boldsymbol{q}_1^{GGT} \\ \vdots \\ \boldsymbol{q}_{N_G}^{GGT} \\ \dot{\boldsymbol{q}}^{T} \\ \vdots \\ \dot{\boldsymbol{q}}_1^{GGT} \\ \dot{\boldsymbol{q}}_{N_G}^{GGT} \end{Bmatrix}$$

$$\boldsymbol{A} = \begin{bmatrix} \boldsymbol{0} & \boldsymbol{0} & \cdots & \boldsymbol{0} & \boldsymbol{I} \\ \frac{\Omega_1}{C_1}\boldsymbol{\Lambda}^{-1}\boldsymbol{R}^T & -\Omega_1 \boldsymbol{I} & \boldsymbol{0} & \boldsymbol{0} & \boldsymbol{0} \\ \vdots & \vdots & \ddots & \vdots & \vdots \\ \frac{\Omega_{N_A}}{C_{N_A}}\boldsymbol{\Lambda}^{-1}\boldsymbol{R}^T & \boldsymbol{0} & \cdots & -\Omega_{N_A} \boldsymbol{I} & \boldsymbol{0} \\ -\boldsymbol{M}^{-1}\boldsymbol{K}_T & \boldsymbol{M}^{-1}\boldsymbol{R} & \cdots & \boldsymbol{M}^{-1}\boldsymbol{R} & \boldsymbol{0} \end{bmatrix}, \quad \boldsymbol{x} = \begin{Bmatrix} \boldsymbol{q}^{T} \\ \boldsymbol{q}_1^{AAT} \\ \boldsymbol{q}_{N_A}^{AAT} \\ \vdots \\ \dot{\boldsymbol{q}}^{T} \end{Bmatrix} \text{ and } \boldsymbol{K}_T = \boldsymbol{K}_e + \boldsymbol{K}_v^{\infty}.$$

Since A is a non-symmetric matrix, the following eigenvalue problems are formulated:

$$\boldsymbol{A}\boldsymbol{R}_r = \boldsymbol{\Lambda}\boldsymbol{R}_r \ , \quad \boldsymbol{A}^T\boldsymbol{R}_l = \boldsymbol{\Lambda}\boldsymbol{R}_l \tag{25}$$

where Λ is the spectral matrix composed by the complex eigenvalues, and R_r and R_l are the modal matrices whose columns contain the right- and left-hand-side eigenvectors, respectively, normalized by the relation, $R_l^T R_r = I$. Thus, the left- and right-hand-side eigenvectors can be separated into a structural and dissipative eigenvectors represented, respectively, by the subscripts e and d according to the following expressions:

$$\boldsymbol{R}_r = [\boldsymbol{R}_{re}, \boldsymbol{R}_{rd}], \boldsymbol{R}_l = [\boldsymbol{R}_{le}, \boldsymbol{R}_{ld}] \tag{26}$$

The eigenvectors related to the internal variables are overdamped with a small contribution to the dynamic behavior. As result, the state of the system can be approached by the contributions of the elastic eigenvectors, $\boldsymbol{x} \approx \boldsymbol{R}_{re}\boldsymbol{x}_e$, so that the Eq. (24) can be expressed under the following form [24], $\dot{x}_e = A_r x_e + B_r u$ and $\boldsymbol{y} = \boldsymbol{C}_r\boldsymbol{x}_e$, where $A_r = R_{le}^T A R_{re}$, $B_r = R_{le}^T B$ and $\boldsymbol{C}_r = \boldsymbol{C}\boldsymbol{R}_{re}$ are input and output state reduced matrices of the system.

This modal reduction technique is very simple to implement since the choice of eigenmodes to be retained is based only on the frequency band of interest. However, for some viscoelastic systems, where the elastic modes may be overdamped, care must be taken, and an eigenfrequency analysis *a priori* must be performed in order to identify these eigenmodes. Also, the main disadvantage of the reduced state-space system, intended to apply control techniques, is that the matrices are complex. However, since all overdamped (relaxation/dissipative) modes were neglected, all elements of $\dot{x}_e$ are composed of complex conjugates pairs, such that they can be rewritten as follows:

$$\boldsymbol{A}_r = \begin{bmatrix} \ddots & & & \\ & \lambda_i & & \\ & & \bar{\lambda}_i & \\ & & & \ddots \end{bmatrix}, \quad \boldsymbol{B}_r = \begin{bmatrix} \vdots \\ \varphi_i \\ \bar{\varphi}_i \\ \vdots \end{bmatrix}, \quad \boldsymbol{C}_r = [\cdots \quad \psi_i \quad \bar{\psi}_i \quad \cdots] \tag{27}$$

where λ_i and $\bar{\lambda}_i$ are the retained elastic eigenvalues and their complex conjugates.

According to Friot and Bouc [40], to construct a real representation of the state-space system represented by Eq. (24), one can use a state transformation matrix such as $x_e = T\bar{x}_e$, where T is defined as:

$$\boldsymbol{T} = \begin{bmatrix} \ddots & & & \\ & -\frac{1}{2}\frac{j}{\mathrm{Im}(\lambda_i)} & \frac{1}{2}\frac{j}{\mathrm{Im}(\lambda_i)} & \\ & \frac{1}{2} - \frac{1}{2}\frac{j\mathrm{Re}(\lambda_i)}{\mathrm{Im}(\lambda_i)} & \frac{1}{2} + \frac{1}{2}\frac{j\mathrm{Re}(\lambda_i)}{\mathrm{Im}(\lambda_i)} & \\ & & & \ddots \end{bmatrix} \tag{28}$$

Thus, the real state-space system can be constructed, $\dot{\bar{x}}_e = \bar{A}_r\bar{x}_e + \bar{B}_r u$ and $y = \bar{C}_r\bar{x}_e$, where $\bar{\boldsymbol{A}}_r = \boldsymbol{T}\boldsymbol{A}_r\boldsymbol{T}^{-1}$, $\bar{B}_r = T B_r$, and $\bar{C}_r = C_r T^{-1}$.

6. Internal balancing method

The internal balancing method is another interesting reduction method of viscoelastic systems. However, it does not guarantee that the reduced coordinates are a subset of the original coordinates of the system. This method is based on the controllability and observability of each balanced state.

Based on the work by Moore [24], the balanced internal system is defined such that the *grammians* are diagonal and equal. For a viscoelastic system expressed by Eq. (24), the controllability and observability grammians, denoted by $\boldsymbol{W}_c$ and $\boldsymbol{W}_0$, are defined in order to satisfy the Lyapunov stability equations, $\boldsymbol{A}\boldsymbol{W}_c + \boldsymbol{W}_c\boldsymbol{A}^T = -\boldsymbol{B}\boldsymbol{B}^T$ and $\boldsymbol{A}^T\boldsymbol{W}_0 + \boldsymbol{W}_0\boldsymbol{A} = -\boldsymbol{C}^T\boldsymbol{C}$, where the Cholesky decomposition of matrix $\boldsymbol{W}_c$ is performed by the relation, $W_c = L_c L_c^T$. Also, a transformation matrix $\boldsymbol{P}$ is introduced, $\boldsymbol{P} = \boldsymbol{L}_c\boldsymbol{U}\boldsymbol{\Lambda}^{-1/2}$, where $\boldsymbol{\Lambda}$ and $\boldsymbol{U}$ are the eigenvalues and eigenvectors of the eigenvalue problem, $L_c W_0 L_c^T$, in such a way that the internal balanced model is given as:

$$\dot{\boldsymbol{x}}_r = \boldsymbol{A}_r\boldsymbol{x}_r + \boldsymbol{B}_r\boldsymbol{u}, \quad \boldsymbol{y} = \boldsymbol{C}_r\boldsymbol{x}_r \tag{29}$$

where $\boldsymbol{A}_r = \boldsymbol{P}^{-1}\boldsymbol{A}\boldsymbol{P}$, $\boldsymbol{B}_r = \boldsymbol{P}^{-1}\boldsymbol{B}$, $\boldsymbol{C}_r = \boldsymbol{C}\boldsymbol{P}$, and $\boldsymbol{x}_r = \boldsymbol{P}^{-1}\boldsymbol{x}$.

According to the definition of an internally balanced system, $\boldsymbol{W}_c(\boldsymbol{P}) = \boldsymbol{W}_0(\boldsymbol{P}) = diag(\sigma_i)$, where $\sigma_i(1, 2, \ldots, N)$ is the controllability of a state i, and N is the number of DOFs. Thus, according to the high and small values of states, σ_i, the internally balanced system (29) can be partitioned into retained states, $\boldsymbol{x}_{rr}$, and reduced states, $\boldsymbol{x}_{rd}$, as follows:

$$\begin{bmatrix} \dot{\boldsymbol{x}}_{rr} \\ \dot{\boldsymbol{x}}_{rd} \end{bmatrix} = \begin{bmatrix} \boldsymbol{A}_{rr} & \boldsymbol{A}_{rd} \\ \boldsymbol{A}_{dr} & \boldsymbol{A}_{dd} \end{bmatrix} \begin{bmatrix} \boldsymbol{x}_{rr} \\ \boldsymbol{x}_{rd} \end{bmatrix} + \begin{bmatrix} \boldsymbol{B}_{rr} \\ \boldsymbol{B}_{dd} \end{bmatrix} \boldsymbol{u} \tag{30}$$

Hence, the undesirable states, $\boldsymbol{x}_{rd}$, must be removed by performing a simple static reduction and retaining only the contributions of states, $\boldsymbol{x}_{rr}$.

7. Robust condensation strategy of viscoelastic systems

For real-world systems incorporating viscoelastic elements, it is practically impossible to perform time and frequency analyses using the GHM, ADF or FDM models, owing to the prohibitive computation times and storage memory required to evaluate the augmented equations of motion. This fact motivates the use of alternatives strategies with the aim of diminishing the model dimensions (and the computational burden, as a result), while keeping a reasonable predictive capacity of the numerical models. This can be done based on the assumption that the exact responses, given by the resolution of Eq. (12), can be approached by solutions in a reduced space as follows [18]:

$$\boldsymbol{Q} = \boldsymbol{T}\boldsymbol{Q}_r, \tag{31}$$

where $\boldsymbol{T} \in C^{N \times NR}$ is the vector basis (or Ritz basis) and $\boldsymbol{Q}_r \in C^{NR}$ where $NR \ll N$ is the number of vectors in the basis. Hence, the transfer function can be rewritten as follows:

$$\boldsymbol{H}_r(\omega, T) = \boldsymbol{c}\boldsymbol{Z}_r(\omega, T)^{-1}\boldsymbol{b} \tag{32}$$

where $Z_r(\omega, T) = T^T K_e T + G(\omega, T) T^T \overline{K}_v T - \omega^2 T^T M T$ is reduced dynamic stiffness matrix.

It can be seen that for viscoelastic systems, the construction of the basis is an issue, since the stiffness matrix is frequency- and temperature-dependent. Thus, three solutions are possible: (a) one can neglect this dependence by considering the stiffness matrix as independent from frequency and temperature [15, 19]. In this case, the basis is also constant; (b) one can use a reduction basis obtained by the resolution of the nonlinear eigenvalue problem [22, 23]; (c) it is possible to use an iterative method, which allows the re-actualization of the basis according to frequency [15, 16].

The strategy adopted here consists in using a constant reduction basis computed by using the tangent stiffness matrix representing the static behavior of the viscoelastic material. As can be seen in **Figure 2**, on the low frequency range, by prolonging the modulus and loss factor curves by asymptotes, the extrapolation leads to the real values G_0 and $\eta_0 = 0$. The tangent stiffness matrix can thus be calculated by the relation, $K_0 = K_e + G_0 \overline{K}_v$ [15].

The nominal basis containing the first retained modes of the viscoelastic system can thus be obtained by the resolution of the following eigenvalue problem [18]:

$$\begin{aligned} &(\boldsymbol{K}_0 - \lambda_i \boldsymbol{M})\boldsymbol{\phi}_i = \boldsymbol{0} \qquad i = 1, \dots, N \\ &\boldsymbol{\phi}_0 = [\boldsymbol{\phi}_1 \quad \boldsymbol{\phi}_2 \quad \cdots \quad \boldsymbol{\phi}_{NR}], \quad \boldsymbol{\Lambda}_0 = diag(\lambda_1, \dots, \lambda_{NR}) \end{aligned} \tag{33}$$

where ϕ_0 contains the mode shapes of the conservative associated system, which is further enriched by introducing the following residues associated with the external loads and the viscoelastic damping forces, respectively:

$$\boldsymbol{R} = \boldsymbol{K}_0^{-1}\boldsymbol{b}, \quad \boldsymbol{R}_v^0 = \boldsymbol{K}_0^{-1}\overline{\boldsymbol{K}}_v\boldsymbol{\phi}_0 \tag{34}$$

Thus, the enriched basis of reduction is represented as follows:

$$\boldsymbol{T}_0 = [\boldsymbol{\phi}_0 \quad \boldsymbol{R} \quad \boldsymbol{R}_v^0] \tag{35}$$

The basis (35) can be used to reduce viscoelastic systems with reasonable accuracy, but it is not *"robust"* enough to taking into account parametric uncertainties and to be used in conjunction with optimization and/or model updating processes. This means that the basis given by Eq. (35) should in principle be updated successively to guarantee a satisfactory accuracy of model reduction during iterations. However, according to the authors in reference [18], the strategy consists in using a fixed reduction basis evaluated from the initial model [as given by Eq. (35)], to be further enriched by a set of residual vectors depending on the parametric modifications introduced by the viscoelastic treatment. In most practical applications, the viscoelastic surface treatments are not applied to the entire structure, but only in specific zones. Thus, based on Eq. (12), one can write the following modified system:

$$[\boldsymbol{\Delta Z}(\omega, T) + \boldsymbol{Z}(\omega, T)]\,\boldsymbol{Q} = \boldsymbol{F} \tag{36}$$

where $\boldsymbol{\Delta Z}(\omega, T) = \boldsymbol{\Delta K}_v(\omega, T) - \omega^2 \boldsymbol{\Delta M}_v$ is the dynamic stiffness matrix due the perturbations.

Hence, Eq. (36) can be interpreted as the equilibrium equation of the nominal model, subjected to forces of modifications, $\boldsymbol{Z}(\omega, T)\boldsymbol{Q} = \boldsymbol{F} + \boldsymbol{f}_\Delta(\omega, T)$, where $\boldsymbol{f}_\Delta(\omega, T) = -\boldsymbol{\Delta Z}(\omega, T)\boldsymbol{Q}$, $\Delta M_v = \sum_{i=1}^{n\_mp} \Delta m_i M_v$ and $\Delta \bar{K}_v = \sum_{i=1}^{n\_mp} \Delta k_i \bar{K}_v$. Δm_i and Δk_i are the mass and stiffness variations, $M_v = \bigcup_i M_{v_i}$ and $\bar{K}_v = \bigcup_i \bar{K}_{v_i}$ are the mass and stiffness matrices of the zones, M_{v_i} and $\bar{K}_{v_i}$, are the elementary viscoelastic mass and stiffness matrices, and $\boldsymbol{\Delta M}_v$ and $\Delta \bar{K}_v$ are the matrices to be reduced, which are in general sparse and nonlinear functions of the design parameters.

7.1. Basis of displacements associated with the structural modifications

The vector of forces of modifications, $\boldsymbol{f}_\Delta(\omega, T)$, depends on the response of the modified system, $\boldsymbol{Q}$. Since this response is unknown *a priori*, the forces associated with the modifications cannot be computed exactly. The strategy is to generate the forces due to the small modifications by $\boldsymbol{f}_\Delta(\omega, T)$ using a truncated basis of normal modes of the mean model according to Eq. (40); next, the static responses of the mean model can be obtained by using the estimated forces of modifications. These two steps must be performed for each design variable subjected to small modifications.

$$\boldsymbol{f}_\Delta(\omega, T) \approx -\boldsymbol{\Delta Z}(\omega, T)\boldsymbol{\phi}_0 \boldsymbol{Q}_r \tag{37}$$

Hence, after obtaining the basis of forces, one can calculate a series of static responses of the system based on the tangent stiffness matrix according to the following form:

$$\boldsymbol{R}_\Delta = \boldsymbol{K}_0^{-1} \boldsymbol{f}_\Delta \tag{38}$$

and the final robust basis of reduction taking into account the small modifications is as follows:

$$\boldsymbol{T} = \begin{bmatrix} \boldsymbol{T}_0 & \boldsymbol{R}_\Delta \end{bmatrix} \tag{39}$$

Figure 3 illustrates a cycle of optimization process by using the robust basis, where the standard Ritz basis is increased by the static residues associated with the external loadings and the forces associated with the viscoelastic modifications. This procedure is used to approximate the behavior of the modified viscoelastic system without the re-actualization of the nominal basis, leading to a drastic reduction of the time required for computing the response of the large-scale viscoelastic systems.

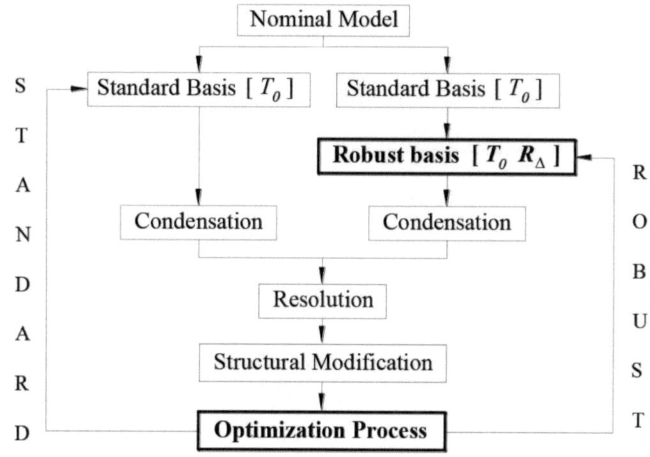

Figure 3. Block diagram showing the standard and the robust optimization procedures.

8. Review of FE modeling of passive constraining layer damping

One type of structure of particular interest in terms of practical viscoelastic applications is the three-layer sandwich plate illustrated in **Figure 4**. In the present work, the FE modeling procedure is summarized based on the original contribution made by Khatua and Cheung [41]

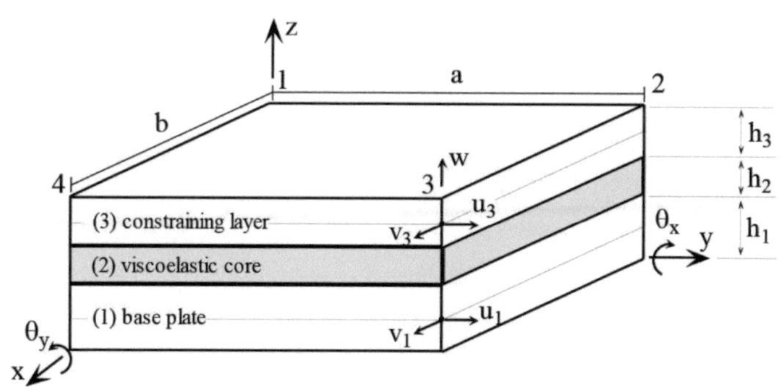

Figure 4. Illustration of the three-layer sandwich plate FE.

and implemented by de Lima et al. [42]. The sandwich element is composed by four nodes and seven DOFs per node, as depicted in the figure, where u and v are the in-plane displacements in directions x and y, respectively, w is the transverse displacement, and θ_x and θ_y are the rotations.

In the development of the theory, it has been assumed that the three layers are perfectly bounded and the materials involved are considered to be isotropic with linear behavior. These assumptions are reasonable [18], since, in practice, the most commercially available viscoelastic materials for vibration attenuation are self-adhesive. The analysis also assumes the Kirchhoff's theory for the base plate and constraining layer with the same rotations, and only for the viscoelastic core, the transverse shear is included (Mindlin's theory). The transverse displacement is assumed to be same for all the layers.

A number of approaches have considered in the open literature to describe with reasonable accuracy the shear behavior of constrained-layer damping treatments. However, the assumptions adopted herein are often used to model moderately thin sandwich beam and plate structures with reasonable accuracy [43].

The displacements are discretized by using linear shape functions for the in-plane displacements of the base plate and constraining layer, and a cubic shape function for the transverse displacement, by the expression, $\boldsymbol{u}(x, y, t) = \boldsymbol{N}(x, y)\boldsymbol{u}_{(e)}(t)$, where $\boldsymbol{N}(x, y)$ represents the matrix of the interpolation functions, and $u_{(e)}(t)=\left[u_1^i \; v_1^i \; u_3^i \; v_3^i \; w^i \; \theta_x^i \; \theta_y^i\right]^T$ with $i = 1$ to 4 is vector formed by the mechanical nodal DOFs. According to the theory of elasticity, the strain–displacement relations are formulated, $\boldsymbol{\varepsilon}(x, y, z, t) = \boldsymbol{B}(x, y, z)\, \boldsymbol{u}_{(e)}(t)$, where the strains for elastic layers, $\varepsilon_k=\left[\varepsilon_x^{(k)} \; \varepsilon_y^{(k)} \; \gamma_{xy}^{(k)}\right]^T$ with $(k = 1, 3)$, and for the viscoelastic core, $\varepsilon_2=\left[\varepsilon_x^{(2)} \; \varepsilon_y^{(2)} \; \gamma_{xy}^{(2)} \; \gamma_{xz}^{(2)} \; \gamma_{yz}^{(2)}\right]^T$, are obtained. Thus, based on the stress states assumed for each layer and the stress–strain relations, the stress responses of the system can be obtained as follows:

$$\boldsymbol{M}^{(e)} = \sum_{k=1}^{3} \rho_k h_k \int_{x=0}^{b} \int_{y=0}^{a} \boldsymbol{N}^T(x,y)\boldsymbol{N}(x,y)\,dydx \tag{40.a}$$

$$\boldsymbol{K}_e^{(e)} = \sum_{k=1,3} \int_{z=0}^{h_k} \int_{x=0}^{b} \int_{y=0}^{a} \boldsymbol{B}_k^T(x,y,z)\boldsymbol{C}_k\boldsymbol{B}_k(x,y,z)\,dydxdz \tag{40.b}$$

$$\boldsymbol{K}_2^{(e)}(\omega,T) = \int_{z=0}^{h_2} \int_{x=0}^{b} \int_{y=0}^{a} \boldsymbol{B}_2^T(x,y,z)\boldsymbol{C}_2(\omega,T)\boldsymbol{B}_2(x,y,z)\,dydxdz \tag{40.c}$$

where b and a designate, respectively, the dimensions of the rectangular plate element in directions x and y, and h_k and ρ_k represent the thickness and the mass density of the kth layer, respectively. The stiffness matrices, $K_e^{(e)}=K_1^{(e)} + K_3^{(e)}$ and $K_v^{(e)}(\omega, T)=K_2^{(e)}(\omega, T)$, give, respectively, the contributions of the purely elastic and viscoelastic parts of the sandwich structure. Hence, the elementary equations of motion are given as follows:

$$M^{(e)}\ddot{q}_{(e)}(t)+\left[K_e^{(e)}+K_v^{(e*)}\right]q_{(e)}(t)=f^{(e)}(t) \tag{41}$$

where $M^{(e)}\in R^{N_e\times N_e}$ is the mass (symmetric, positive definite) matrix, and $K_e^{(e)}\in R^{N_e\times N_e}$ and $K_v^{(e*)}=K_v^{(e)}(\omega,\ T)\in R^{N_e\times N_e}$ are the stiffness matrices (symmetric, non-negative definite) corresponding to the purely elastic and viscoelastic substructures, respectively. $q_{(e)}(t)\in R^{N_e}$ and $f^{(e)}(t)\in R^{N_e}$ are displacements and load vectors, respectively.

9. FE modeling of discrete viscoelastic dampers

From the practical standpoint, the use of viscoelastic materials in mounts and joints is an interesting alternative [1, 31]. **Figure 5a** illustrates the two mostly used configurations of viscoelastic mounts with the corresponding geometrical parameters. The placement of those mounts in structures is illustrated in **Figure 5b**. The mounts can be conveniently represented by springs, meaning that a translation mount produces damping forces while a rotational mount generates damping moments. In the same figure, the translational and rotational stiffness coefficients, $K^t(s)$ and $K^r(s)$, are given.

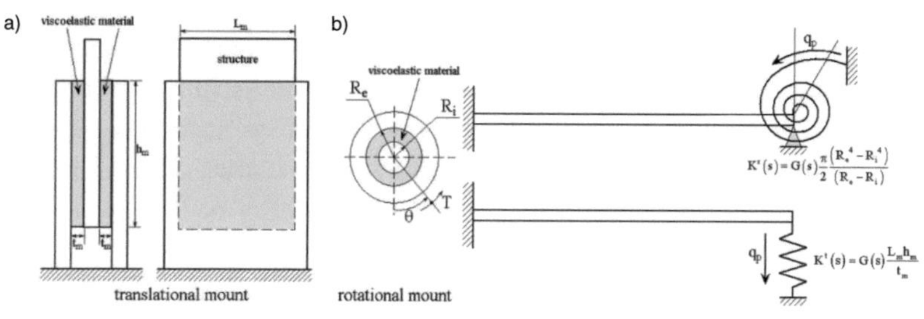

Figure 5. Sketches (**a**) and springs representation (**b**) of discrete viscoelastic devices.

Designating by p the order of the coordinate following the which the viscoelastic mount works, the inclusion of the viscoelastic effect into the equations of motion can be easily done by using the concept of dyadic structural modifications [44]. Thus, the equations of motion of the structural system with viscoelastic mount can be written as follows:

$$M\ddot{q}(t)+K_e q(t)+G(s)\overline{K}_v q(t)=f(s) \tag{42}$$

where $\mathbf{K}_v(s)=\mathbf{K}^t(s)I_p^T I_p$ for a translation mount, and $\mathbf{K}_v(s)=\mathbf{K}^r(s)I_p^T I_p$ for a rotation mount, and I_p designates the pth column of the identity matrix of order N.

Hence, the global system of equations of motion can be expressed under the form:

10. Numerical examples

The purpose of this section is to perform numerical examples in order to illustrate the main features and capabilities of the viscoelastic modeling procedures intended to design and performance analysis of the viscoelastic damping treatments presented herein. In addition, experimental investigations with a freely suspended rectangular plate were performed, where frequency-response functions (FRFs) and modal analysis have been performed to demonstrate the accuracy of the viscoelastic models and to confirm the effectiveness of the viscoelastic materials applied in the context of vibration attenuation.

10.1. Curve fitting of the viscoelastic model parameters

In the simulations that follow the viscoelastic characteristics of commercially available ISD112 manufactured by 3M [33] have been used. The material data provided by the manufacturer, in terms of storage and loss moduli, at 25°C in the frequency band [8–8000 Hz], have been used to identify the parameters for each viscoelastic model. Eq. (3) was used to form the objective function, which was minimized with respect to the unknown set of model parameters. Such objective function is symbolically defined as follows:

$$F = \sum_{i=1}^{N_f} \frac{\left| \mathrm{Re}\left[G_{model}(\omega_i)\right] - \mathrm{Re}\left[G_{data}(\omega_i)\right] \right|}{\mathrm{Re}\left[G_{data}(\omega_i)\right]} + \frac{\left| \mathrm{Im}\left[G_{model}(\omega_i)\right] - \mathrm{Im}\left[G_{data}(\omega_i)\right] \right|}{\mathrm{Im}\left[G_{data}(\omega_i)\right]} \tag{43}$$

Optimization was carried out by using genetic algorithms [45], with populations of 800 individuals, allowing for 200 generations and using side constraints. For illustration, **Figure 6** shows the storage modulus, loss modulus, and loss factor functions reconstructed from the identified parameters only for the GHM model with five mini-oscillators, superimposed to the experimental counterparts. As can be seen, good quality of the curve fitting could be achieved. The same quality could be obtained for the FDM model and the ADF model with five anelastic fields. Negligible improvement was obtained by increasing the order of those models. The values of the parameters obtained for the models are defined in **Table 1**.

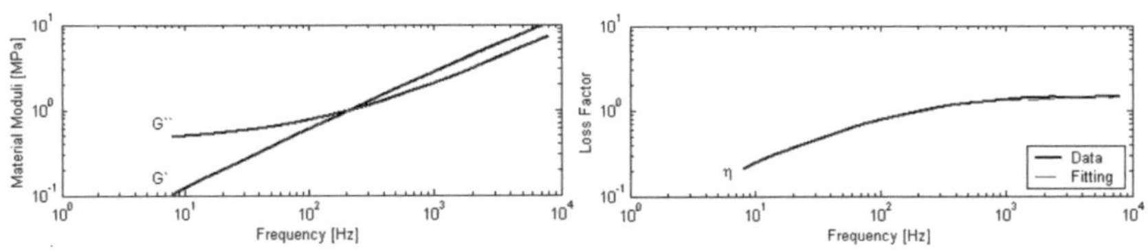

Figure 6. Curve fitting of modulus functions for 3M ISD112 according to the GHM model.

10.2. Model validation

To verify the model summarized in Section 8, experimental tests were performed on a freely suspended plate made of aluminum with a constraining damping layer made of a thin ISD112 viscoelastic material and an outer thin aluminum sheet. The experiments consisted in obtaining a set of 20 FRFs corresponding to point I, indicated in **Figure 7**. Only the average FRF is shown here. The number of elements used to generate the model is shown in the same figure, formed by 378 elastic DOFs, and the anelastic displacements are computed according to the mini-oscillator terms defined in **Table 1**.

Table 2 provides the physical and geometrical properties used to generate the FE model.

GHM				ADF				FDM
G_r (MPa)	α_i	ω_i (*rad/s*)	ζ_i	G_r (MPa)	Δ_i	Ω_i (rad/s)	C_i	G_0=0.428 (MPa)
0.4623	0.26	991.33	4.575	0.4680	0.205	103.48	460.7	G_1=0.0088 (MPa)
	0.95	6986.15	4.27		0.682	103.48	138.2	α=0.67
	2.04	103,437.5	2.702		1.942	638.33	48.5	β=0.41
G_u(MPa)	3.69	22,950.1	1.923	G_u(MPa)	7.062	3054.43	13.4	
28.49	53.7	266,466.6	1.299	44.08	83.37	17,583.2	1.13	

Table 1. Identified parameters for the GHM, FDM, and ADF models.

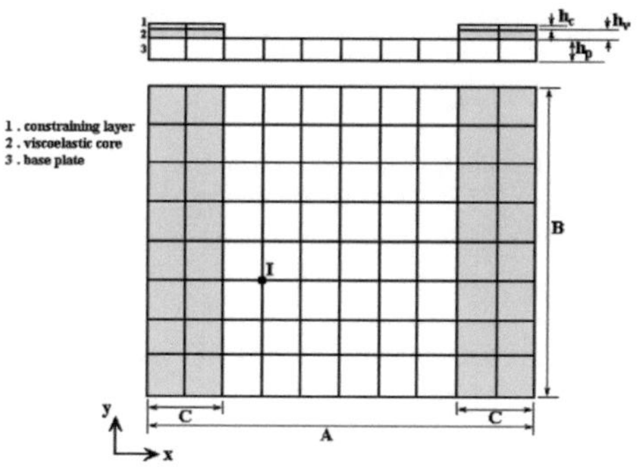

Figure 7. Illustration of the FE model for the plate with partial viscoelastic treatment.

Base plate	Viscoelastic core	Constrained layer
A = 20 × 10^{-2} m	C = 2 × 10^{-2} m	C = 2 × 10^{-2} m
B = 25 × 10^{-2} m		
h_p = 5 × 10^{-4} m	h_v = 20 × 10^{-5} m	h_c = 5 × 10^{-4} m
E = 70.3 × 10^9 N/m	ϱ = 1099.5 kg/m³	E = 70.3 × 10^9 N/m
ϱ = 2750 kg/m³	ν = 0.5	ϱ = 2750 kg/m³

Table 2. Physical and geometrical characteristics of the plate FE model.

Figure 8 shows the amplitudes of the average FRFs calculated from the experiments, compared to the numerically acquired counterparts. It can be seen the efficiency of the surface damping treatment in mitigating the amplitudes of vibrations in the frequency band of interest. Also, it can be noted the accuracy of the model in predicting the dynamic response of the viscoelastic system.

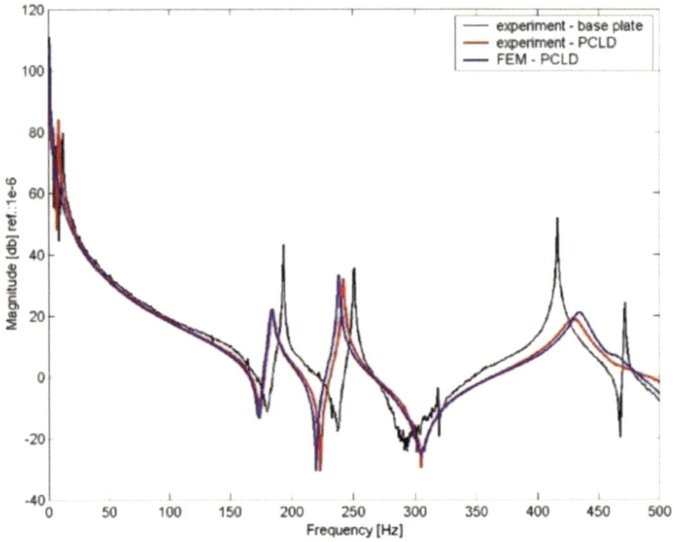

Figure 8. FE and experimental FRFs of the system with and without treatment.

Mode	1		2		3	
Modal parameters	ω_1 (Hz)	ζ_1	ω_2 (Hz)	ζ_2	ω_3 (Hz)	ζ_3
Experimental	184.37	8.1×10^{-3}	242.47	1.9×10^{-3}	429.38	18.2×10^{-3}
FE prediction	184.38	7.5×10^{-3}	238.75	1.8×10^{-3}	434.4	14.4×10^{-3}
Deviations (%)	0.00	8.00	1.60	5.50	1.30	26.40

Table 3. Experimental natural frequencies and modal damping factors of the plate.

Table 3 compares numerical and experimental frequencies and damping factors obtained by applying the half-power bandwidth method [1]. It can be noted that the two sets are reasonably close to each other. However, the differences observed are mostly due to the identification procedure of the mini-oscillator parameters from the experimental data for the ISD112 material; the theory adopted in the FE model such as the perfectly bounded conditions; the variations on the temperature during the tests; and the boundary conditions.

10.3. Two-dimensional truss with a translational viscoelastic mount

Figure 9 shows the two-dimensional truss FE model in which a translational mount is applied on node 7 with the direction indicated on the same figure.

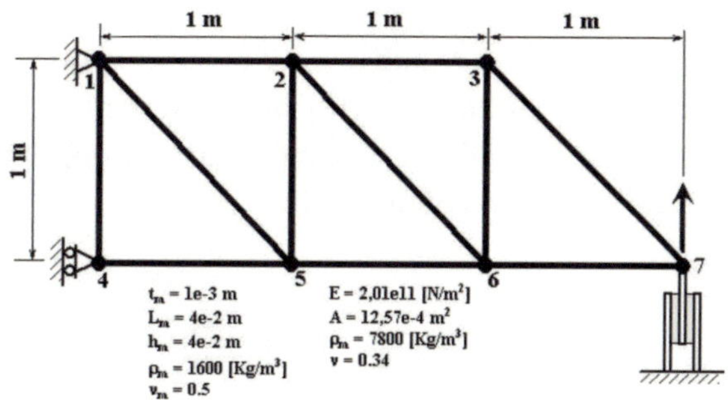

Figure 9. Two-dimensional truss with a translational viscoelastic mount.

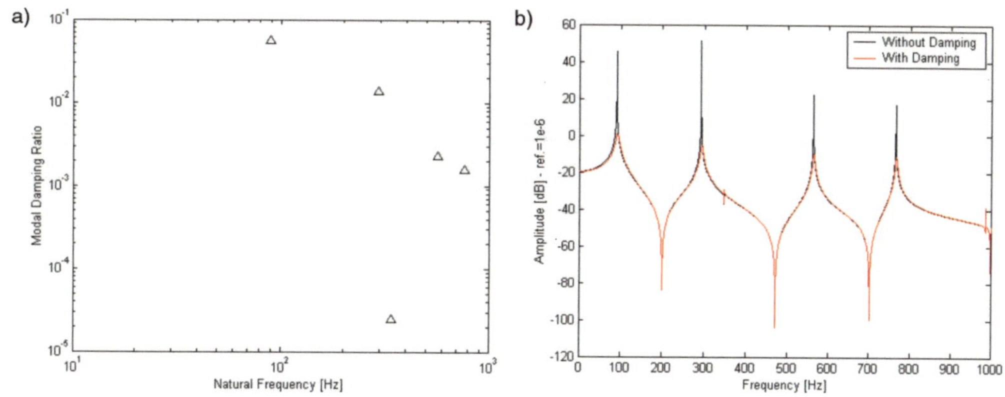

Figure 10. Natural frequencies and damping ratios (a) and FRFs (b) for the truss.

Using the theory presented in Section 9 combined with the FDM model, the complex eigenvalue problem was performed to obtain the natural frequencies and damping ratios. The results corresponding to the five vibration modes in the frequency band [80–800 Hz] are presented in **Figure 10a**. In **Figure 10b**, the amplitudes of the FRFs of the systems with and without viscoelastic damper are compared. Again, it is possible to evaluate the influence of the damping on the response amplitudes and the influence of the frequency on damping and stiffness of the structure. The FRFs are related to the vertical displacement of node 7 indicated in **Figure 9**.

10.4. Internally balanced method

Figure 12 shows the results for a beam-like structure partially treated with constrained-layer damping, as illustrated in **Figure 11**, in terms of the controllability and observability grammians in the balanced realization. It can be noted that W_c and W_0 must be equal and diagonal, as predicted by the theory of internally balanced reduction method detailed in Section 6. Also, since the retained states must be composed by the states with major controllability and observability indices, the first five modes will be considered in the model reduction system.

The time response to an impulse excitation applied at point P (indicated in **Figure 11**), and the amplitudes of the FRFs of the beam before and after reduction are depicted in **Figure 13**. It can be clearly perceived the efficiency of the internally balanced method in predicting both time and frequency responses of the viscoelastic system by the appropriate choice of the major controllability and observability indices of the states.

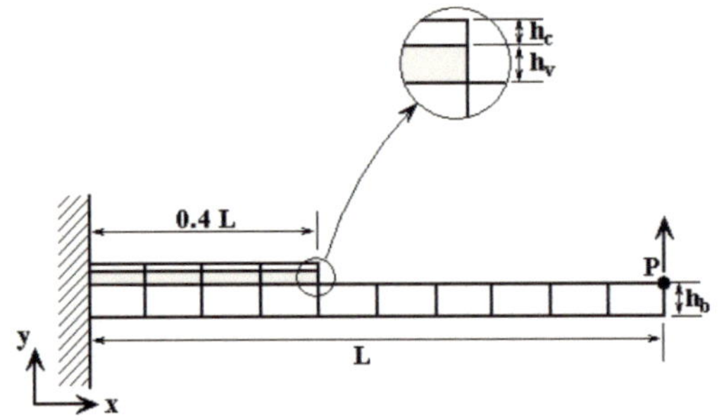

Figure 11. Illustration of the beam partially treated with constraining viscoelastic layer.

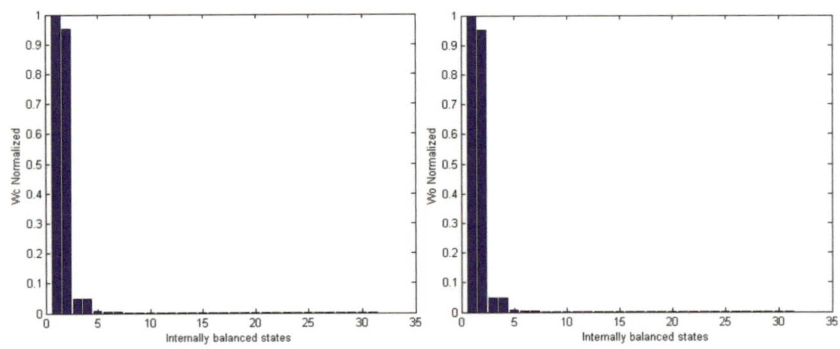

Figure 12. W_c and W_0 versus internally balanced states for the viscoelastic beam.

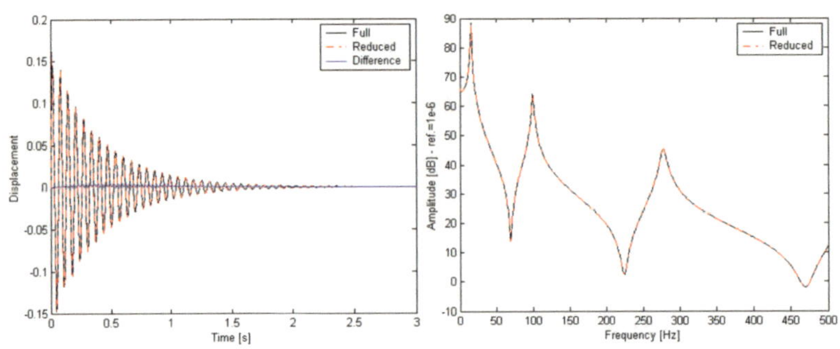

Figure 13. Time and FRFs for the full and reduced systems—internally balanced method.

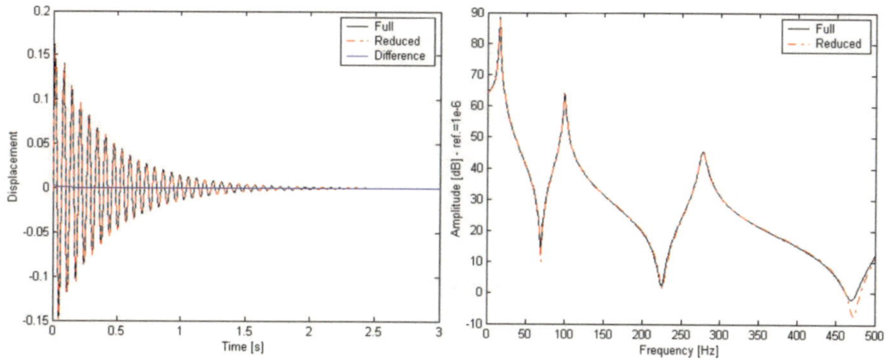

Figure 14. Time and FRFs for the full and reduced systems—constant enriched basis.

10.5. Modal reduction method

Figure 14 shows the time and frequency domain responses of the reduced beam system by using the enriched modal reduction method compared with the full system. In this case, it has been considered only the first five modes of the associated conservative viscoelastic system, ϕ_0, enriched with the static residues associated with the external loads, $\boldsymbol{R}$, and the viscoelastic damping forces, R_v^0. As can be seen, both impulse responses and FRFs appear as expected when compared with the time and frequency responses obtained by the internally balanced method, leading to conclude that the reduction method by applying the constant enriched basis (35) is also a viable method to reduce viscoelastic systems.

11. The self-heating phenomenon

The good damping performance and inherent stability of viscoelastic materials in relatively broad frequency bands, besides cost-effectiveness, offers many possibilities for practical engineering applications. However, some drawbacks must be dealt with, such as ageing and chemical instability in the presence of some substances, the mass added and the fact that in most traditional design procedures of viscoelastic dampers subjected to cyclic loadings, uniform and constant temperature is generally assumed and does not take into account the self-heating phenomenon. Also, for viscoelastic dampers subjected to dynamic loadings superimposed on static preloads, especially when good isolation characteristics are required at high frequencies, traditional design guidelines can lead to poor designs or even severe failures, since it is observed a rapidly increasing rate of temperature change and an accompanying stiffness reduction.

The self-heating can cause temperature increases in viscoelastic materials, affecting significantly their damping capacity [26–28]. Thus, in applications in which the viscoelastic materials are subjected to cyclic loadings superimposed on static preloads, such as engine mounts and tall buildings, the interest to obtain high isolation characteristics becomes essential, since the vibration amplitudes are directly related to fatigue and, consequently, to structural integrity [4, 29, 35]. Moreover, depending on the magnitude of the applied loadings, the vibration energy of the viscoelastic material is converted to heat at a rate faster than the heat is conducted away, leading to a rapidly increasing rate of local temperature change known as *thermal runaway* phenomenon [26]. Thus, it is expected that it can have a strong influence on the stiffness and damping properties of viscoelastic materials, leading to unexpected damping performance or even severe failures of viscoelastic damping devices.

Figure 15 shows the experimental results obtained for a viscoelastic damper subjected to a vertical cyclic loadings during 3396 s, superimposed on different values of static displacement applied to the specimen by the screws shown in the same figure.

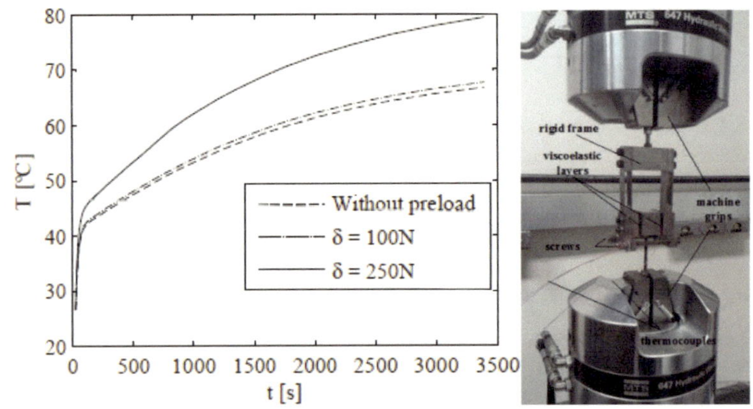

Figure 15. Time evolution of the temperature inside the viscoelastic material and the experimental setup.

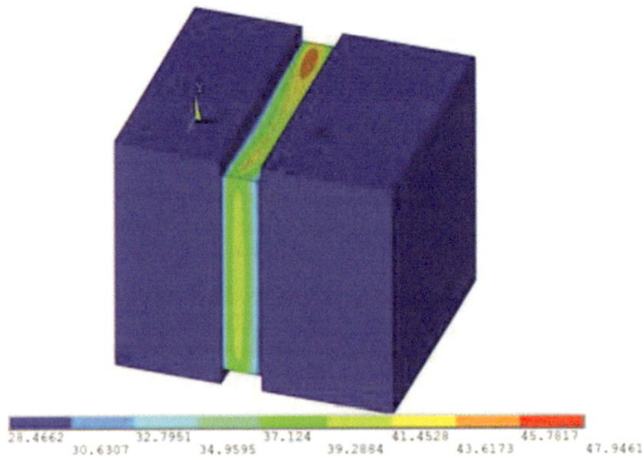

Figure 16. Temperature contours for one half of the damper at $t = 100$ s for $\delta = 250$ N.

One can conclude that as the static preload increases, the self-heating becomes more pronounced. As a result, an increasing in the temperature values of the viscoelastic material is observed, leading to a significantly reduction of its damping capacity or even its complete failure in practical engineering applications. Moreover, it is not possible to identify a progressive stabilization of the temperatures in the loading phase, indicating the occurrence of the so-called thermal runaway phase [29].

Figure 16 enables to conclude that the assumption of assuming a constant and uniform temperature distribution for viscoelastic materials subjected to cyclic loadings is not correct, since the temperatures are not constant and vary from one point to another.

12. Concluding remarks

A comprehensive review of the modeling strategies of engineering structures incorporating viscoelastic materials has been showed. The FE modeling procedure of two-dimensional sandwich plates treated with viscoelastic materials as a passive constrained-layer damping and a modeling strategy of discrete viscoelastic damping devices including translational and rotational mounts have been also implemented. As can be noted, the modeling of viscoelastic materials was conceived so as to encompass different designs, regarding the type of treatment applied as surface or discrete viscoelastic vibration dampers. The GHM, ADF, and FDM models were used to include the frequency- and temperature-dependent viscoelastic behavior into FE matrices, in spite of the significant increase in the order of the system's augmented matrices, entailed by the inclusion of internal variables especially for the GHM and ADF models. Moreover, the separation of the material modulus function of each viscoelastic model into real and imaginary parts to enable the identification of the material modulus parameters from experimental data has also been addressed and illustrated for the ISD112 viscoelastic material as detailed in the examples.

The ongoing work aims at developing a user-friendly computer code incorporating various modeling tools available to date to be used for the design, performance analysis, and optimization of different types of viscoelastic vibration dampers taking into account the self-heating phenomenon, as can be available in numerical examples. Also, the implementation of efficient numerical procedures as model reduction methods for the resolution of the equations of motion for modal and frequency-domain analyses of more complex engineering systems incorporating viscoelastic materials was addressed.

In general, the numerical simulations presented enabled to illustrate the application of the modeling procedure as a tool to evaluate the damping effectiveness in terms of eigenvalue and frequency response analysis. Based on the obtained results, one can conclude about the convenience of using more elaborate viscoelastic models in combination with FE models of complex medium- to large-scale structural systems.

Currently, the modeling procedure is being extended to include other types of structural elements, such as three-dimensional beams, plates, and shells. Also, the implementation of efficient numerical and experimental procedures of the self-heating phenomenon and the thermal runaway phase in viscoelastic materials is a topic under investigation.

Acknowledgements

The authors are grateful to the state agencies FAPEMIG and FAPEG for research funding of their research projects. A.M.G. de Lima is thankful to CNPq and CAPES for the financial support to their research activities and for the grant of doctorate and postdoctorate scholarships.

Author details

Antonio Marcos G. de Lima[1*], Luiz Fernando F. Rodovalho[1] and Romes A. Borges[2]

*Address all correspondence to: amglima@mecanica.ufu.br

1 School of Mechanical Engineering, Campus Santa Mônica, Federal University of Uberlândia, Uberlândia, MG, Brazil

2 Department of Applied Mathematics, Federal University of Goiás, Catalão, GO, Brazil

References

[1] Nashif AD, Jones DIG, Henderson JP. Vibration Damping. John Wiley & Sons, New York, 1985.

[2] Mead DJ. Passive Vibration Control. Wiley, Canada, 1998, pp. 554.

[3] Palmeri A, Ricciardelli F. Fatigue analyses of buildings with viscoelastic dampers. Journal of Wind Engineering and Industrial Aerodynamics, 94 (2006): 377–395.

[4] de Lima AMG, Lambert S, Rade DA, Pagnacco E, Khalij L. Fatigue reliability analysis of viscoelastic structures subjected to random loads. Mechanical Systems and Signal Processing. 43 (2014): 305–318.

[5] Rao MD. Recent applications of viscoelastic damping for noise control in automobiles and commercial airplanes. In: USA Symposium on Emerging Trends in Vibration and Noise Engineering, India, 2001.

[6] Samali B, Kwok KCS. Use of viscoelastic dampers in reducing wind- and earthquake-induced motion of building structures. Engineering Structures. 17 (1995): 639–654.

[7] Bagley RL, Torvik PJ. A generalized derivative model for an elastomer damper. Shock and Vibration Bulletin. 49 (1979): 135–143.

[8] Bagley RL, Torvik PJ. Fractional calculus—A different approach to the analysis of viscoelastically damped structures. AIAA Journal. 21(1983): 741–748.

[9] Bagley RL, Torvik PJ. Fractional calculus in the transient analysis of viscoelastically damped structures. AIAA Journal. 23 (1985): 918–925.

[10] Golla DF, Hughes PC. Dynamics of viscoelastic structures—A time domain finite element formulation. Journal of Applied Mechanics. 52 (1985): 897–906.

[11] MacTavish D, Hughes PC. Modeling of linear viscoelastic space structures. Journal of Vibration and Acoustics. 115 (1993): 103–115.

[12] Lesieutre GA. Finite element for dynamic modelling of uniaxial rods with frequency dependent material properties. International Journal of Solids and Structures. 29 (1992): 1567–1579.

[13] Lesieutre GA, Bianchini E. Time domain modeling of linear viscoelasticity using anelastic displacement fields. Journal of Vibration and Acoustics, Transactions of the ASME. 117 (1995): 424–430.

[14] Lesieutre GA, Lee U. A finite element for beams having segmented active constrained layers with frequency-dependent viscoelastics. Smart Materials and Structures. 5 (1996): 615–627.

[15] Balmès E. Model reduction for systems with frequency dependent damping properties. In: 15th International Modal Analysis Conference (IMAC), Orlando, Florida, USA, 1997.

[16] Balmès E. Damping and complex modes. In: Proceedings of the 21st International Modal Analysis Conference (IMAC), USA, 2003.

[17] de Lima AMG, Rade DA. Modeling of structures supported on viscoelastic mounts using FRF substructuring. In: Proceedings of the Twelfth International Congress on Sound and Vibration, ICSV12, Lisbon, Portugal, 2005.

[18] de Lima AMG, da Silva AR, Rade DA, Bouhaddi N. Component mode synthesis combining robust enriched Ritz approach for viscoelastically damped structures. Engineering Structures. 32 (2010) 1479–1488.

[19] Yae KH, Inman DJ. Control-oriented order reduction of finite element model. Dynamic Systems and Control. 115 (1993): 708–711.

[20] Trindade MA, Benjeddou A. Hybrid active-passive damping treatments using viscoelastic and piezoelectric materials: Review and assessment. Journal of Vibration and Control. 8 (2002): 699–746.

[21] Park CH, Baz A. Vibration control of bending modes of plates using active constrained layer damping. Journal of Sound and Vibration. 227 (1999): 711–734.

[22] Salmanoff J. A finite element, reduced order, frequency dependent model of viscoelastic damping. MSc. Thesis, Faculty of the Virginia Polytechnic Institute, Blacksburg, VA, USA, 1997.

[23] Lam MJ, Hybrid active/passive models with frequency dependent damping. Ph. D. Thesis, Faculty of the Virginia Polytechnic Institute, Blacksburg, VA, USA, 1997.

[24] Moore BC. Principal component analysis for linear systems: Controllability, observability, and model reduction. Institute of Electronic Engineers Transactions on Automatic Control. AC-26 (1981) 17–31.

[25] de Lima AMG, Rade DA, Bouhaddi N. Optimization of viscoelastic systems combining robust condensation and metamodeling. Journal of the Brazilian Society of Mechanical Sciences and Engineering. 32 (2010) 485–495.

[26] de Lima, AMG, Rade DA, Lacerda HB, Araújo CA. An investigation of the self-heating phenomenon in viscoelastic materials subjected to cyclic loading accounting for prestress. Mechanical Systems and Signal Processing. 58–59 (2015) 115–127.

[27] Schapery RA. Effect of cyclic loading on the temperature in viscoelastic media with variable properties. AIAA Journal. 2 (1964) 827–835.

[28] Lesieutre GA, Mingori DL. Finite element modelling of frequency-dependent material damping using augmenting thermodynamic fields. Journal of Guidance Control Dynamics. 13 (1990) 1040–1050.

[29] Lesieutre GA, Govindswamy K. Finite element modelling of frequency-dependent and temperature-dependent dynamic behaviour of viscoelastic materials in simple shear. International Journal of Solids and Structures. 33 (1996) 419–432.

[30] Christensen RM. Theory of Viscoelasticity: An Introduction. Academic Press Inc., New York, 2nd edition, 1982.

[31] de Lima AMG, Rade DA, Lépore-Neto FP. An efficient modelling methodology of structural systems containing viscoelastic dampers based on frequency response function substructuring. Journal of Mechanical System and Signal Processing. 23 (2009) 1272–1281.

[32] Drake ML, Soovere J. A design guide for damping of aerospace structures. In: AFWAL Vibration Damping Workshop Proceedings 3, 1984.

[33] http://www.3m.com/ (Accessed 01 April 2015)

[34] Daya EM, Poitier-Ferry M. A numerical method for nonlinear eigenvalue problems. Applications to vibrations of viscoelastic structures. Computers and Structures. 79 (2001): 533–541.

[35] Palmeri A, Ricciardelli F. Fatigue analyses of buildings with viscoelastic dampers. Journal of Wind Engineering and Industrial Aerodynamics. 94 (2006): 377–95.

[36] Palmeri A, Ricciardelli F, Muscolino G, De Luca A. Effects of viscoelastic memory on the buffeting response of tall buildings. Wind Structures. 7 (2004): 89–106.

[37] Yuan L, Agrawal OP. A numerical scheme for dynamic systems containing fractional derivatives. AME Journal of Vibration and Acoustics. 124 (2002): 321–324.

[38] Wagner N, Adhikari S. Symmetric state-space method for a class of nonviscously damped systems. AIAA Journal. 41 (2003): 951–956.

[39] Cunha-Filho AG, de Lima AMG, Donadon MV, Leão LS. Flutter suppression of plates subjected to supersonic flow using passive constrained viscoelastic layers and Golla-Hughes-McTavish method. Aerospace Science and Technology. 52 (2016): 70–80.

[40] Friot E, Bouc R. Large frequency band localized control of a plate subjected to a wind force. In: 2nd GDR shock and vibration symposium, Marseille, France, 1996.

[41] Khatua TP, Cheung YK. Bending and vibration of multilayer sandwich beams and plates. International Journal for Numerical Methods in Engineering. 6 (1973): 11–24.

[42] de Lima AMG, Stoppa MH, Rade DA, Steffen Jr. V. Sensitivity analysis of viscoelastic structures. Shock and Vibration. 13 (2006): 545–558.

[43] Austin EM. Variations on modelling of constrained-layer damping treatments. Shock and Vibration Digest. 31 (1999): 275–280.

[44] Montalvão D, Maia NMM, Ribeiro AMR. A review of vibration-based structural health monitoring with special emphasis on composite materials. Shock and Vibration Digest. 38 (2006): 295–324.

[45] Srinivas N., Deb K. Multiobjective using non-dominated sorting in genetic algorithms. Technical Report, Department of Mechanical Engineering Institute of Technology, India, 1993.